£39.99

CW00421726

Analyzing Sensory Data with R

Chapman & Hall/CRC
The R Series

Series Editors

John M. Chambers
Department of Statistics
Stanford University
Stanford, California, USA

Torsten Hothorn
Division of Biostatistics
University of Zurich
Switzerland

Duncan Temple Lang
Department of Statistics
University of California, Davis
Davis, California, USA

Hadley Wickham
RStudio
Boston, Massachusetts, USA

Aims and Scope

This book series reflects the recent rapid growth in the development and application of R, the programming language and software environment for statistical computing and graphics. R is now widely used in academic research, education, and industry. It is constantly growing, with new versions of the core software released regularly and more than 5,000 packages available. It is difficult for the documentation to keep pace with the expansion of the software, and this vital book series provides a forum for the publication of books covering many aspects of the development and application of R.

The scope of the series is wide, covering three main threads:
- Applications of R to specific disciplines such as biology, epidemiology, genetics, engineering, finance, and the social sciences.
- Using R for the study of topics of statistical methodology, such as linear and mixed modeling, time series, Bayesian methods, and missing data.
- The development of R, including programming, building packages, and graphics.

The books will appeal to programmers and developers of R software, as well as applied statisticians and data analysts in many fields. The books will feature detailed worked examples and R code fully integrated into the text, ensuring their usefulness to researchers, practitioners and students.

Published Titles

Stated Preference Methods Using R, *Hideo Aizaki, Tomoaki Nakatani, and Kazuo Sato*

Using R for Numerical Analysis in Science and Engineering, *Victor A. Bloomfield*

Event History Analysis with R, *Göran Broström*

Computational Actuarial Science with R, *Arthur Charpentier*

Statistical Computing in C++ and R, *Randall L. Eubank and Ana Kupresanin*

Reproducible Research with R and RStudio, *Christopher Gandrud*

Introduction to Scientific Programming and Simulation Using R, Second Edition, *Owen Jones, Robert Maillardet, and Andrew Robinson*

Displaying Time Series, Spatial, and Space-Time Data with R, *Oscar Perpiñán Lamigueiro*

Programming Graphical User Interfaces with R, *Michael F. Lawrence and John Verzani*

Analyzing Sensory Data with R, *Sébastien Lê and Theirry Worch*

Analyzing Baseball Data with R, *Max Marchi and Jim Albert*

Growth Curve Analysis and Visualization Using R, *Daniel Mirman*

R Graphics, Second Edition, *Paul Murrell*

Multiple Factor Analysis by Example Using R, *Jérôme Pagès*

Customer and Business Analytics: Applied Data Mining for Business Decision Making Using R, *Daniel S. Putler and Robert E. Krider*

Implementing Reproducible Research, *Victoria Stodden, Friedrich Leisch, and Roger D. Peng*

Using R for Introductory Statistics, Second Edition, *John Verzani*

Dynamic Documents with R and knitr, *Yihui Xie*

Analyzing Sensory Data with R

Sébastien Lê
Thierry Worch

CRC Press
Taylor & Francis Group
Boca Raton London New York

CRC Press is an imprint of the
Taylor & Francis Group, an **informa** business

A CHAPMAN & HALL BOOK

CRC Press
Taylor & Francis Group
6000 Broken Sound Parkway NW, Suite 300
Boca Raton, FL 33487-2742

© 2015 by Taylor & Francis Group, LLC
CRC Press is an imprint of Taylor & Francis Group, an Informa business

No claim to original U.S. Government works

Printed on acid-free paper
Version Date: 20140819

International Standard Book Number-13: 978-1-4665-6572-2 (Hardback)

Visit the Taylor & Francis Web site at
http://www.taylorandfrancis.com

and the CRC Press Web site at
http://www.crcpress.com

Printed and bound by CPI Group (UK) Ltd, Croydon, CR0 4YY

Contents

Foreword

Back in the 1980s when I was still a consultant statistician supporting clients from sensory and other disciplines, I dreamed of the day when there would be free, easy-to-use statistical software for everyone to use. It seems that in sensory science that day is approaching fast with the arrival first of SensoMineR from Sébastien and colleagues at Agrocampus Ouest, Rennes, France, and now with a textbook backed up by online exercise data files.

The benefits to sensory scientists will be that they can undertake specialised sensory analyses such as MFA or GPA that are not widely available in other statistical packages. However, perhaps of far more benefit is that students and practitioners around the world can undertake the full range of standard and specialised sensory analyses at zero cost (as well as saving them time with automated routines)! I know that Sébastien and Thierry had this fully in mind as they have labored to finish this text.

All practicing sensory scientists will also find much to benefit from as there are sections throughout the book for the experienced (advanced) user. In fact, some of the examples tackled in this book are quite advanced (sorted Napping, comparison of various profiles, multivariate analyses with confidence ellipses to evaluate the panel performance, *etc.*) and very close to research.

The authors are aiming to produce a hands-on style reference text for the sensory scientist and with the teaching experience of Sébastien and the fact that both are connected to Agrocampus Ouest in Rennes, sharing the Jérôme Pagès vision of the statistics in the sensory area, it is clear that this will be a practical text with a strong multivariate emphasis.

Finally I must note that Sensometrics is still a dynamic and fast-moving discipline—but the beauty of the R software is that new methods can be quickly programmed and added to the R portfolio. Actually, the analysis of sensory data within R has been extended with the useful algorithms available for instance in the sensR package emanating from Per Brockhoff and Rune Christensen in Denmark.

I congratulate these two young scientists on their enthusiasm and energy and I support this project. Now I am looking forward to becoming an R user for the analysis of my sensory data!

Hal MacFie
Hal MacFie Training, Food Quality and Preference

Preface

Writing is definitely a foolish act or an act of great bravery, in which an author reveals their own thoughts to others. Writing a textbook could be interpreted as arrogant and immodest, as by definition a textbook is a "book used as a *standard* work for the study of a particular subject." Why would someone take the risk to reveal their thoughts and consider themself as knowledgeable enough to write a standard?

When we first started on this project, our main objective was simply to help—to fill a need, by providing to sensory scientists a textbook that would or could make their everyday life easier. We came up with this idea by realizing that in practice it is very common for (sensory) scientists to entrust their experimental data to statisticians for their analyses. Similarly, it is common to see companies externalizing their data treatment by asking statistical consultancy agencies to perform the analyses for them.

With the desire of understanding why this way of proceeding is so common, we went back to the *source*, and went through a large number of existing teaching programs and syllabuses throughout worldwide universities, to finally discover that sensory evaluation is literally taught as the evaluation[1] of the senses (*i.e.*, evaluation in the sense of measure). Although these sensory evaluation courses are often attached to statistical courses, the statistics learned are meant to be general to all types of data, and not specific to sensory data.

Due to this disconnection during the teaching process between the sensory measurement and the statistical analysis of the resulting data, it is not so surprising that sensory scientists, whether they are rookies or experts, have difficulties sometimes fully understanding sensory evaluation in its big picture: the reasons why sensory data are collected, the way they are collected, the intrinsic meaning of these data, the way they can be analyzed, and last but not least, the way results issued from the analysis of such data can be interpreted.

It is precisely to reconnect all these different dimensions of sensory evaluation that we decided to develop this textbook. We conceived it regarding two main objectives that seem apparently irreconcilable: to become a reference book[2] and, at the same time, to provide a more integrated vision of the analysis of sensory data. We hope that by consulting the book, jumping from one chapter to another, sensory scientists will find all the answers they are

---

[1] The making of a judgment about the amount, number, or value of something.

[2] A book is intended to be consulted for information on specific matters rather than read from beginning to end.

looking for in terms of sensory data analysis. We also hope that by reading the book linearly, from beginning to end, scientists will acquire more than a good background in statistics. In other words, we hope that they will get the fundamentals of statistics that will render them autonomous for the analysis of sensory data. In summary, from a *sensory* point of view, each chapter was conceived to be self-sufficient but, as a whole, chapters were arranged to be coherent, from a *statistical* point of view.

To achieve this, we tried to provide a book that aims to be didactic, the main objective being to get the readers familiar with the general concept of analysis, and with the analysis of sensory data in particular. The spirit of this book is thus to present the data gathered from different sensory evaluations, and for each evaluation, the corresponding sensory implications. In practice, we are asking ourselves what is possible to extract from the data in each case. By fully understanding what is possible to do with the data gathered, we provide tailored statistical analysis. In each case, we attach great importance to a good understanding of the sensory evaluation in terms of objectives, so we systematically provide, for each objective, the notions of statistical unit, of variables–in brief the notion of data set. Once the data set is presented, we show how to perform the required analysis by using the statistical software R, and more especially (but not only!) the SensoMineR package. Besides this essential practical aspect, we present the results from each analysis, as well as the keys for the interpretation.

For these reasons, we called this textbook *Analyzing Sensory Data with R*, which summarizes well the three main aspects of the book: 1) the act of performing analyses on 2) sensory data[3] by using 3) R as practical tool. With this in mind, the book intends to be methodological, applied to a particular scope, and pragmatic. We intend here to learn how to tackle methodically a sensory problem in order to provide concrete solutions that facilitate decision making. The analysis of this problem ends up in the statistical treatment of the data related to that problem.

Writing a book involves making firm choices regarding its contents, and this textbook is no exception. As mentioned previously, we tried to be as progressive as possible in terms of statistical methods, with the constraint of being as coherent and independent as possible in terms of sensory problems. The book is divided into three main parts: 1) the quantitative approaches; 2) the qualitative approaches; and 3) the affective approaches. Here, the notions of "quantitative" and "qualitative" are not specifically linked to the statistical nature of the data, but rather to their sensory nature. In the first part "Quantitative descriptive approaches", products are evaluated in details according to different sensory attributes, while in the second part, "Qualitative descriptive approaches", products are evaluated as a whole. In the third part, "Affective descriptive approaches", products are evaluated with respect to a

---

[3]It can be noted that the second point incorporates two words ("sensory" and "data") which are inseparable from our point of view.

very special measure, *i.e.*, the hedonic measure, and associated with the vast notion of "ideal products."

This organization of the parts may seem counterintuitive. Indeed, qualitative approaches are usually used to screen products and/or attributes, or when a quick overview of the products is needed (such as in quality control or overall product evaluation). *De facto* they are often used prior to quantitative approaches, which aim to describe how products are perceived through the human senses, and to provide the so-called product profiles. Still, to be as progressive and pedagogic as possible in terms of statistics, we fully assume that choice for two reasons. First, from a practical point of view, starting with the quantitative approach gives the scientists the chance to get familiar with the book while reading something they can (most probably) already relate to. Second, from a statistical point of view, it gives us the chance to gradually increase difficulty throughout the book. Indeed, in the first part, we introduce the notion of a model with Analysis of Variance (ANOVA), as well as fundamental[4] exploratory multivariate methods on quantitative data, such as Principal Component Analysis (PCA) and Multiple Factor Analysis (MFA), in its most common usage. In the second part, models are getting more complex with the Bradley–Terry and Thurstonian models, as well as multivariate methods, with Correspondence Analysis (CA), Multiple Correspondence Analysis (MCA), and MFA applied to contingency tables. Finally, the third part is probably the most complex one, since to get meaningful information, methods have to be combined. It is, for example, the case with linear (or PLS) regression and PCA to create external preference maps.

Nevertheless, within each part, this textbook is driven by sensory issues: as soon as a sensory issue is put forward, the most appropriate statistical methods to answer that issue are presented. To facilitate the reading, their explanations and outputs are limited to the required essentials. Consequently, to fully appreciate the richness of a statistical method and its outputs, the reader must often refer to different chapters.

However, our choices should not go against scientists' needs. For that reason, we made the different parts as independent as possible so they are able to draw their own reading path. To do so, we decided to be as consistent as possible in the writing of the different chapters. Throughout this textbook, each chapter corresponds to one main sensory topic. Chapters are all structured in a similar way: they all start with a first section called *Data, sensory issues, notations*, which presents the nature of the sensory evaluation and its objectives. This section also develops the sensory particularities/issues related to the sensory evaluation, provides details about the data set thus obtained (in terms of statistical individuals, variables, *etc.*), and introduces the statistical analyses required. In the second section called *In practice*, the tailored statistical analyses mentioned, as well as their key outputs and interpretation, are presented in more detail. More precisely, the way such analyses are performed

---

[4] From our point of view, "forming a necessary base or core."

in R is presented step by step through real life examples. In the third section called *For experienced users*, variants or extensions of the methods presented are proposed. These variants or extensions can be related to the sensory task itself, to the statistical methodology, or (in most cases) to both. Finally, since the book is meant to be pedagogic, various exercises and recommended readings are proposed in sections four and five of each chapter. These exercises and readings aim at getting readers more accustomed to the sensory evaluations, the sensory data, the statistical analyses, and with R. These readings and exercises are used to feed the readers' curiosity, as they are useful complements to the chapters.

Although R is the tool used for the statistical analysis, we do not assume that readers are fluent in this programming language. Hence, the R code used in the various chapters is detailed and commented on throughout the book. For R-beginners, an additional section introducing R is given in the Appendix. Without the pretense of making readers R-experts, this book still should give them enough knowledge to 1) be familiar with this programming language, 2) provide a good reference list of useful functions, 3) give them the aptitude to analyze and interpret sensory data, and 4) readapt the code to their own needs.

In addition, there is a blog dedicated to the analysis of sensory data with R at http://www.sensorywithr.org. This is the website we are making reference to throughout the book, as it contains all the data sets used in the different exercises, as well as the solutions.

Acknowledgments

This book would not have been possible without the help of others. Amongst these people, we are grateful to Rob Calver, Sarah Gelson, Robin Lloyd-Starkes, and Shashi Kumar for their support, their help, and their dedication.

We are happy to thank all the students who collected almost all the data sets used in this textbook.

And, of course, we are more than happy to thank Jérôme Pagès, François Husson, and Marine Cadoret: without these main contributors in terms of methodology and development, it would have been impossible to write the book.

Last but not least, the authors would like to thank their families, in particular Betina and Virginie, for their understanding and their endless support.

Part I

Quantitative descriptive approaches

Quantitative descriptive approaches

By definition, *quantitative* means "measured by the quantity of something rather than its quality." Literally, and considering our context, the title of this first part "Quantitative descriptive approaches" encompasses all these sensory methods that seek to describe a set of products (or stimuli) by using panelists[5] as the instrument to gauge the intensity of some sensory attributes. In other words, the data that are considered are obtained from panelists who rate products (*i.e.*, assign a value according to a particular scale) according to sensory attributes. The object of this exercise is to understand the set of products regarding the sensory attributes involved in the experiment, and to obtain a description of each product: the so-called sensory profile.

Within that framework, two main situations are considered: 1) a reference situation, where the same set of sensory attributes is used by all the panelists to describe the products, and 2) an alternative situation, also quite common in the industry, where panelists use their own set of attributes to describe the products. Depending on the situation, the sensory scientist has different expectations from the data, and is confronted with different problematic and different strategies for its analysis.

For the most common situation (*i.e.*, the first one), a list of sensory attributes is imposed to the panelists. Although they are usually trained, panelists may not feel comfortable with using and rating all the attributes. Due to the uniqueness of the list of attributes, the quality of the data provided can (quite) easily be assessed, at least from a statistical point of view. In other words, the performance of the panelists (as a whole and individually) can be assessed. By doing so, we ensure that the information obtained on the products is reliable. And if we trust the panelists, we ultimately trust the data, and we also trust the information regarding the products.

In the second situation, panelists choose their own list of attributes. In that case, it is reasonable to think that they only choose attributes they feel comfortable with. Consequently, it is rational to assume that they rate properly the products regarding the attributes that they have chosen. However, this assumption may still be questionable. But due to the diversity of the lists of attributes, it is structurally (almost) impossible to assess the quality of the data, at least from a statistical point of view. Still, this situation is of utmost importance from an application point of view, and being able to solve it helps

---

[5]A member of a panel, *e.g.*, an expert, a trained panelist, a consumer.

us solve related problems, at least in terms of data structure (the so-called multiple data tables).

This first part is divided into three chapters. Two thirds of this part - the first two chapters - are dedicated to the first situation, *i.e.*, the case where panelists describe quantitatively products according to a same set of attributes. Such emphasis is justified as this situation is the reference case that every sensory scientist should master. As is the reference situation, the statistical methods (also considered as a reference) used to analyze the data are carefully explained. More precisely, Chapter 1 deals with the quality of the data, in other words, the assessment of the performance of the panelists (as a whole and individually), while Chapter 2 deals with the study of the product space. Finally, Chapter 3 deals with the case where panelists use their own set of attributes. As we will see, the statistical method used for the analysis of data presented in the last chapter also allows comparing data provided from different panels (*e.g.*, a panel made of experts and a panel made of consumers, or two panels from two different countries).

From a statistical point of view, as explained in the preamble, the most simple techniques are used and we try to present the best of them. In Chapter 1, for the assessment of the performance of the panelists, mostly Analysis of Variance (ANOVA), and particularly the so-called *omnibus* test (F-test), is used. In Chapter 2, for the representation and the understanding of the product space, exploratory multivariate methods (such as Principal Component Analysis [PCA] and Hierarchical Ascendant Classification [HAC]) are used, but not only as the t-test is also considered for the statistical significance of the estimations of the coefficients in an ANOVA model. Finally, in Chapter 3, Multiple Factor Analysis (MFA) is used for the comparison of different sensory profiles.

1

When panelists rate products according to a single list of attributes

CONTENTS

"...it is not only possible, but also recommended, to assess the performance of the panelists, as a whole and individually. By doing so, one ensures the quality of the data used for understanding the product space (cf. Chapter 2). This procedure can be done quite easily using the **panelperf** *and the* **paneliperf** *functions of the SensoMineR package."*

1.1 Data, sensory issues, and notations

Descriptive analysis methods refers to a large family of methods. It goes from the Flavor profile method to Free Choice Profiling (FCP), through Quantitative Descriptive Analysis® (QDA®) and Spectrum®. All these methods have been conceived to provide a sensory description of products that is as accurate as possible.

This sensory description is obtained by rating a set of products according to a list of sensory attributes. This evaluation is done by a peculiar measuring instrument, the human being, usually referred to as panelist or subject. To obtain the most accurate sensory description possible, the panel must be as efficient as possible, and *de facto* panelists are usually screened and trained.

In practice, with QDA (certainly the most emblematic method used in the industry), the panel is composed of 10 to 12 panelists. The training of these panelists, which aims at harmonizing and calibrating them, can take months. During the training sessions, a strategic phase (which can take up to a week) is taking place: the language development. During this phase, the panelists work together with a panel leader to generate a list of attributes used to describe the category of products they are studying. Hence, all together, the obtainment of a fine sensory description for a set of products costs money and time. Therefore, it is important to assess the performance of our measuring instrument, as a whole and individually. It is also very important (and this issue is linked to the previous one) to provide insightful and actionable information to the panel leader on the one hand, and to the panelists on the other hand. Should the panel or a panelist in particular, be further trained on an attribute?

The statistical assessment of the performance of our panelists is essentially based on three important qualities: their ability to differentiate the products (discrimination) consistently (repeatable) and consensually (in agreement). The indicators used to measure these three qualities must be actionable for the panel leader, quick to report to the panelists, and also easy to understand for them. Indeed, beyond the physical qualities good panelists must possess, they must feel implicated in the process and therefore need to be motivated. A good way to do so is to provide to panelists feedback on their assessments.

More formally, let's consider a set of I products being evaluated by K panelists in monadic sequence (*i.e.*, each panelist tests the products one by one according to a well-defined experimental design) and rated on J attributes. Without loss of generality, let's consider that all the panelists have evaluated all the I products (*i.e.*, full design) S times. In that case, the experiment generates $I \times K \times J \times S$ scores. For this experiment, the statistical unit of interest is one assessment, in other words, one of the $I \times K \times S$ different combinations of product, panelist, and session. Here, each statistical unit is associated with a combination of J scores, each score being related to one sensory attribute and to three categorical values, indicating which product is tested by which panelist, in which session. These scores are structured in a table comprising $I \times K \times S$ rows and $3 + J$ columns, one for the *product* information, one for the *panelist* information, one for the *session* information, and J for the sensory attributes. Such a way of presenting the data is very common in statistics, and we shall encounter it on many other occasions.

To assess the performance of our panel, essentially ANOVA is used. The ANOVA is a reference method applied to understand the influence of experimental factors[1] on a quantitative dependent variable[2]. In our case, the dependent variables are the sensory attributes; the experimental factors, also

---

[1] A factor of an experiment is a controlled independent variable; a variable whose levels are set by the experimenter.

[2] The dependent variables are variables to characterize or predict from other independent variables.

called independent variables, are the factors associated with the product effect, the panelist effect, the session effect, and eventually all their first-order interactions. Within this chapter, amongst others, the notions of histogram, box plot, sum of squares decomposition, hypothesis testing, *p-value*, model, interaction, and *F*-test are evoked.

1.2   In practice

The data set considered here consists of a test involving 12 perfumes, rated by 12 trained panelists twice on 12 attributes. These data were collected by Mélanie Cousin, Maëlle Penven, Mathilde Philippe, and Marie Toularhoat as part of a larger master's degree project. Data are stored in the file entitled *perfumes_qda_experts.csv*, which can be downloaded from the book website[1].

Let's import the data into R with the **read.table** function. This function is the principal means of reading tabular data into R. Its main parameters are:

- `file`, the name of the file which the data are to be read from;

- `header`, a logical value indicating whether the file contains the names of the variables in the first row;

- `sep`, the field separator character;

- `dec`, the character used in the file for decimal points;

- and eventually `row.names` which indicates the position of the vector containing the row names.

```
> experts <- read.table(file="perfumes_qda_experts.csv",header=TRUE,
+ sep=",",dec=".",quote="\"")
```

Once imported, it is important to validate the importation. To do so, the **summary** function is used. This function is a generic function that can be used in many different contexts: its primary use is for producing summaries of the results of various model fitting functions, but it is also used for getting an overview of a data set. In the latter case, for the categorical variables, it shows the different categories and their respective number of occurrences, and for the continuous variables, it shows their minimum and maximum, as well as their quartiles and means. If missing data are detected (which is not the case here), an additional row called NA (for "Not Available") mentions the number of missing values detected for the given variable.

```
> summary(experts)
```

---

[1]http://www.sensorywithr.org

```
      Panelist        Session          Rank                      Product
CM      : 24    Min.    :1.0    Min.    : 1.00   Angel              : 24
CR      : 24    1st Qu.:1.0    1st Qu.: 3.75   Aromatics Elixir   : 24
GV      : 24    Median :1.5    Median : 6.50   Chanel N5          : 24
MLD     : 24    Mean    :1.5    Mean    : 6.50   Cinéma             : 24
NMA     : 24    3rd Qu.:2.0    3rd Qu.: 9.25   Coco Mademoiselle: 24
PR      : 24    Max.    :2.0    Max.    :12.00   J'adore EP         : 24
(Other):144                                    (Other)            :144
       Spicy            Heady            Fruity            Green
Min.    : 0.000   Min.    : 0.000   Min.    : 0.000   Min.    :0.0000
1st Qu.: 0.000   1st Qu.: 0.300   1st Qu.: 0.500   1st Qu.:0.0000
Median : 0.500   Median : 1.900   Median : 2.300   Median :0.2000
Mean    : 2.249   Mean    : 3.963   Mean    : 3.487   Mean    :0.8458
3rd Qu.: 3.625   3rd Qu.: 8.125   3rd Qu.: 6.125   3rd Qu.:0.7000
Max.    :10.000   Max.    :10.000   Max.    :10.000   Max.    :9.8000

      Vanilla           Floral           Woody            Citrus
Min.    : 0.000   Min.    : 0.000   Min.    :0.000   Min.    :0.000
1st Qu.: 0.200   1st Qu.: 3.775   1st Qu.:0.000   1st Qu.:0.000
Median : 1.200   Median : 6.800   Median :0.500   Median :0.200
Mean    : 3.113   Mean    : 6.119   Mean    :1.252   Mean    :1.049
3rd Qu.: 6.000   3rd Qu.: 9.000   3rd Qu.:1.525   3rd Qu.:1.125
Max.    :10.000   Max.    :10.000   Max.    :9.700   Max.    :7.800

      Marine           Greedy           Oriental         Wrapping
Min.    :0.0000   Min.    : 0.000   Min.    : 0.000   Min.    : 0.000
1st Qu.:0.0000   1st Qu.: 0.100   1st Qu.: 0.100   1st Qu.: 2.775
Median :0.0000   Median : 1.000   Median : 1.200   Median : 6.300
Mean    :0.4674   Mean    : 2.924   Mean    : 3.699   Mean    : 5.737
3rd Qu.:0.2000   3rd Qu.: 5.100   3rd Qu.: 8.000   3rd Qu.: 8.900
Max.    :9.5000   Max.    :10.000   Max.    :10.000   Max.    :10.000
```

A first overview of the data shows that the variables *Session* and *Rank* are considered as continuous variables. Since these columns correspond, respectively, to the session information and to the order of presentation, they should be considered as factors.

To transform these columns from numerical to categorical, the **as.factor** function is applied on these two variables. To do so, we use the $ symbol, which is very important in R, as it allows specifying a particular component of an R object (*cf.* Appendix). In our case, `experts$Session` refers to the variable *Session* of the *experts* data set.

```
> experts$Session <- as.factor(experts$Session)
> experts$Rank <- as.factor(experts$Rank)
> summary(experts)
      Panelist    Session    Rank                   Product            Spicy
CM      : 24    1:144    1    : 24   Angel              : 24   Min.    : 0.000
CR      : 24    2:144    2    : 24   Aromatics Elixir : 24   1st Qu.: 0.000
GV      : 24             3    : 24   Chanel N5          : 24   Median : 0.500
MLD     : 24             4    : 24   Cinéma             : 24   Mean    : 2.249
NMA     : 24             5    : 24   Coco Mademoiselle: 24   3rd Qu.: 3.625
PR      : 24             6    : 24   J'adore EP         : 24   Max.    :10.000
(Other):144           (Other):144   (Other)            :144
```

```
        Heady                Fruity               Green               Vanilla
Min.    : 0.000    Min.    : 0.000    Min.     :0.0000    Min.    : 0.000
1st Qu.: 0.300    1st Qu.: 0.500    1st Qu.:0.0000    1st Qu.: 0.200
Median : 1.900    Median : 2.300    Median :0.2000    Median : 1.200
Mean    : 3.963    Mean    : 3.487    Mean     :0.8458    Mean    : 3.113
3rd Qu.: 8.125    3rd Qu.: 6.125    3rd Qu.:0.7000    3rd Qu.: 6.000
Max.    :10.000    Max.    :10.000    Max.     :9.8000    Max.    :10.000

        Floral               Woody               Citrus              Marine
Min.    : 0.000    Min.    :0.000    Min.     :0.000    Min.     :0.0000
1st Qu.: 3.775    1st Qu.:0.000    1st Qu.:0.000    1st Qu.:0.0000
Median : 6.800    Median :0.500    Median :0.200    Median :0.0000
Mean    : 6.119    Mean    :1.252    Mean     :1.049    Mean     :0.4674
3rd Qu.: 9.000    3rd Qu.:1.525    3rd Qu.:1.125    3rd Qu.:0.2000
Max.    :10.000    Max.    :9.700    Max.     :7.800    Max.     :9.5000

        Greedy               Oriental            Wrapping
Min.    : 0.000    Min.    : 0.000    Min.     : 0.000
1st Qu.: 0.100    1st Qu.: 0.100    1st Qu.: 2.775
Median : 1.000    Median : 1.200    Median : 6.300
Mean    : 2.924    Mean    : 3.699    Mean     : 5.737
3rd Qu.: 5.100    3rd Qu.: 8.000    3rd Qu.: 8.900
Max.    :10.000    Max.    :10.000    Max.     :10.000
```

1.2.1   What basic information can I draw from the data?

From the output of the **summary** function, we can see that the experimental design used to collect the data seems balanced: in the data set, there are 24 occurrences for each panelist and each product, and there are 144 occurrences for each session.

To be more precise we can cross, for instance, the two variables *Panelist* and *Product* to make sure that each panelist has tested twice each of the 12 products. To do so, we use the **table** function as follows, to produce a contingency table:

```
> table(experts$Product,experts$Panelist)
```

```
                   CM CR GV MLD NMA PR RL SD SM SO SQ ST
Angel               2  2  2   2   2  2  2  2  2  2  2  2
Aromatics Elixir    2  2  2   2   2  2  2  2  2  2  2  2
Chanel N5           2  2  2   2   2  2  2  2  2  2  2  2
Cinéma              2  2  2   2   2  2  2  2  2  2  2  2
Coco Mademoiselle   2  2  2   2   2  2  2  2  2  2  2  2
J'adore EP          2  2  2   2   2  2  2  2  2  2  2  2
J'adore ET          2  2  2   2   2  2  2  2  2  2  2  2
L'instant           2  2  2   2   2  2  2  2  2  2  2  2
Lolita Lempicka     2  2  2   2   2  2  2  2  2  2  2  2
Pleasures           2  2  2   2   2  2  2  2  2  2  2  2
Pure Poison         2  2  2   2   2  2  2  2  2  2  2  2
Shalimar            2  2  2   2   2  2  2  2  2  2  2  2
```

The output speaks for itself: each panelist has evaluated twice each product.

Beyond those numerical indicators, data can also be visualized graphically. To visualize how a sensory attribute was used by the panel, we can plot a histogram of this attribute. To generate histograms, the **hist** function is used. The main input of this function is the vector of numbers corresponding to the variable to represent. Let's consider the sensory attribute *Wrapping*: to generate its histogram, the **hist** function is called and applied on `experts$Wrapping`.

To enhance the histogram, the estimation of the distribution of the attribute in question is added. This is done by using the **lines** function combined with the **density** function.

```
> hist(experts$Wrapping,col="lightgray",border="gray",proba=TRUE)
> lines(density(experts$Wrapping),lwd=2,col="darkgray"))
```

These commands produce Figure 1.1.

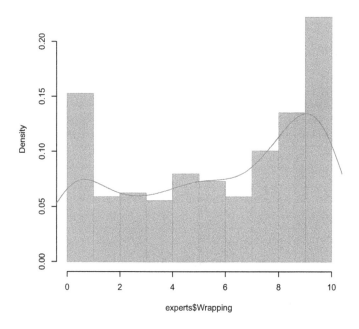

FIGURE 1.1
Histogram combined with the estimation of the distribution of the attribute *Wrapping* obtained with the **hist** and **density** functions (*experts* data set).

Similarly, to plot the box plot of a given sensory attribute, the **boxplot** function is used. This function also takes as input a vector of numerical values. This following command produces Figure 1.2.

```
> boxplot(experts$Wrapping,col="lightgray",border="gray",main="Wrapping")
```

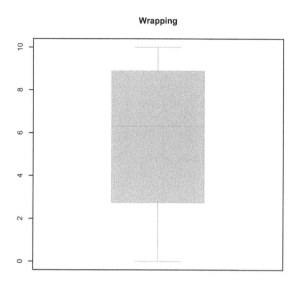

FIGURE 1.2
Box plot of the attribute *Wrapping* with the **boxplot** function (*experts* data set).

Alternatively, the box plots of a given sensory attribute can be generated conditionally to a categorical variable. This requires the introduction of the concept of model that will be used throughout the book, in which a dependent variable is explained by several independent variables. In that case, the command `Wrapping~Product` expresses that the analysis of the dependent variable *Wrapping* is a function of the independent variable *Product*. In other words, this is very similar to explaining the attribute *Wrapping* by the factor associated with the *Product* effect.

```
> boxplot(Wrapping~Product,data=experts,col="lightgray",main="Wrapping")
```

This command produces Figure 1.3, where differences amongst the products can be seen for the attribute *Wrapping*: the panelists have certainly differentiated the products for that sensory attribute.

The summary and the graphical representations of the data allow assessing the overall quality of the data. Note that the graphical representations proposed here are minimal with respect to R possibilities. Many more graphics exist in R, and some more examples are provided in the Appendix. To

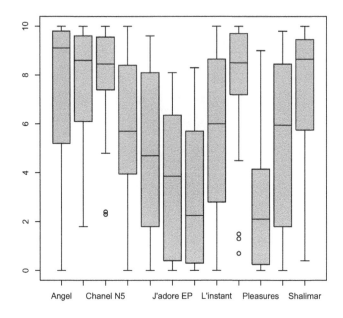

FIGURE 1.3
Box plots of the attribute *Wrapping* per product obtained with the **boxplot**
function (*experts* data set).

evaluate the performance of the panelists, as a whole and individually, some
more advanced methods are required. These methods and their results are
presented in the following sections.

1.2.2 How can I assess the performance of my panel?

The performance of the panel is assessed for each sensory attribute separately.
To do so, an Analysis of Variance (ANOVA) is performed on each sensory
attribute.

ANOVA is a statistical method "for partitioning the overall variability
in a set of observations into components due to specified influences and to
random error (...). The resulting ANOVA table provides a concise summary
of the structure of the data and a descriptive picture of the different sources
of variation" (Chatfield, 1992).

In our example, we consider $I = 12$ perfumes, $K = 12$ panelists, and
$S = 2$ sessions. In other words, each panelist has rated each perfume twice on

12 sensory attributes. We want to understand the variability of each sensory attribute (dependent variable) regarding different sources, such as the fact that different products were evaluated, by different panelists, during different sessions (independent variables). Let's first consider the following ANOVA model that we apply systematically to all the sensory attributes later on:

$$Y_{iks} \sim \mu + \alpha_i + \beta_k + \gamma_s + \alpha\beta_{ik} + \alpha\gamma_{is} + \beta\gamma_{ks} + \epsilon_{iks},$$

where

- α_i is the ith coefficient associated with the *Product* effect;

- β_k is the kth coefficient associated with the *Panelist* effect;

- γ_s is the sth coefficient associated with the *Session* effect;

- $\alpha\beta_{ik}$ is the ik^{th} coefficient associated with the *Product-Panelist* interaction;

- $\alpha\gamma_{is}$ is the is^{th} coefficient associated with the *Product-Session* interaction;

- $\beta\gamma_{ks}$ is the ks^{th} coefficient associated with the *Panelist-Session* interaction;

- and ϵ_{iks} denotes the error term.

In the model, the errors are assumed to be normally distributed with mean zero and constant variance, σ^2, and also to be independent:

- $\forall(i, k, s), \epsilon_{iks} \sim N(0, \sigma^2)$;

- $\forall(i, k, s) \neq (i', k', s'), \mathrm{Cov}(\epsilon_{iks}, \epsilon_{i'k's'}) = 0$.

Let's have a look at the meaning of the different effects introduced in the model (main effects and interaction effects):

- The *Product* effect indicates whether the products are perceived as different on that attribute. In other words, if the *Product* effect is significant, the panel has discriminated the products with respect to the sensory attribute of interest (the one we want to explain in the model). This effect is of main interest when assessing the performance of a panel: it corresponds to the discrimination ability of the panel.

- The *Panelist* effect indicates whether panelists use the scale of notation similarly or not. This effect is of less interest.

- The *Session* effect indicates whether the scale of notation is used consistently from one session to the other. This effect is also of less interest.

- The *Product-Panelist* interaction indicates whether products are perceived similarly by the different panelists. In other words, it indicates whether there is a consensus amongst the panelists while rating the products on the attribute of interest. If the *Product-Panelist* interaction is significant, no consensus amongst the panelists (within the panel) is observed: the panelists do not have the same perception of the products with respect to the sensory attribute of interest. This interaction is hence of main interest while assessing the performance of a panel: it corresponds to the agreement between panelists.

- The *Product-Session* interaction indicates whether products are perceived similarly from one session to the other. For a given sensory attribute of interest, it indicates whether the attribute is used similarly from one session to the other. If the *Product-Session* interaction is significant, the panel is not repeatable from one session to the other. This interaction is also of main interest while assessing the performance of a panel: it measures the repeatability of the panelists.

- The *Panelist-Session* interaction indicates whether some panelists use the scale of notation differently, from one session to the other. This effect is of less interest.

Let's consider the sensory attribute *Marine*, and let's fit on that attribute an ANOVA model, by testing the *Product* effect, *Panelist* effect, *Session* effect, and their respective 2-way interactions (also called first-order interactions). To do so, the **aov** function is used. The results of this ANOVA model thus obtained are shown using the **summary.aov** function (or its generic form **summary**).

```
> res.aov <- aov(Marine~Product+Panelist+Session+Product:Panelist
+ +Product:Session+Panelist:Session,data=experts)
> summary.aov(res.aov)
                  Df Sum Sq Mean Sq F value   Pr(>F)
Product           11  35.41   3.219   2.219 0.017453 *
Panelist          11  67.12   6.102   4.206 2.84e-05 ***
Session            1  21.02  21.017  14.485 0.000223 ***
Product:Panelist 121 172.08   1.422   0.980 0.543772
Product:Session   11  35.56   3.233   2.228 0.016950 *
Panelist:Session  11  79.22   7.202   4.964 2.46e-06 ***
Residuals        121 175.56   1.451
---
Signif. codes:  0 '***' 0.001 '**' 0.01 '*' 0.05 '.' 0.1 ' ' 1
```

These results are of utmost importance as they indicate which sources of variability are important to consider to explain the overall variability. The importance of a source of variability is assessed by the F-test. The test statistic in an F-test is the ratio of two scaled sums of squares[2] associated with different

---

[2]Sums of squares divided by their respective degrees of freedom. With respect to the output, the elements of the third column (Mean Sq) are obtained by dividing those from the second column (Sum Sq) by those from the first column (Df).

sources of variability. For example, the test statistic of the *Product* effect is calculated using the following formula:

$$F_{statistic} = \frac{SS_{Product}/df_{Product}}{SS_{Residual}/df_{Residual}}.$$

Under the null hypothesis (*i.e.*, the source of variability considered does not explain the overall variability), the test statistic follows an F-distribution. To each source of variability, we can associate a *p-value* that indicates whether the source should be considered to explain the overall variability.

As we can see, for the sensory attribute *Marine* and for a significance level of 0.05 (=5%):

- With a *p-value* of 0.017, the *Product* effect is significant; the panel differentiated between products, which is what we should expect from a trained panel.

- With a *p-value* of 0.544, the *Product-Panelist* interaction is not significant; within the panel, there's a consensus amongst panelists for that sensory attribute, which is also what we should expect from a sensory panel.

- With a *p-value* of 0.017, the *Product-Session* interaction is significant; the panel, as a whole, is not repeatable from one session to the other, which is not what we should expect from a sensory panel.

For a better understanding of the notion of interaction, let's use the **graphinter** function of the SensoMineR package. For a given couple of factors $A = (\alpha_i)_{i \in I}$ and $B = (\beta_j)_{j \in J}$, this function plots the means for each level $\alpha\beta_{ij}$ of the interaction effect AB in an intuitive graphical output (*cf.* Figure 1.4). This function takes as arguments the data set of interest (*experts*), the positions of the factors of interest within the data set (4 for the *Product* effect, 2 for the *Session* effect), as well as the position of the sensory attribute of interest, in our case *Marine* (13). Additional graphical arguments (related to the number of graphical output per window to plot, here 1) can be set. As **graphinter** belongs to the SensoMineR package, the latter needs to be loaded using the **library** function.

```
> library(SensoMineR)
> graphinter(experts,col.p=4,col.j=2,firstvar=13,lastvar=13,numr=1,numc=1)
```

The x-axis of this output represents the average values of the products calculated over all sessions, whereas the y-axis represents the average values of the products calculated for each session. When there is no interaction effect between products and sessions, the two broken lines are parallel (one line per session). On the contrary, when there is a significant interaction, the two lines split and/or cross; in our case they split.

As mentioned previously, the fact that the *Product-Session* interaction is significant expresses a lack of repeatability of the panelists for the attribute

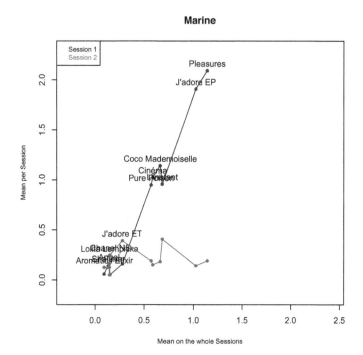

FIGURE 1.4
Visualization of the *Product-Session* interaction for the attribute *Marine* with the **graphinter** function (*experts* data set).

Marine. In our case, the previous graphical output shows which products have contributed the most to this interaction. The important notion of interaction can also be understood with the **interact** function of the SensoMineR package (for more information concerning this function, *cf.* Section 1.4).

Now that the performance of the panel is assessed for one attribute in particular, let's see how those results can be automatically generated for all the other attributes simultaneously. A simple solution consists in using the **panelperf** function of the SensoMineR package, as it performs automatically an ANOVA on all the continuous variables of a given data set. In such case, it is mandatory to specify one unique model for all the attributes. Here, the model used is the one defined previously. Before applying **panelperf** to the *experts* data set, other additional arguments are required. Amongst these arguments, the position of the first sensory attribute of interest (`firstvar = 5`) needs to be informed.

Once set, let's run the function and store the results in an object called arbitrarily `res.panelperf`. In order to understand the structure of the results provided by **panelperf**, the **names** function is applied on `res.panelperf`. This function is a generic function to get or set the names of an object.

```
> res.panelperf <- panelperf(experts,firstvar=5,formul="~Product+Panelist+
+ Session+Product:Panelist+Product:Session+Panelist:Session")
> names(res.panelperf)
[1] "p.value"     "variability" "res"         "r2"
```

The most important output of **panelperf** is in its first component, called *p.value*. This output contains a matrix in which the rows correspond to the sensory attributes of the data set, the columns to the effects tested in the ANOVA model, and in which each element corresponds to the *p-value* associated with the F-test of an effect for a given attribute.

```
> res.panelperf$p.value
```

	Product	Panelist	Session	Product:Panelist
Spicy	1.996468e-22	1.648251e-12	0.891091356	6.576338e-04
Heady	4.307737e-31	1.702560e-15	0.644266574	1.437929e-07
Fruity	2.769452e-13	3.703762e-09	0.936755834	8.841554e-05
Green	2.585917e-09	8.402604e-09	0.066405749	1.502685e-04
Vanilla	9.030883e-23	8.552791e-10	0.230450643	7.707797e-07
Floral	1.075929e-19	4.324579e-16	0.026866379	3.138436e-04
Woody	1.622612e-06	6.315267e-11	0.382461653	3.666210e-02
Citrus	8.308177e-02	1.798865e-09	0.155682764	2.213355e-06
Marine	1.519361e-02	2.842584e-05	0.115613651	5.437725e-01
Greedy	1.146599e-30	6.125002e-05	0.490598620	2.958274e-07
Oriental	2.682067e-12	5.350357e-18	0.142287536	1.473964e-12
Wrapping	4.458847e-11	3.471630e-17	0.007782676	1.794984e-07

	Product:Session	Panelist:Session
Spicy	0.21284234	1.516094e-01
Heady	0.35287310	1.001730e-01
Fruity	0.90330674	7.592500e-01
Green	0.87265815	7.346610e-01
Vanilla	0.56817260	2.814368e-02
Floral	0.03750445	1.845189e-01
Woody	0.55521775	5.019211e-02
Citrus	0.56432959	5.177717e-01
Marine	0.01694954	2.457826e-06
Greedy	0.61492619	9.706534e-04
Oriental	0.98061658	9.238796e-04
Wrapping	0.40350391	4.647913e-01

For a better visualization of the results, the **coltable** function of the SensoMineR package is used. This function represents graphically a table, and colors the cells with respect to thresholds (and colors) that can be specified within the function. By default, the cells are highlighted in pink when values are lower that 0.05 (significance level); this can be changed using the `col.lower` argument. Figure 1.5 corresponds to the result of **coltable** applied to the matrix `res.panelperf$p.value`. Moreover, results of the ANOVAs have been sorted according to the significance of the *Product* effect. To do so, the **order** function is applied to the column of the matrix that needs to be sorted.

```
> coltable(res.panelperf$p.value[order(res.panelperf$p.value[,1]),],
+ col.lower="gray")
```

	Product	Panelist	Session	Product:Panelist	Product:Session	Panelist:Session
Heady	4.308e−31	1.703e−15	0.6443	1.438e−07	0.3529	0.1002
Greedy	1.147e−30	6.125e−05	0.4906	2.958e−07	0.6149	0.0009707
Vanilla	9.031e−23	8.553e−10	0.2305	7.708e−07	0.5682	0.02814
Spicy	1.996e−22	1.648e−12	0.8911	0.0006576	0.2128	0.1516
Floral	1.076e−19	4.325e−16	0.02687	0.0003138	0.0375	0.1845
Fruity	2.769e−13	3.704e−09	0.9368	8.842e−05	0.9033	0.7593
Oriental	2.682e−12	5.35e−18	0.1423	1.474e−12	0.9806	0.0009239
Wrapping	4.459e−11	3.472e−17	0.007783	1.795e−07	0.4035	0.4648
Green	2.586e−09	8.403e−09	0.06641	0.0001503	0.8727	0.7347
Woody	1.623e−06	6.315e−11	0.3825	0.03666	0.5552	0.05019
Marine	0.01519	2.843e−05	0.1156	0.5438	0.01695	2.458e−06
Citrus	0.08308	1.799e−09	0.1557	2.213e−06	0.5643	0.5178

FIGURE 1.5
Assessment of the performance of the panel with the **panelperf** and the **coltable** functions (*experts* data set).

Figure 1.5 shows that the panel discriminates between the products for all the sensory attributes, except *Citrus*. It shows also that panelists have particularly well differentiated the products, considering the singular values taken by the *p-values* of the *Product* effect (remarkably small). Problematically, for the *Product-Panelist* interaction, there is no consensus amongst the panelists, except for *Marine*. When looking at the *Product-Session* interaction, the panel is repeatable from one session to the other for all the sensory attributes, except for *Floral* and *Marine*.

One main debate amongst sensory scientists is the way the *Panelist* effect should be considered: should it be considered as *fixed* or as *random*? When the *Panelist* effect is considered as fixed, it means that the effect of this factor is totally linked to the levels of the factor. In other words, the focus is on these (particular) panelists specifically. When the *Panelist* effect is considered as random, it means that the effect of this factor is not strictly linked to the levels of the factor. In other words, the attention is in what panelists, in general, can express, and not in these (particular) panelists, specifically.

By default, in the SensoMineR package, *p-values* are calculated by considering the *Panelist* effect as random. This explains why the *p-values* associated with the attribute *Marine* are not the same when running the **aov** function (in that case the *Product* effect is considered as fixed) and when running **panelperf**. This option can be easily modified by changing the **random** argument to FALSE instead of TRUE. It is important to keep in mind that the interpretation of the results is different from one case to the other, as the intrinsic nature of the test is different. Beyond those considerations, the calculation of the *p-values* depends, as well, on the nature of the effect, whether it is considered as random or not. For instance, when the *Panelist* effect is considered as random, the statistic test for the *Product* effect is calculated the following way (for the statistical test of the *Session* effect, replace *Product* by *Session* in the formula):

$$F_{statistic} = \frac{SS_{Product}/df_{Product}}{SS_{Product:Panelist}/df_{Product:Panelist}}.$$

Tips: Random effects and mixed models

In **panelperf**, it is possible to define whether the *Panelist* and *Session* effects should be considered as fixed or random (through the **random** argument). This latter situation is the most frequent one in sensory analysis. However, it is not possible to consider, for instance, the *Panelist* effect as random and the *Session* effect as fixed.

In order to gain in flexibility, other R functions can be used. The main package proposing functions that allows integrating mixed models is the lme4 package (Bates, Maechler, Bolker, & Walker, 2013). The **lmer** function is then used. Although this function is more flexible, this package has the drawback not to provide all useful results. For instance, the *p-values* associated with the different effects are not provided. For that reason, we prefer to use the **lmer** function from the lmerTest package (Kuznetsova, Brockhoff, & Christensen, 2013). This function is an extension of the **lmer** function of the lme4 package, but with additional results. To set a factor to random, the code (1|random factor) should be used. The results of the analysis are provided by the **anova** function (fixed effects) and by the **rand** function (random effects).

```
> library(lmerTest)
> mixedmodel <- lmer(Fruity~Product+Session+Product:Session+
+ (1|Panelist)+(1|Product:Panelist)+(1|Panelist:Session),data=experts)
> anova(mixedmodel)
Analysis of Variance Table of type 3 with
Satterthwaite  approximation for degrees of freedom
```

```
            Df Sum Sq Mean Sq F value  Denom    Pr(>F)
Product         11 503.43  45.767 10.4719 120.98 2.773e-13 ***
Session          1   0.02   0.020  0.0046 131.99    0.9462
Product:Session 11  24.46   2.223  0.5087 131.99    0.8947
---
Signif. codes:  0 '***' 0.001 '**' 0.01 '*' 0.05 '.' 0.1 ' ' 1
> rand(mixedmodel)
Analysis of Random effects Table:
                        Chi.sq Chi.DF p.value
(1 | Panelist)            9.74      1   0.002 **
(1 | Product:Panelist)   17.53      1   3e-05 ***
(1 | Panelist:Session)    0.00      1   1.000
---
Signif. codes:  0 '***' 0.001 '**' 0.01 '*' 0.05 '.' 0.1 ' ' 1
```

A selection of the best model can be obtained through the **step** function. In this function, it is possible to decide whether the nonsignificant fixed and/or random effects should be discarded from the model through the `reduce.fixed` and `reduce.random` arguments.

```
> mm.step <- step(mixedmodel,reduce.fixed=TRUE,reduce.random=TRUE)
> mm.step
Random effects:
                        Chi.sq Chi.DF elim.num p.value
(1 | Panelist)            12.5      1        0 4e-04 ***
(1 | Product:Panelist)    17.5      1        0 3e-05 ***
(1 | Panelist:Session)     0.0      1        1     1
---
Signif. codes:  0 '***' 0.001 '**' 0.01 '*' 0.05 '.' 0.1 ' ' 1

Fixed effects:
                Sum Sq Mean Sq NumDF DenDF F.value elim.num    Pr(>F)
Product         503.43   45.77    11   121 10.47188        0 2.77e-13 ***
Session           0.02    0.02     1   143  0.00476        2    0.945
Product:Session  24.46    2.22    11   132  0.50874        1    0.895
---
Signif. codes:  0 '***' 0.001 '**' 0.01 '*' 0.05 '.' 0.1 ' ' 1
```

Using the **plot** function on the outputs of the **step** function (*e.g.*, `plot(mm.step)`), the results of the paired comparisons can be graphically represented (not shown here). This graphic is color-coded based on the significance of the difference between each pair.

1.2.3   How can I assess the performance of my panelists?

Similar results can be obtained at the panelist level. To do so, the same strategy as in the assessment of the panel performance is used. However, since the objective is slightly different, the ANOVA model needs to be adapted to the panelist. Let's consider the following simple ANOVA model that we apply systematically to all panelists, and for each sensory attribute:

$$Y_{is} \sim \mu + \alpha_i + \gamma_s + \epsilon_{is},$$

where α_i is the ith coefficient associated with the product effect, γ_s is the sth coefficient associated with the session effect, and ϵ_{is} denotes the residual. Moreover, we suppose that:

- $\epsilon_{is} \sim N(0, \sigma^2)$;

- $\forall (i, s) \neq (i', s'), Cov(\epsilon_{is}, \epsilon_{i's'}) = 0$.

In the same manner as for **panelperf**, the **paneliperf** function of the SensoMineR package produces the results of two ANOVA models applied to all the continuous variables of a given data set: the first ANOVA applies the first model to the panel, while the second ANOVA applies the second model to each panelist. Hence, the first model considered is the same one as in the previous part, whereas the second model is `formul.j="~Product+Session"`. Like for **panelperf**, **paneliperf** requires additional information about the data: the argument `col.j` informs about the position of the panelist, whereas `firstvar` locates the position of the first continuous variable within the data set.

Let's apply **paneliperf** to our data set. We store the results of this analysis in an object arbitrarily called `res.paneliperf`. As previously, the names of the objects stored in `res.paneliperf` are printed by using the **names** function.

```
> res.paneliperf <- paneliperf(experts,formul="~Product+Panelist+Session+
+ Product:Panelist+Product:Session+Panelist:Session",
+ formul.j="~Product+Session",col.j=1,firstvar=5,synthesis=TRUE)
> names(res.paneliperf)
 [1] "prob.ind"   "vtest.ind"   "res.ind"    "r2.ind"      "signif.ind"
 [6] "agree.ind"  "complete"    "p.value"    "variability" "res"
```

The most important outputs of **paneliperf** are stored in its first, third, and sixth components respectively called `prob.ind`, `res.ind`, and `agree.ind`. These three elements contain matrices that present in rows the panelists and in columns the sensory attributes.

In the matrix `prob.ind`, the *p-values* associated with the *F*-test of the *Product* effect of the ANOVAs performed on each panelist are stored. By construction, the test statistic of the *Product* effect evaluates the global performance of each panelist, as

$$F_{statistic} = \frac{SS_{Product}/df_{Product}}{SS_{Residual}/df_{Residual}}.$$

When the numerator is high, the panelist has well discriminated the products. When the denominator is low, the panelist is repeatable. In other words, when the numerator is high and the denominator is low, then the ratio is high; consequently, the *p-value* associated with the *F*-test is small, and the panelist is performing well.

In the matrix `res.ind`, the residuals associated with the ANOVAs performed at the panelist level are stored. There's no possible test for analyzing the repeatability of the panelists individually. Still, we can compare them attribute per attribute to the standard deviation of the residual.

In the matrix `agree.ind`, the correlation coefficients between the adjusted means for the *Product* effect calculated for each panelist, on the one hand, and for the panel on the other hand are stored. This matrix corresponds obviously to the notion of consensus between each panelist and the whole panel.

Those matrices can be sorted by rows and by columns, simultaneously, using the **magicsort** function of the SensoMineR package. By using the argument `method` set to `"median"`, this function (1) calculates the median over each row and each column and (2) uses this information to sort the rows and the columns of the table.

```
> res.magicsort <- magicsort(res.paneliperf$res.ind,method="median")
> round(res.magicsort,3)
```

	Marine	Green	Citrus	Woody	Greedy	Spicy	Vanilla
CM	0.176	0.895	0.928	0.512	1.036	1.347	1.852
SM	1.101	1.119	0.685	0.668	0.864	0.963	1.182
CR	0.866	0.133	1.192	1.112	1.590	0.965	0.934
SO	1.322	0.828	1.991	1.124	1.046	1.091	1.103
PR	0.415	0.919	1.255	1.237	1.037	1.322	1.561
ST	0.000	1.421	0.041	1.740	0.898	1.384	2.842
SQ	0.564	0.376	1.578	0.800	1.731	1.580	1.598
SD	0.051	2.183	0.795	1.703	1.615	1.675	1.230
RL	0.055	1.229	1.687	1.402	1.124	2.287	2.605
GV	1.231	0.145	2.009	1.626	2.737	2.152	2.076
MLD	2.266	0.608	0.845	0.329	1.712	2.261	1.030
NMA	2.882	1.759	0.082	2.444	2.215	3.367	2.165
median	0.715	0.907	1.060	1.180	1.357	1.482	1.580

	Floral	Heady	Oriental	Wrapping	Fruity	median
CM	0.667	0.613	0.817	1.895	1.963	0.911
SM	1.547	0.693	0.968	1.652	2.479	1.034
CR	1.535	2.348	1.989	2.025	1.121	1.157
SO	0.814	1.625	2.329	2.346	1.968	1.223
PR	0.904	1.116	2.199	1.329	1.744	1.246
ST	2.843	2.003	0.843	1.659	1.437	1.429
SQ	1.661	1.610	0.529	1.865	1.156	1.579
SD	2.357	1.556	0.196	2.795	2.057	1.645
RL	3.173	1.281	2.427	1.651	3.060	1.669
GV	2.298	1.813	1.394	1.804	2.529	1.911
MLD	2.147	2.411	2.354	2.157	0.755	1.930
NMA	1.078	2.095	3.570	2.088	3.081	2.190
median	1.604	1.618	1.691	1.880	1.966	1.504

For a better visualization of the results, the **coltable** function is applied to the table stored in `res.magicsort`, and rounded to 2 decimals using the **round** function. In this case, since the threshold of 0.05 is only meaningful for the `prob.ind` table, other arbitrary thresholds need to be defined. Here, we consider 1 as lower level (`level.lower=1`) and 2 as higher level (`level.upper=2`). In other words, if the standard deviation of the residual is lower than 1, cells are colored in "gainsboro"; if the standard deviation of the residual is higher than 2, cells are colored in "dark gray." Those thresholds can be determined by plotting the distribution of all standard deviations (based on all possible ANOVAs, for all panelists and all sensory attributes).

```
> coltable(round(res.magicsort,2),level.lower=1,level.upper=2,
+ col.lower="gainsboro",col.upper="gray")
```

Figure 1.6 corresponds to the result of **coltable** applied to the matrix `res.paneliperf$res.ind`. It shows clearly which panelists are repeatable (*e.g.*, *CM* or *SM*) and which are not (*e.g.*, *MLD* or *NMA*). It also shows which sensory attributes are easy to assess, from a repeatability point of view (*e.g.*, *Marine* or *Green*), and which are not (*e.g.*, *Wrapping* or *Fruity*).

	Marine	Green	Citrus	Woody	Greedy	Spicy	Vanilla	Floral	Heady	Oriental	Wrapping	Fruity	median
CM	0.18	0.89	0.93	0.51	1.04	1.35	1.85	0.67	0.61	0.82	1.9	1.96	0.91
SM	1.1	1.12	0.68	0.67	0.86	0.96	1.18	1.55	0.89	0.97	1.65	2.48	1.03
CR	0.87	0.13	1.19	1.11	1.59	0.96	0.93	1.53	2.35	1.99	2.03	1.12	1.16
SO	1.32	0.83	1.99	1.12	1.05	1.09	1.1	0.81	1.63	2.33	2.35	1.97	1.22
PR	0.42	0.92	1.25	1.24	1.04	1.32	1.56	0.9	1.12	2.2	1.33	1.74	1.25
ST	0	1.42	0.04	1.74	0.9	1.38	2.84	2.84	2	0.84	1.66	1.44	1.43
SQ	0.56	0.38	1.58	0.8	1.73	1.58	1.8	1.66	1.81	0.53	1.86	1.16	1.58
SD	0.06	2.18	0.79	1.7	1.61	1.68	1.23	2.36	1.56	0.2	2.79	2.06	1.64
RL	0.06	1.23	1.69	1.4	1.12	2.29	2.61	3.17	1.28	2.43	1.65	3.06	1.67
GV	1.23	0.14	2.91	1.63	2.74	2.15	2.06	2.3	1.81	1.39	1.8	2.53	1.91
MLD	2.27	0.61	0.84	0.33	1.71	2.26	1.03	2.15	2.41	2.35	2.16	0.76	1.93
NMA	2.88	1.76	0.08	2.44	2.21	3.37	2.17	1.08	2.1	3.57	2.09	3.08	2.19
median	0.72	0.91	1.06	1.18	1.36	1.48	1.58	1.6	1.62	1.69	1.88	1.97	1.5

FIGURE 1.6
Assessment of the repeatability of each panelist with the **paneliperf** and the **coltable** functions (*experts* data set).

1.3 For experienced users: Measuring the impact of the presentation order on the perception of the products?

Sensory tasks, in which products are presented to the panelists in monadic sequence (*e.g.*, QDA), require specific designs. Indeed, it is often assumed that the first product is perceived as more intense than the other products. This phenomena is also known as *first-order* effect. Additionally, if the product space presents some *strong* products, the perception of the products tested right after these products can be influenced by them. This phenomena is also known as *carry-over* effect. To balance these undesired effects, each product should be tested equally at each position, and should follow equally each of the other products. Such particularities should be considered in the experimental design, which should equally balance the presentation order and carry-over effects across panelists. It is not under the scope of this chapter to explain how to generate such designs. However, for those interested, it is worth mentioning that some packages in R are dedicated to that (*e.g.*, AlgDesign, crossdes, SensoMineR to a lesser extent).

A methodology that studies the impact of the presentation order and/or carry-over effect on the perception of the perfumes is presented here. This methodology requires the information related to the order of presentation for each panelist. Such information is stored in the third column (called *Rank*) of the data set *experts*. To evaluate the *Rank* effect, let's consider the same ANOVA model as previously, in which the presentation order effect is added:

$$Y_{ikso} \sim \mu + \alpha_i + \beta_k + \gamma_s + \omega_o + \alpha\beta_{ik} + \alpha\gamma_{is} + \beta\gamma_{ks} + \epsilon_{ikso},$$

where

- α_i is the ith coefficient associated with the *Product* effect;

- β_k is the kth coefficient associated with the *Panelist* effect;

- γ_s is the sth coefficient associated with the *Session* effect;

- ω_o is the oth coefficient associated with the *Rank* effect;

- $\alpha\beta_{ik}$ is the coefficient associated with the *Product-Panelist* interaction;

- $\alpha\gamma_{is}$ is the coefficient associated with the *Product-Session* interaction;

- $\beta\gamma_{ks}$ is the coefficient associated with the *Panelist-Session* interaction;

- and ϵ_{ikso} denotes the residual.

In the model, we suppose that the residual term has the following properties:

- $\epsilon_{ikso} \sim N(0, \sigma^2)$;

- $\forall (i, k, s, o) \neq (i', k', s', o'), Cov(\epsilon_{ikso}, \epsilon_{i'k's'o'}) = 0.$

In this ANOVA, the effects have the same meaning as previously. Additionally, the *Rank* effect indicates whether the products are perceived as different depending on the position in which they are tested. In other words, if the *Rank* effect is significant, the panel has perceived the products differently depending on the position in which they are tested, for the sensory attribute of interest. For its commodity, let's use the **panelperf** function to evaluate the order effect. The code is the same as previously used, except that the *Rank* effect is added in the model.

```
> res.order <- panelperf(experts,firstvar=5,formul="~Product+Panelist+Session+
+ Rank+Product:Panelist+Product:Session+Panelist:Session")
> coltable(round(res.order$p.value,3),col.lower="gray")
```

The results (*cf.* Figure 1.7) show that 7 out of 12 attributes are associated with a significant *Rank* effect. Such results comfort us in having used a well-balanced design.

In practice, it is often assumed that the first product tested is perceived as more intense compared to the other products. In other words, the first order is a strong contributor to the significance of that effect. To avoid such effect, some practitioners use a *dummy* product, *i.e.*, a warm-up product that is always tested first (this product is not used in the analysis). To assess the importance of the first product tested, let's add to the *experts* data set a column named *First*. This column takes a "yes" if the product is tested first, a "no" otherwise. The previous analysis is then performed by substituting *Rank* with *First* in the previous model.

```
> library(car)
> experts$First <- recode(experts$Rank,"1='yes'; else='no'",as.factor=TRUE)
> res.first <- panelperf(experts,firstvar=5,lastvar=16,formul="~Product+
+ Panelist+Session+First+Product:Panelist+Product:Session+Panelist:Session")
> coltable(round(res.first$p.value,3),col.lower="gray")
```

Five of the 7 previous attributes have a significant first-order effect (*cf.* Figure 1.8). In other words, for 5 out of 12 sensory attributes, the products were perceived as differently when tested first or when tested at another position. To understand which products in particular, we refer to Exercise 1.5 as a methodology is presented there.

Next, let's assess the impact on the perception of an attribute, for a given product, of the product tested previously (carry-over effect). To assess the carry-over effect, a column with the information related to the previous column is required. This additional column, named *Previous*, takes the name of the

	Product	Panelist	Session	Rank	Product:Panelist	Product:Session	Panelist:Session
Spicy	0	0	0.891	0.006	0	0.173	0.119
Heady	0	0	0.644	0.012	0	0.373	0.111
Fruity	0	0	0.937	0.153	0	0.91	0.774
Green	0	0	0.066	0.069	0	0.856	0.707
Vanilla	0	0	0.23	0.003	0	0.56	0.027
Floral	0	0	0.027	0.26	0	0.036	0.179
Woody	0	0	0.382	0.12	0.041	0.544	0.047
Citrus	0.076	0	0.156	0.002	0	0.504	0.456
Marine	0.016	0	0.116	0.6	0.546	0.018	0
Greedy	0	0	0.491	0.001	0	0.57	0.001
Oriental	0	0	0.142	0.027	0	0.977	0.001
Wrapping	0	0	0.008	0.003	0	0.303	0.361

FIGURE 1.7
Assessment of the impact of the presentation order on the perception of the
products with the **panelperf** and **coltable** functions (*experts* data set).

previous product tested. For the product tested first, since no other product
was tested previously, a 0 is set. Such column can be generated using the
following code lines (please be very careful as these code lines require the data
set to be sorted according to the presentation order within each panelist to
run correctly):

```
> Previous <- matrix("0",nrow(experts),1)
> colnames(Previous) <- "Previous"
> for (i in 1:nrow(experts)){
>     if (!experts$Rank[i]==1){
>         Previous[i,1] <- as.character(experts$Product[i-1])
>     }
> }
> experts <- cbind(experts,Previous)
```

	Product	Panelist	Session	First	Product:Panelist	Product:Session	Panelist:Session
Spicy	0	0	0.891	0.001	0.001	0.189	0.132
Heady	0	0	0.644	0.223	0	0.341	0.094
Fruity	0	0	0.937	0.03	0	0.906	0.764
Green	0	0	0.066	0.798	0	0.876	0.74
Vanilla	0	0	0.23	0.284	0	0.557	0.026
Floral	0	0	0.027	0.058	0	0.037	0.182
Woody	0	0	0.382	0.044	0.039	0.548	0.048
Citrus	0.074	0	0.156	0.003	0	0.551	0.505
Marine	0.016	0	0.116	0.534	0.525	0.017	0
Greedy	0	0	0.491	0.049	0	0.571	0.001
Oriental	0	0	0.142	0.163	0	0.979	0.001
Wrapping	0	0	0.008	0.075	0	0.329	0.388

FIGURE 1.8
Assessment of the first-order effect with the **panelperf** and **coltable** functions (*experts* data set).

The **panelperf** function is then applied to this data set with the same model as previously used, except for *Rank* now substituted by *Previous*.

```
> res.previous <- panelperf(experts,firstvar=5,lastvar=16,formul="~Product+
+ Panelist+Session+Previous+Product:Panelist+Product:Session+Panelist:Session")
> coltable(round(res.previous$p.value,3),col.lower="gray")
```

From this analysis, the perception of six sensory attributes is impacted by the product tested previously (*cf.* Figure 1.9). Here again, details regarding the assessment of the influence of each product on the perception of the following product are given in the Exercise 1.5.

	Product	Panelist	Session	Previous	Product:Panelist	Product:Session	Panelist:Session
Spicy	0	0	0.893	0.034	0.001	0.197	0.122
Heady	0	0	0.647	0.231	0	0.322	0.079
Fruity	0	0	0.935	0.001	0	0.888	0.768
Green	0	0	0.076	0.055	0	0.854	0.646
Vanilla	0	0	0.207	0.002	0	0.482	0.031
Floral	0	0	0.025	0.018	0.002	0.042	0.212
Woody	0	0	0.398	0.649	0.02	0.527	0.029
Citrus	0.075	0	0.163	0.108	0	0.555	0.477
Marine	0.026	0	0.102	0.167	0.213	0.009	0
Greedy	0	0	0.498	0.017	0	0.589	0.001
Oriental	0	0	0.145	0.382	0	0.979	0.001
Wrapping	0	0	0.005	0.004	0	0.299	0.487

FIGURE 1.9
Assessment of the carry-over effect with the **panelperf** and **coltable** functions
(*experts* data set).

1.4　Exercises

Exercise 1.1 *Apprehending the notion of random effect with the* pan-elperf *function*

The aim of this exercise is to apprehend the notion of random effect in an
ANOVA model. By default, in the **panelperf** function, the factor associated
with the panelist effect is considered as random.

- Import the *perfumes_qda_experts.csv* from the book website, into a data
  frame named *experts*.

- Run the **panelperf** function on the imported data set using the default options. Run an ANOVA using the same model as the one used in the previous step. Compare the results.

- Now run the **panelperf** function on the *experts* data by switching the `random` option to `FALSE`. Compare the results.

Exercise 1.2 *Understanding the notion of interaction with the* graphinter *and the* interact *functions*

The notion of interaction is a concept of utmost importance in statistics. Alas, it is as important as it is hard to apprehend and, consequently, hard to interpret practically. The idea of this exercise is to explore this concept by focusing on the "product by session" interaction and the "product by panelist" interaction.

- Import the *perfumes_qda_experts.csv* from the book website.

- Run the **panelperf** function on the imported data set using the default options. For which attributes do you observe a significant product by session interaction?

- For these attributes, apply the **graphinter** function to "visualize" that interaction. Which product has not been assessed the same way from one session to the other?

- Apply the **interact** function to confirm what you have just observed. For this function, the contribution of a product i to the product by session interaction is calculated the following way:

$$Contribution(i) = \frac{\sum_s \alpha\gamma_{is}^2}{\sum_i \sum_s \alpha\gamma_{is}^2},$$

where $\alpha\gamma_{is}$ denotes the interaction coefficient for product i and session s (*cf.* notations in Section 1.2.2).

- Apply the same procedure for the product by panelist interaction. Comment on your results.

Exercise 1.3 *Understanding performance from a univariate point of view with the* paneliperf *function*

The aim of this exercise is to look at two important features when assessing the performance at a panelist level: the fact that the panelist is in agreement with the rest of the panel, and the capability for a panelist to discriminate between products.

- Import the *perfumes_qda_experts.csv* from the book website, into a data frame named *experts*.

- Apply and comment line by line the following lines of code:

```
> res <- paneliperf(experts,formul="~Product+Panelist+Session+Product:Panelist+
+ Product:Session+Panelist:Session",formul.j="~Product+Session",col.j=1,
+ firstvar=5,synthesis=TRUE)
> resprob <- magicsort(res$prob.ind,method="median")
> coltable(resprob,level.lower=0.05,level.lower2=0.01,level.upper=1,main.title=
+ "P-value of the F-test (by panelist)")
> dev.new()
> hist(resprob,main="Histogram of the P-values",xlab="P-values",xlim=c(0,1),
+ nclass=21,col=c("red","mistyrose",rep("white",19)))
```

- Apply and adapt the same procedure for the object named **res$agree.ind**. To do so, you will create a new object called **resagree**. Comment on your results by re-adjusting the colors to the type of values stored in **resagree**.

The following exercise requires knowledge on exploratory multivariate analyses in general; on Principal Component Analysis (PCA) and Multiple Factor Analysis (MFA) in particular. If you already have knowledge about these techniques and want to give a try, you can do the following exercise. If you are not familiar with them, or need some refresher, we advise you to read the following chapters (as Chapter 2 presents the PCA and Chapter 3 the MFA) and to come back later to do this exercise.

Exercise 1.4 *Understanding performance from a multivariate point of view with the* paneliperf *function*

The aim of this exercise is to evaluate the performance of the panel from a multivariate point of view. This exercise is a good complement to this first chapter.

- Import the *perfumes_qda_experts.csv* from the book website.

- Run the **paneliperf** function on the imported data set using the default options, and save the results in an object named **res**. Run the **PCA** function of the FactoMineR package on the matrix named **res$agree.ind**. Comment on the results.

- Run the **PCA** function of the FactoMineR package on the matrix named **res$vtest.ind**. Comment on the results.

- Finally, use the **cbind** function to merge the two matrices **res$agree.ind** and **res$vtest.ind** into a new matrix named **performance**. Run the **MFA** function of the FactoMineR package on **performance** by considering two groups of 12 variables each, and by using the default options. Comment on the results.

- Run the **paneliperf** function on the imported data set by setting the `graph` argument to `TRUE`.

The following exercise requires knowledge on the t-test and on the notion of V-test. If you already have knowledge about these notions, and want to give a try, you can do the following exercise. If you are not familiar with them, or need some refresher, we advise you to read the following chapter (Chapter 2) and to come back later to do this exercise.

Exercise 1.5 *Introduction to the notion of t-test and V-test*

In Section 1.3 (*For experienced users*), the first-order effect and the carry-over effect are evaluated. If it has been shown for which attributes a significant first-order effect and a carry-over effect are observed, additional details regarding which products are playing a role have not been given. The aim of this exercise is to provide a methodology to assess this in more details.

- Import the *perfumes_qda_experts.csv* file from the book website. Verify the importation and make sure that all the variables are set properly.

- Add a column to the data set called *First*. This column should take a "yes" if the product has been seen in first position, a "no" otherwise. To do so, the **recode** function from the car package can be used.

- As in Section 1.3, evaluate the first-order effect. To do so, the **panelperf** function of the SensoMineR package can be used.

- Using the **decat** function of the SensoMineR package, evaluate the impact of being tested first on the evaluation of each sensory attribute. Note that since the order of the effects plays an important role in the **decat** function, the *First* effect should be positioned first in the ANOVA model.

- Using the `tabT` outputs of the **decat** function, conclude.

- Add a column to the data set called *Previous*. This column should take the name of the product tested prior to the one evaluated. For the product tested first, since no products were tested previously, a "0" can be used.

- As in Section 1.3, evaluate the carry-over effect. To do so, the **panelperf** function of the SensoMineR package can be used.

- Using the **decat** function of the SensoMineR package, evaluate the carry-over on the evaluation of each sensory attribute. Here again, the *Previous* effect should be positioned first in the ANOVA model.

- Using the `tabT` outputs of the **decat** function, conclude.

1.5 Recommended readings

- Bates, D., Maechler, M., Bolker, B., & Walker, S. (2013). lme4: linear mixed-effects models using Eigen and S4. *R package* version 1.4.

- Bi, J. (2003). Agreement and reliability assessments for performance of sensory descriptive panel. *Journal of Sensory Studies*, 18, (1), 61-76.

- Brockhoff, P. M. (1998). Assessor modelling. *Food Quality and Preference*, 9, (3), 87-89.

- Brockhoff, P. (2001). Sensory profile average data: combining mixed model ANOVA with measurement error methodology. *Food Quality and Preference*, 12, (5), 413-426.

- Brockhoff, P. B. (2003). Statistical testing of individual differences in sensory profiling. *Food Quality and Preference*, 14, (5), 425-434.

- Brockhoff, P. M., Hirst, D., & Naes, T. (1996). Analysing individual profiles by three-way factors analysis. In T. Naes & E. Risvik (Eds.), Multivariate analysis of data in sensory science. Elsevier Science Publishers.

- Brockhoff, P. M., & Skovgaard, I. M. (1994). Modelling individual differences between assessors in sensory evaluations. *Food Quality and Preference*, 5, (3), 215-224.

- Chatfield, C. (1992). Problem Solving. A statistician's guide. Chapman & Hall.

- Couronne, T. (1997). A study of assessors' performance using graphical methods. *Food Quality and Preference*, 8, (5), 359-365.

- Danzart, M. (1983). Evaluation of the performance of panel judges. In *Food research and data analysis: proceedings from the IUFoST Symposium*, September 20-23, 1982, Oslo, Norway. Edited by H. Martens and H. Russwurm, Jr.

- Dijksterhuis, G. (1995). Assessing panel consonance. *Food Quality and Preference*, 6, (1), 7-14.

- Kermit, M., & Lengard, V. (2005). Assessing the performance of a sensory panel-panellist monitoring and tracking. *Journal of Chemometrics*, 19, (3), 154-161.

- King, M. C., Hall, J., & Cliff, M. A. (2001). A comparison of methods for evaluating the performance of a trained sensory panel. *Journal of Sensory Studies*, 16, (6), 567-581.

- Kuznetsova, A., Brockhoff, P. B., & Christensen, R. H. B. (2013). lmerTest: tests for random and fixed effects for linear mixed effect models (lmer objects of lme4 package). *R package* version 2.0-3.

- Labbe, D., Rytz, A., & Hugi, A. (2004). Training is a critical step to obtain reliable product profiles in a real food industry context. *Food Quality and Preference*, 15, (4), 341-348.

- Lawless, H. (1998). Commentary on random vs. fixed effects for panelists. *Food Quality and Preference*, 9, (3), 163-164.

- Lea, P., Naes, T., & Rødbotten, M. (1997). Analysis of variance of sensory data. J. Wiley and Sons.

- Lea, P., Rødbotten, N., & Naes, T. 1995. Measuring validity in sensory analysis. *Food Quality and Preference*, 6, (4), 321-326.

- Ledauphin, S., Hanafi, M., & Qannari, E. M. (2006). Assessment of the agreement among the subjects in fixed vocabulary profiling. *Food Quality and Preference*, 17, (3), 277-280.

- MacFie, H. J. H., Bratchell, N., Greenhoff, K., & Vallis, L.V. (1989). Designs to balance the effect of order of presentation and first-order carry-over effects in hall tests. *Journal of Sensory Studies*, 4, (2), 129-148.

- McEwan, J.A., Hunter, E.A., Gemert, L.J., & Lea, P. (2002). Proficiency testing for sensory profile panels: measuring panel performance. *Food Quality and Preference*, 13, (3), 181-190.

- Naes, T. (1991) Handling individual differences between assessors in sensory profiling. *Food Quality and Preference*, 2, (3), 187-199.

- Naes, T. (1998). Detecting individual differences among assessors and differences among replicates in sensory profiling. *Food Quality and Preference*, 9, (3), 107-110.

- Naes, T., & Langsrud, Ø. (1998). Fixed or random assessors in sensory profiling? *Food Quality and Preference*, 9, (3), 145-152.

- Naes, T., & Solheim, R. (1991). Detection and interpretation of variation within and between assessors in sensory profiling. *Journal of Sensory Studies*, 6, (3), 159-77.

- Pagès, J., & Périnel, E. (2004). Panel performance and number of evaluations in a descriptive sensory study. *Journal of Sensory Studies*, 19, (4), 273-291.

- Peltier, C., Brockhoff, P.B., Visalli, M., & Schlich, P. (2014). The MAM-CAP table: a new tool for monitoring panel performances. *Food Quality and Preference*, 32, 24-27.

- Périnel, E., & Pagès, J. (2003). Optimal nested cross-over designs in sensory analysis. *Food Quality and Preference*, 15, (5), 439-446.

- Pineau, N., Chabanet, C., & Schlich, P. (2007). Modeling the evolution of the performance of a sensory panel: a mixed-model and control chart approach. *Journal of Sensory Studies*, 22, (2), 212-241.

- Rossi, F. (2001). Assessing sensory panelist performance using repeatability and reproducibility measures. *Food Quality and Preference*, 12, (5), 467-479.

- The PanelCheck project. `http://www.panelcheck.com`.

- Wakeling, I., Hasted, A., & Buck, D. (2001). Cyclic presentation order designs for consumer research. *Food Quality and Preference*, 12, (1), 39-46.

- Wakeling, I., & MacFie, H. J. H. (1995). Designing consumer trials balanced for first and higher orders of carry-over effects. *Food Quality and Preference*, 6, (4), 299-308.

2

When products are rated according to a single list of attributes

CONTENTS

"... the classical representation of the product space can be enhanced with confidence ellipses around the products. Such ellipses represent the variability around each product, in other words, the different positions that may take each product every time the composition of the panel is modified. These virtual panels are obtained by bootstrap, and using the "raw" data is indispensable. Similarly, the representation of the sensory attributes can be enhanced by plotting the variability of each attribute. Such graphical outputs can be obtained using the **panellipse** *function of the SensoMineR package."*

2.1   Data, sensory issues, and notations

This chapter is a logical follow-up of the previous one. Once the quality of the data provided by the panelists has been checked, it makes sense to finally adopt a product perspective, as ultimately, data were collected for that purpose.

As mentioned previously, descriptive analysis methods aim at obtaining an accurate description of the products regarding a list of sensory attributes. Such description, also known as the so-called sensory profile of a product, has two main purposes. First, it is used to compare a new product to existing and/or competitors' products, and positions it within the product space. Second, it is used to identify the sensory attributes that are the most important to define that product space, and reveals their relationships. Indeed, these sensory attributes are the ones that constitute the sensory dimensions of the product space.

In this chapter, depending on the objectives, two data sets are considered. The first one is the one described in Chapter 1 and used for ANOVA. Let's recall that, in this data set, a statistical unit of interest consists of a sensory evaluation. In other words, it corresponds to an assessment in which a given subject k $(k = 1, \ldots, K)$ gives scores to a given product i $(i = 1, \ldots, I)$ during a given session s $(s = 1, \ldots, S)$ according to J attributes. Consequently, the data set consists of $I \times K \times S$ rows and $3 + J$ columns - one for the *product*, one for the *panelist*, one for the *session*, and J for the sensory attributes. The second data set can be seen as a "contraction" of the first one, and hence results directly from it. In this second data set, the statistical units of interest are the products, and the variables are the J sensory attributes.

Similarly to Chapter 1, the attributes that differentiate the products are identified through ANOVA (performed attribute per attribute) applied on the first data set. Besides to its classical outputs, ANOVA is also used to compute adjusted means, which defines the "contracted" sensory profiles of the products, constituting the second data set. The analysis of this second data set (and more precisely its visualization) is done by using one of the most important exploratory multivariate methods: Principal Component Analysis (PCA).

Throughout this chapter, particular emphasis is made on the t-test in ANOVA, and on the not-so-well-known (but still very useful) geometrical point of view in PCA. More precisely, the notion of contrasts as well as the notion of supplementary information (also known as illustrative information) are presented. In practice, the FactoMineR package (a package dedicated to exploratory multivariate analysis in R) and some of its functions dedicated to the analysis of multivariate data sets are presented and used.

2.2   In practice

The data set used to illustrate the methodology presented here is the same as in Chapter 1. It consists of 12 panelists, testing and rating 12 luxurious women perfumes on 12 attributes. Each panelist rated each product twice.

First, let us quickly recall the *experts* data set that has been used in Chapter 1, by using the **summary** function.

```
> summary(experts)
   Panelist   Session      Rank                 Product        Spicy
 CM     : 24  1:144   1      : 24  Angel            : 24  Min.   : 0.000
 CR     : 24  2:144   2      : 24  Aromatics Elixir : 24  1st Qu.: 0.000
 GV     : 24          3      : 24  Chanel N5        : 24  Median : 0.500
 MLD    : 24          4      : 24  Cinéma           : 24  Mean   : 2.249
 NMA    : 24          5      : 24  Coco Mademoiselle: 24  3rd Qu.: 3.625
 PR     : 24          6      : 24  J'adore EP       : 24  Max.   :10.000
 (Other):144          (Other):144  (Other)          :144
     Heady           Fruity           Green           Vanilla
 Min.   : 0.000  Min.   : 0.000  Min.   :0.0000  Min.   : 0.000
 1st Qu.: 0.300  1st Qu.: 0.500  1st Qu.:0.0000  1st Qu.: 0.200
 Median : 1.900  Median : 2.300  Median :0.2000  Median : 1.200
 Mean   : 3.963  Mean   : 3.487  Mean   :0.8458  Mean   : 3.113
 3rd Qu.: 8.125  3rd Qu.: 6.125  3rd Qu.:0.7000  3rd Qu.: 6.000
 Max.   :10.000  Max.   :10.000  Max.   :9.8000  Max.   :10.000

     Floral           Woody           Citrus          Marine
 Min.   : 0.000  Min.   :0.000   Min.   :0.000   Min.   :0.0000
 1st Qu.: 3.775  1st Qu.:0.000   1st Qu.:0.000   1st Qu.:0.0000
 Median : 6.800  Median :0.500   Median :0.200   Median :0.0000
 Mean   : 6.119  Mean   :1.252   Mean   :1.049   Mean   :0.4674
 3rd Qu.: 9.000  3rd Qu.:1.525   3rd Qu.:1.125   3rd Qu.:0.2000
 Max.   :10.000  Max.   :9.700   Max.   :7.800   Max.   :9.5000

     Greedy          Oriental        Wrapping
 Min.   : 0.000  Min.   : 0.000  Min.   : 0.000
 1st Qu.: 0.100  1st Qu.: 0.100  1st Qu.: 2.775
 Median : 1.000  Median : 1.200  Median : 6.300
 Mean   : 2.924  Mean   : 3.699  Mean   : 5.737
 3rd Qu.: 5.100  3rd Qu.: 8.000  3rd Qu.: 8.900
 Max.   :10.000  Max.   :10.000  Max.   :10.000
```

As previously explained, the first four columns are categorical: these are the independent factors of our experiment. The other variables are continuous: these are the sensory attributes.

2.2.1 How can I get a list of the sensory attributes that structure the product space?

Similarly to Chapter 1, ANOVA is used to define the sensory attributes that structure the product space. However, here, the interpretation of the results is not panelist oriented, but product oriented.

To explain the sensory attribute *Citrus* (dependent variable) with respect to the main effects *Product*, *Panelist*, *Session*, and their first-order interactions (independent variables), the following ANOVA model is considered:

$$Citrus_{iks} \sim \mu + \alpha_i + \beta_k + \gamma_s + \alpha\beta_{ik} + \alpha\gamma_{is} + \beta\gamma_{ks} + \epsilon_{iks},$$

where

- α_i is the ith coefficient associated with the *Product* effect;

- β_k is the kth coefficient associated with the *Panelist* effect;

- γ_s is the sth coefficient associated with the *Session* effect;

- $\alpha\beta_{ik}$ is the ik^{th} coefficient associated with the *Product-Panelist* interaction;

- $\alpha\gamma_{is}$ is the is^{th} coefficient associated with the *Product-Session* interaction;

- $\beta\gamma_{ks}$ is the ks^{th} coefficient associated with the *Panelist-Session* interaction;

- and ϵ_{iks} denotes the error term.

As seen in Chapter 1, the errors are assumed to be normally distributed, with mean zero, constant variance σ^2, and independent:

- $\epsilon_{iks} \sim N(0, \sigma^2)$;

- $\forall (i, k, s) \neq (i', k', s'), Cov(\epsilon_{iks}, \epsilon_{i'k's'}) = 0$.

By focusing on the *Product* effect, the question we want to answer is: For the sensory attribute *Citrus*, is my *Product* effect *statistically* significant? This question can be rephrased the following way: Are my products significantly different regarding the sensory attribute *Citrus*? If it is the case, this sensory attribute *Citrus* structures[1] the product space, and should play an important role in the explanation of the sensory dimensions.

As explained in Chapter 1, let us have a look at the ANOVA table, and more particularly at the results of the F-test. To do so, alternatively to the **aov** function presented in Chapter 1, the **lm** function is used to fit the ANOVA model. The **anova** function is then applied on the result issued from the **lm** function to generate the ANOVA table.

```
> citrus.lm <- lm(Citrus~Product+Panelist+Session+Product:Panelist+
+ Product:Session+Panelist:Session,data=experts)
> anova(citrus.lm)
Analysis of Variance Table

Response: Citrus
                   Df Sum Sq Mean Sq F value    Pr(>F)
Product            11  69.59  6.3263  3.9512 6.520e-05 ***
Panelist           11 128.70 11.7002  7.3076 1.799e-09 ***
Session             1   3.45  3.4453  2.1518    0.1450
Product:Panelist  121 452.60  3.7405  2.3362 2.213e-06 ***
Product:Session    11  15.45  1.4048  0.8774    0.5643
Panelist:Session   11  16.31  1.4830  0.9263    0.5178
Residuals         121 193.73  1.6011
---
Signif. codes:  0 '***' 0.001 '**' 0.01 '*' 0.05 '.' 0.1 ' ' 1
```

---

[1]Construct or arrange according to a plan; give a pattern or organization to.

With a significance threshold at 5% (=0.05) and a *p-value* of 6.52e-05, the ANOVA table shows a highly significant *Product* effect: the products have been differentiated regarding the sensory attribute *Citrus*. Based on this result, we can expect *Citrus* to play a role in the structure of the product space.

To get the list of attributes that structures the product space, the **decat** function of the SensoMineR package is used. This function systematically performs ANOVA on each sensory attribute using a given model. The main feature of the **decat** function is to produce result summaries that are specific to one particular effect (here the *Product*). For the function to know on which effect to focus on, it is of utmost importance to position that effect (here *Product*) in the first place when specifying the ANOVA model.

To use the function, first load the SensoMineR package if it has not already been done. The main arguments of the **decat** function to specify are: the data set on which the analyses are performed, the ANOVA model, and the positions of the first and last sensory attributes (by default, the position of the last sensory attribute corresponds to the last column of the data set). The **decat** function produces a list of results that we store here in an object called `res.decat`.

```
> library(SensoMineR)
> res.decat <- decat(experts,formul="~Product+Panelist",firstvar=5,
+ lastvar=ncol(experts),graph=FALSE)
```

The names of the different components stored in `res.decat` are obtained using the **names** function.

```
> names(res.decat)
[1] "tabF"    "tabT"    "coeff"   "adjmean" "resF"    "resT"
```

Amongst the different results provided by the **decat** function, the one we are directly interested in here is `res.decat$resF`, as it stores the results associated with the *F*-test.

```
> res.decat$resF
           Vtest        P-value
Heady    15.843938 7.740772e-57
Greedy   15.504561 1.615578e-54
Vanilla  13.142580 9.385764e-40
Spicy    12.488234 4.327667e-36
Floral   11.455297 1.106016e-30
Oriental  9.821087 4.567635e-23
Fruity    9.619339 3.312781e-22
Wrapping  8.856596 4.124750e-19
Green     7.755499 4.399839e-15
Woody     5.510370 1.790403e-08
Citrus    2.508168 6.067947e-03
```

This output highlights the sensory attributes for which products are differentiated at a significance threshold of 0.05 (this threshold can be changed using the argument `proba` in the **decat** function). This list of attributes is sorted from the most significant (*Heady* with a *p-value* of 7.74e-57) to the less (but still) significant (*Citrus* with a *p-value* of 6.07e-03).

As expressed by the *p-values* that are singularly small, products have been extremely differentiated by the panelists: it seems that some attributes, such as *Heady* or *Greedy*, are really specific to some products.

2.2.2   How can I get a sensory profile for each product?

Now that the list of sensory attributes differentiating the products has been defined, the natural continuity consists in defining which products are specific for those attributes. In other words, rather than focusing on the main effects, we are interested in the effects of the levels associated with the factors and their interactions. This new question to answer can be rephrased as: For the sensory attribute *Citrus*, which product can I consider as significantly different ("positively" or "negatively", in a sense that will be specified latter) from some kind of an average product?

The answer to that question lies in the analysis of the coefficients α_i ($i = 1, \ldots, I$) associated with the *Product* effect. Such an analysis of the coefficients is done through the Student's t-test, in which the following hypotheses are tested for each product, *i.e.*, for each level of the *Product* effect:

$H_0 : \alpha_i = 0$ *versus* $H_1 : \alpha_i \neq 0$.

To get a unique estimate for each α_i, constraints need to be set on them. These constraints are also called *contrasts*, in the statistical jargon. Different contrasts exist, and the one we are choosing here is a very natural one, in the sense that no *a priori* on the products is considered:

$$\sum_{i=1}^{I} \alpha_i = 0.$$

This contrast consists in testing each product with respect to some kind of an average product, and not with respect to a specific product. Incidentally, the point of view adopted on the contrast also fits with the point of view adopted by PCA, as we will see further on.

To set contrasts, the **options** function is used. This function allows setting up general options, which affect the way R computes and displays its results.

```
> options(contrasts=c("contr.sum","contr.sum"))
```

To get the results of the t-test, the **summary.lm** function (or more generically, the **summary** function) is applied to the results of the **lm** function. In our case, this corresponds to applying the **summary.lm** function to `citrus.lm`:

```
> summary.lm(citrus.lm)
```

```
Call:
lm(formula = Citrus ~ Product + Panelist + Session + Product:Panelist +
    Product:Session + Panelist:Session, data = experts)

Residuals:
    Min      1Q  Median      3Q     Max
-3.3135 -0.2703  0.0000  0.2703  3.3135

Coefficients:
               Estimate Std. Error t value Pr(>|t|)
(Intercept)    1.048958   0.074561  14.068  < 2e-16 ***
Product1      -0.640625   0.247292  -2.591 0.010759 *
Product2      -0.444792   0.247292  -1.799 0.074567 .
Product3      -0.119792   0.247292  -0.484 0.628968
Product4       0.001042   0.247292   0.004 0.996646
Product5       0.192708   0.247292   0.779 0.437339
Product6       1.117708   0.247292   4.520 1.45e-05 ***
Product7       0.526042   0.247292   2.127 0.035433 *
Product8      -0.340625   0.247292  -1.377 0.170925
Product9      -0.236458   0.247292  -0.956 0.340882
Product10      0.559375   0.247292   2.262 0.025482 *
Product11     -0.448958   0.247292  -1.815 0.071924 .
(...)
---
Signif. codes:  0 '***' 0.001 '**' 0.01 '*' 0.05 '.' 0.1 ' ' 1

Residual standard error: 1.265 on 121 degrees of freedom
Multiple R-squared:  0.7798,Adjusted R-squared:  0.4777
F-statistic: 2.581 on 166 and 121 DF,  p-value: 4.003e-08
```

The previous output is impossible to interpret, unless we have the correspondence between `Product1`, ..., `Product11` and the levels of the *Product* effect. To get this correspondence, the **levels** function is applied to the variable *Product*:

```
> levels(experts$Product)
 [1] "Angel"             "Aromatics Elixir" "Chanel N5"     "Cinéma"
 [5] "Coco Mademoiselle" "J'adore EP"       "J'adore ET"    "L'instant"
 [9] "Lolita Lempicka"   "Pleasures"        "Pure Poison"   "Shalimar"
```

Now we know that `Product1` corresponds to *Angel*, `Product2` to *Aromatics Elixir*, ..., and `Product11` to *Pure Poison*, so the results of our ANOVA can be interpreted. To do so, the products associated with a *p-value* higher than 0.05 are separated from the products with a *p-value* lower than 0.05.

For the first ones (*p-value* > 0.05), the products are not significantly different from the average product regarding the sensory attribute *Citrus*. This is the case for *Aromatics Elixir, Chanel N5, Cinéma, Coco Mademoiselle, L'instant,* and *Lolita Lempicka*. On the contrary, the second ones are significantly different from the average product regarding the attribute *Citrus*. In this case, a distinction between the products that have been perceived with a high intensity of *Citrus* (at least higher than the average product regarding that attribute) and the products that have been perceived with a low intensity of *Citrus* (at least lower than the average product regarding that attribute) should be made. Such a distinction is made using the sign of the estimates: the products that have a positive estimate (first column) and a "small" *p-value* (< 0.05, *cf.* last column) are significantly more intense in *Citrus* than the average product. This is the case for *J'adore EP*, *J'adore ET*, and *Pleasures*. Inversely, the products associated with a negative estimate and a "small" *p-value* (< 0.05, *cf.* last column) are significantly less intense in *Citrus* than the average product. This is the case for *Angel* and *Pure Poison*.

In practice, looking at the results of the *t*-tests for all sensory attributes can quickly become tedious. As evoked previously, the **decat** function aims at running systematically all possible ANOVAs, using a given model, and summarizes the results in different matrices. This function is designed to point out the sensory attributes that are the most characteristic of a set of products as a whole, as well as product by product.

The results of the *t*-tests are stored in **res.decat$resT**. This list is composed of as many objects as there are products (more precisely, as there are levels in the *Product* effect). For instance, for *Angel* and *Pleasures*, the following results are obtained:

```
> res.decat$resT$Angel
              Coeff Adjust mean       P-value
Greedy     4.9638889   7.8875000 5.706734e-27
Heady      3.8788194   7.8416667 9.571760e-17
Vanilla    4.0701389   7.1833333 1.023563e-16
Spicy      1.6506944   3.9000000 9.545053e-05
Wrapping   1.8131944   7.5500000 4.422015e-04
Citrus    -0.6406250   0.4083333 4.194330e-02
Green     -0.7333333   0.1125000 6.804738e-03
Fruity    -1.5659722   1.9208333 1.659853e-03
Floral    -3.6270833   2.4916667 7.646973e-15

> res.decat$resT$Pleasures
              Coeff Adjust mean       P-value
Green      2.4041667   3.2500000 6.710969e-17
Floral     2.1312500   8.2500000 2.149873e-06
Marine     0.6743056   1.1416667 1.119319e-02
Fruity     0.9756944   4.4625000 4.874226e-02
Spicy     -1.7576389   0.4916667 3.370726e-05
Oriental  -2.6576389   1.0416667 7.409649e-06
Greedy    -2.2527778   0.6708333 1.038211e-07
Wrapping  -3.1118056   2.6250000 3.606021e-09
Vanilla   -2.8381944   0.2750000 2.236329e-09
Heady     -3.0503472   0.9125000 2.199124e-11
```

Each sensory profile is structured according to three components:

- in the first column, the estimate of α_i (*i.e.*, $\hat{\alpha}_i$);

- in the second column, the estimate of $\mu + \alpha_i$ (*i.e.*, $\hat{\mu} + \hat{\alpha}_i$);

- in the third column, the *p-value* associated with the test $H_0 : \alpha_i = 0$ versus $H_1 : \alpha_i \neq 0$.

For each product, the attributes that are associated with *p-values* lower than the predefined threshold (the significance level can be changed using the `proba` option, set by default at $\alpha = 0.05$) are shown. These attributes are then sorted according to two key parameters: the sign of the estimate of the coefficient α_i and the value of the *p-value*.

Based on these results, it can be concluded that *Angel* has been perceived as *Greedy*, *Heady*, to a lesser degree as *Spicy*; on the contrary, it has not been perceived much as *Fruity*, nor as *Floral*. Similarly, *Pleasures* has been perceived as *Green*, to a lesser extent as *Floral*; it has not been perceived much as *Wrapping*, nor as *Heady*. This constitutes the major information of the sensory profiles of these two products. Such information is extremely useful to understand the product space and the differences between products. Still there is a need for a more global understanding through graphical representations.

Tips: Contrasts in Analysis of Variance

Analyses of Variance are used to evaluate the significance of one or more factors on a continuous variable. The global significance of each factor is evaluated through the F-test. Additionally, the significance of the different levels within each factor is evaluated through the t-test. Usually, this additional step tests whether the coefficients α_i, associated with the ith level of the categorical variable of interest, are significantly different from 0. Without loss of generality, let us consider the simplest ANOVA model in which one continuous variable is explained by one categorical variable:

$$Y_{ij} \sim \mu + \alpha_i + \epsilon_{ij}.$$

The estimate of μ and the different levels α_i depends on so-called contrasts. Mainly, three types of contrasts are used:

- $\alpha_1 = 0$, the intercept μ corresponds to the average score for the level 1 of that factor, and the coefficient α_i corresponds to the deviation between level i and level 1 $(i = 2, \ldots, I)$;

- $\alpha_I = 0$, the intercept μ corresponds to the average score for the level I of that factor, and the coefficient α_i corresponds to the deviation between level i and level I $(i = 1, \ldots, I - 1)$;

- $\sum_{i=1}^{I} \alpha_i = 0$, the intercept μ corresponds to the average score of that factor, and the coefficient α_i corresponds to the deviation of level i from the average.

When there is no *a priori* on the levels, or in other words when there is no specified level of reference, the third contrast is used. However, R uses by default the second one, *i.e.*, $\alpha_I = 0$. To change the contrast to $\sum_i \alpha_i = 0$, the following code should be used in R. This code should be run before actually performing the ANOVA.

```
> options(contrasts=c("contr.sum","contr.sum"))
```

Note that, by changing the contrast, the estimate of the coefficients change (by definition), but the global analysis does not. However, since the default contrast in R uses the last level as reference ($\alpha_I = 0$), it only displays the results for the $(I - 1)$ first coefficients. When the contrast $\sum_{i=1}^{I} \alpha_i = 0$ is used, the last coefficient should then be calculated. This could be done in two different ways:

- calculate manually the last coefficient by summing the $I - 1$ first coefficients, and by multiplying this sum by -1;

- change the order of the levels of the factor (so that α_I is not the last one anymore) using the **relevel** function, and re-run the analysis.

Note that in the first case, only the estimate of α_I is provided, whereas in the second case, the *p-value* is also available, as the test associated with that coefficient is performed.

This second case is automated with the **AovSum** function of the FactoMineR package. This function is similar to the **aov** function except that it considers the $\sum_{i=1}^{I} \alpha_i = 0$ contrast and provides directly the results of the *t*-test for all the modalities of each effect added in the model. These results are stored in the object $Ttest.

2.2.3   How can I represent the product space on a map?

Amongst the different outputs of the **decat** function, one still needs to be mentioned. It is its fourth element called adjmean. The object res.decat$adjmean is a matrix that presents, in rows, the different levels of the categorical variable of interest (*i.e.*, the first effect in the model, here *Product*), and in columns the variables of interest (here the sensory attributes). As stated by its name, this matrix contains the adjusted means by products for the different attributes. For each sensory attribute, such adjusted means are obtained by computing, for each product i, the score obtained using the following formula: $\hat{\mu} + \hat{\alpha}_i$.

```
> round(res.decat$adjmean,3)
```

	Spicy	Heady	Fruity	Green	Vanilla	Floral	Woody
Angel	3.900	7.842	1.921	0.112	7.183	2.492	1.175
Aromatics Elixir	6.304	8.308	0.613	0.517	1.821	4.296	2.638
Chanel N5	3.733	8.213	0.967	0.437	1.788	6.150	0.950
Cinéma	1.083	2.196	5.125	0.212	4.863	5.550	1.017
Coco Mademoiselle	0.912	1.142	5.063	0.779	1.950	7.975	0.804
J'adore EP	0.262	1.179	6.404	1.563	0.467	8.400	0.912
J'adore ET	0.342	1.287	5.625	1.483	0.879	8.179	0.875
L'instant	0.737	2.283	3.842	0.296	4.888	7.383	0.975
Lolita Lempicka	1.400	4.408	3.350	0.492	8.079	3.025	0.708
Pleasures	0.492	0.913	4.463	3.250	0.275	8.250	0.708
Pure Poison	1.663	1.896	3.546	0.629	1.921	7.229	1.346
Shalimar	6.163	7.888	0.925	0.379	3.246	4.496	2.921

	Citrus	Marine	Greedy	Oriental	Wrapping
Angel	0.408	0.142	7.888	4.758	7.550
Aromatics Elixir	0.604	0.092	0.342	7.450	7.721
Chanel N5	0.929	0.150	0.625	6.379	7.846
Cinéma	1.050	0.587	4.375	2.871	5.571
Coco Mademoiselle	1.242	0.662	2.938	3.088	4.796
J'adore EP	2.167	1.025	1.300	1.137	3.567
J'adore ET	1.575	0.275	1.858	0.917	3.325
L'instant	0.708	0.683	3.379	3.050	5.654
Lolita Lempicka	0.813	0.146	9.154	3.683	7.637
Pleasures	1.608	1.142	0.671	1.042	2.625
Pure Poison	0.600	0.571	1.492	2.396	5.217
Shalimar	0.883	0.133	1.062	7.621	7.333

This data set defines the sensory profiles of the products (without the notion of significance), and corresponds to the second data set presented in Section 2.1. Since we are interested in the similarities and differences between products, this table is particularly important. Of course, comparison between products can be directly done by looking at this matrix, but such comparison quickly becomes tedious with the increase of the number of products and/or attributes.

How is it possible to visually highlight the similarities and differences between sensory profiles? A solution could be to position the products on a map (called product space), in which products are close if their sensory profiles are similar, and far apart if their sensory profiles are different. Such a challenge seems all the more difficult as the number of attributes increases, or, in other words, as the data set is multivariate. To get such representation of the products, one of the most important *principal component methods* is used: Principal Component Analysis (PCA).

PCA applies when individuals are described by quantitative variables. In PCA, the data set, $X = (x_{ij})$, to be analyzed is structured the following way: the I rows correspond to individuals (or statistical units) and the J columns to variables. At the intersection of row i and column j, the value x_{ij} corresponds to the score of the individual i for the variable j.

The main objective of PCA is to summarize the information contained in a multivariate data set into graphical representations of individuals and variables. The general idea is to represent the scatter plot of the individuals, N_I, in a low-dimensional subspace (usually two dimensions), that respects as well as possible the distances between individuals. This subspace is the best low-dimensional representation possible of R^J, the vector space formed of vectors of J real numbers (the J variables). Similarly, PCA aims at providing a representation of the scatter plot of the variables, N_J, in a low-dimensional subspace (usually two dimensions), that respects as well as possible the distances between variables. This subspace is the best low-dimensional representation possible of R^I, the vector space formed of vectors of I real numbers (the I statistical units).

Geometrically, PCA simply consists in changing the frame of reference, by representing the cloud of points N_I (*resp.* N_J), usually defined in R^J (*resp.* R^I), into a lower-dimensional subspace. It is worth mentioning here that the first transformation (obligatory) in PCA consists in centering the data by columns. To do so, the average score for a variable j is subtracted from each value x_{ij}. Geometrically, this step positions the average individual into the center of the new lower-dimensional subspace. An additional step (optional), which consists in standardizing the variables (dividing each value x_{ij} by the standard deviation of the corresponding variable j), can also be performed. This step consists in giving the same weight to each variable (*i.e.*, the variance of each variable is then equal to 1). Without loss of generality, and as it is a very common situation, this additional transformation is performed.

In PCA, the way distances are interpreted in R^J (distance between the individuals) and R^I (distance between the variables) is totally different. In R^J, two individuals are all the more close as their values, measured over all variables, are close:

$$d^2(i, i') = \sum_{j=1}^{J} (x_{ij} - x_{i'j})^2.$$

In R^I, two variables are all the more close as their correlation coefficient, calculated over all individuals, is high.

To represent the individuals, PCA looks for a sequence of orthogonal vectors (u_s), such as u_s maximizes the variability of N_I projected on u_s. Let F_s denote the vector of the coordinates of N_I projected on u_s, and λ_s the variance of F_s. The sequence of (u_s) is such that (λ_s) is a decreasing sequence of positive numbers. In other words, the variance of F_1, the vector of the coordinates of N_I projected on u_1, is maximum. Orthogonally to F_1, the vector of the coordinates of N_I projected on u_2, denoted F_2, has the second highest variance, and so on.

This principle is the same for N_J, the scatter plot of the variables. Let's denote by (v_s) the sequence of vectors on which N_J is projected, G_s the vector of the coordinates of the variables of N_J projected on v_s, and μ_s the variance of G_s. The sequence of (v_s) is such that (μ_s) is a decreasing sequence of positive numbers.

As rows and columns are linked through the data set X, so are the representations of N_I and N_J. It can be shown that for each s, $\lambda_s = \mu_s$, and

$$F_s = \frac{1}{\sqrt{\lambda_s}} X G_s,$$

$$G_s = \frac{1}{\sqrt{\lambda_s}} X^t F_s.$$

These previous so-called *transition formulae* suggest that both representations can (and have to) be interpreted conjointly. This is why users tend to represent both representations (*i.e.*, the individuals and the variables) in one unique plot, called *biplot*. As the two spaces are originally different (defined in R^J and R^I), we do not adopt this point of view, and keep the representations separated.

As mentioned previously, the same weight is given to each sensory attribute by standardizing the data. As it is the point of view the most commonly used, it is done by default in the **PCA** function of the FactoMineR package (this can be changed by setting the option `scale.unit` to `FALSE`). To run the **PCA** function on our data set, the FactoMineR package[2] needs to be loaded first. The **PCA** function is then applied directly on the sensory profiles of the products, *i.e.*, `res.decat$adjmean`. We store the results in an object called `res.pca`. As usual, the names of the different objects saved in `res.pca` are obtained using the **names** function.

```
> library(FactoMineR)
> res.pca <- PCA(res.decat$adjmean)
> names(res.pca)
[1] "eig"  "var"  "ind"  "svd"  "call"
```

The most important results provided by the **PCA** function are:

- **eig**, which contains the eigenvalues and consequently the percentage of variability associated with each dimension;

- **var**, which contains the results associated with the variables, *i.e.*, their coordinates on the components, their correlations with the components, their contributions to the construction of the components, and their quality of representation on each component;

- **ind**, which contains the results associated with the individuals, *i.e.*, their coordinates on the components, their contributions to the construction of the components, their quality of representation on each component, and their distance to the center of gravity of the scatter plot N_I.

The table stored in `res.pca$eig` shows that the first two principal components gather 86% of the total variability (also called variance, or inertia) of the scatter plot N_I: by itself, the first component explains 64.2% whereas the second component explains 21.9% of the total variability.

```
> res.pca$eig[1:5,]
       eigenvalue percentage of variance cumulative percentage of variance
comp 1 7.706131876            64.21776563                          64.21777
comp 2 2.623886409            21.86572008                          86.08349
comp 3 0.562201334             4.68501111                          90.76850
comp 4 0.379796487             3.16497072                          93.93347
comp 5 0.373641220             3.11367683                          97.04714
```

---

[2]Since SensoMineR depends on FactoMineR, loading SensoMineR loads automatically FactoMineR. However, the inverse is not true, and loading FactoMineR does not load SensoMineR.

The representation of the perfumes (*cf.* Figure 2.1) is based on the coordinates of the individuals stored in `res.pca$ind$coord`:

```
> res.pca$ind$coord
                        Dim.1       Dim.2       Dim.3       Dim.4       Dim.5
Angel               3.3874304  -2.0620946   0.70855424  -0.10077019   0.21469121
Aromatics Elixir    3.7364394   2.3858030   0.09282138   0.22471652   0.19236189
Chanel N5           2.1005261   1.2336084  -0.12500002  -1.50615298  -0.89853647
Cinéma             -0.5031301  -1.4376269  -0.54931646   0.59255407   0.01427407
Coco Mademoiselle  -1.8422142  -0.2393964  -0.68037120  -0.03712329  -0.17730847
J'adore EP         -4.0120305   0.6275234   0.29301889   0.72175276  -0.63017941
J'adore ET         -2.8131149   0.2099981  -0.09874017   0.11343629  -1.00461126
L'instant          -0.5232300  -1.0366391  -1.06802487  -0.24732524   0.69002535
Lolita Lempicka     1.7740343  -3.2283776   0.82789114   0.05333337  -0.26697309
Pleasures          -4.2541608   1.1576474   1.52135272  -0.52203319   0.98536364
Pure Poison        -0.6583864   0.2549858  -1.09119389  -0.26069607   0.79320863
Shalimar            3.6078368   2.1345686   0.16900823   0.96830795   0.08768391
```

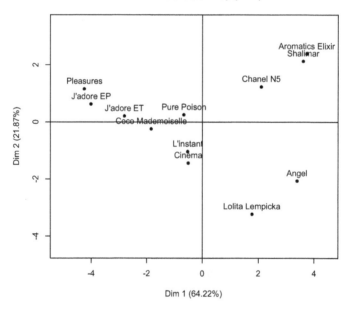

FIGURE 2.1
Representation of the perfumes on the first two dimensions resulting from PCA, using the **PCA** function on `res.decat$adjmean` (*experts* data set).

As hypothesized in Section 2.2.2, Figure 2.1 confirms that the main dimension of variability (*i.e.*, the first component) opposes products such as *Pleasures* to products such as *Angel*. This result is supported by the following output, which highlights the contribution of the products to the construction of the components (note that the sum of the contributions across products equals 100 for a given component):

```
> res.pca$ind$contrib
                      Dim.1       Dim.2       Dim.3       Dim.4        Dim.5
Angel              12.4086087 13.5048849  7.4417160  0.22280863   1.027998522
Aromatics Elixir   15.0972624 18.0776877  0.1277094  1.10799508   0.825281404
Chanel N5           4.7713192  4.8331286  0.2316045 49.77439399  18.006752260
Cinéma              0.2737430  6.5639631  4.4727240  7.70413315   0.004544223
Coco Mademoiselle   3.6699705  0.1820157  6.8614946  0.03023854   0.701169410
J'adore EP         17.4064728  1.2506435  1.2726768 11.42996030   8.857117110
J'adore ET          8.5577043  0.1400568  0.1445156  0.28233964  22.509199923
L'instant           0.2960518  3.4129416 16.9078978  1.34216122  10.619255470
Lolita Lempicka     3.4033478 33.1010957 10.1595239  0.06241168   1.589640605
Pleasures          19.5708691  4.2562424 34.3073135  5.97948852  21.654942619
Pure Poison         0.4687530  0.2064931 17.6494320  1.49120305  14.032622285
Shalimar           14.0758975 14.4708469  0.4233920 20.57286622   0.171476169
```

The main contributors to the construction of the first component are *Pleasures*, *J'adore EP*, *Aromatics Elixir*, *Shalimar*, and *Angel*.

Tips: The plot.PCA function of the **FactoMineR** package

By default, the **PCA** function generates two graphics: the representation of the individuals, and the representation of the variables. Unless stated differently, these graphics are using the first two components. Through different arguments of the **PCA** function, it is possible to adapt your graphics (representing the results using different components, for instance). However, in order to avoid re-running the entire analysis, we suggest the use of the **plot.PCA** function (or the generic **plot** function) on the results of your **PCA**.

By running the following lines of code,

```
> plot.PCA(res.pca,choix="ind",axes=c(2,3))
> plot.PCA(res.pca,choix="ind",select="contrib 7")
> plot.PCA(res.pca,choix="ind",select="cos2 0.8")
> plot.PCA(res.pca,choix="ind",select="cos2 5")
> plot.PCA(res.pca,choix="var",axes=c(2,3))
```

we successively plot:

- the individuals on components 2 and 3;

- the seven individuals that have the highest contribution;

- the individuals that have a cumulated quality of representation on the first two components higher than 0.8;

- the five individuals that have the highest cumulated quality of representation on the first two components;

  - the variables on components 2 and 3.

As expressed by Figure 2.2, this opposition between those two poles of products can be explained in terms of sensory attributes. Figure 2.2 is based on the coordinates of the variables on the components, stored in res.pcavarcoord. When variables are standardized, which is our case, the coordinates are equal to the correlation coefficients of the variables with the components.

```
> res.pca$var$coord
              Dim.1        Dim.2        Dim.3        Dim.4        Dim.5
Spicy     0.8716443   0.45963687   0.107856675   0.08976533   0.033406069
Heady     0.9246736   0.22330561   0.198242266  -0.12056791  -0.126629461
Fruity   -0.8985310  -0.31098223  -0.111280152   0.24776600  -0.129826609
Green    -0.7409308   0.31449573   0.544549889  -0.11199439   0.149363058
Vanilla   0.5518335  -0.80242160   0.064000738   0.10689898   0.117028889
Floral   -0.8727855   0.33940701  -0.295290558  -0.13800792  -0.031290877
Woody     0.6414953   0.60021583  -0.050048095   0.43595209   0.159966247
Citrus   -0.8138241   0.22734986   0.258793055   0.20388280  -0.381013940
Marine   -0.8659626   0.05604599   0.037178995   0.03779279   0.351476971
Greedy    0.2787301  -0.93036189   0.195788226   0.09349221  -0.005996179
Oriental  0.9157147   0.33647512  -0.005797401  -0.01476709  -0.042191781
Wrapping  0.9637400  -0.11909239  -0.042781294  -0.10386606  -0.080858506
```

As shown in Figure 2.2, the main dimension of variability opposes sensory attributes such as *Floral, Citrus, Green, Marine,* and *Fruity,* to attributes such as *Oriental, Heady, Spicy,* and *Wrapping.*

Tips: Automatic description of the dimensions of a multivariate analysis

An automatic description of the dimensions can be obtained using the **dimdesc** function of the FactoMineR package. This function provides the list of variables that are significantly linked to each component. This list is sorted according to two keys: the sign of the correlation and the significance of the *p-value.* This function does not, however, preclude the interpretation of the dimensions. For example, for the first dimension, we have:

```
> res.dimdesc <- dimdesc(res.pca)
> res.dimdesc$Dim.1
$quanti
            correlation       p.value
Wrapping      0.9637400  4.644746e-07
Heady         0.9246736  1.681388e-05
Oriental      0.9157147  2.904051e-05
Spicy         0.8716443  2.203341e-04
Woody         0.6414953  2.454812e-02
```

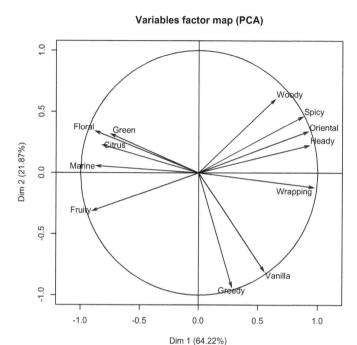

FIGURE 2.2

Representation of the sensory attributes on the first two dimensions resulting from PCA, using the **PCA** function on `res.decat$adjmean` (*experts* data set).

```
Green     -0.7409308 5.834473e-03
Citrus    -0.8138241 1.276815e-03
Marine    -0.8659626 2.709002e-04
Floral    -0.8727855 2.111339e-04
Fruity    -0.8985310 7.128849e-05
```

Note that for each dimension, only the attributes that are significantly different at a certain threshold (defined by the argument **proba** set by default at 0.05) are shown. In this case, the sensory attributes that are the most "positively" linked to the first dimension are *Wrapping* and *Heady* (on the top of the list), whereas the sensory attributes that are the most "negatively" linked are *Fruity* and *Floral* (on the bottom of the list).

Finally, to summarize, a nice combination of the results provided by both the *t*-test and the PCA of the sensory profiles (adjusted means) is given by the **decat** function. It consists in a representation of the sensory profile within a table, in which the significance of the *t*-test is color-coded, and in which the rows and the columns are sorted according to the results of the PCA (*cf.* Figure 2.3).

Such graphical representation is produced automatically by the **decat** function, when the argument `graph` is set to TRUE, or by using the following code:

```
> profiles.sort <- res.decat$adjmean[rownames(magicsort(res.decat$tabT)),
+ colnames(magicsort(res.decat$tabT))]
> coltable(profiles.sort,magicsort(res.decat$tabT),level.lower=-1.96,
+ level.upper=1.96,col.upper="gray",col.lower="gainsboro")
```

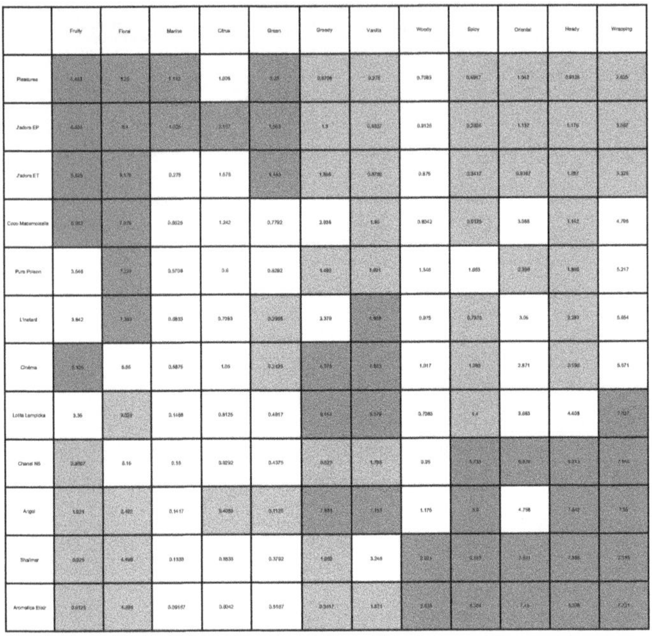

FIGURE 2.3
Representation of the sensory profiles within a sorted color-coded table as provided by the **decat** function (*experts* data set).

2.2.4 How can I get homogeneous clusters of products?

Principal component methods (such as PCA) and clustering methods (such as Hierarchical Ascendant Classification [HAC]) are often complementary. Indeed, clustering methods aim at studying the distance between individuals, whereas PCA aims at visualizing these distances. As PCA represents the scatter plot of individuals N_I on two components only, two individuals that are represented as relatively close on a factorial plane may be actually very distant, hence leading to unfortunate misinterpretations. By combining PCA and HAC, such misinterpretation can be avoided. For example, the cluster information can be added on the factorial plane by using a color code. In this case, our two former individuals, who are represented as relatively close on the two dimensions used, would be associated with different colors. This difference in colors suggests that they are distant on other dimensions, as they belong to two distinct clusters. Similarly, results from HAC can be complemented by PCA, as the notion of distance between clusters can *de facto* be visualized in a factorial plane.

Such a combination between multivariate analysis and clustering is the core of the **HCPC** function (Hierarchical Clustering on Principle Components; FactoMineR package), as it takes as input the results issued from any multivariate methods implemented in FactoMineR, and run HAC on the individual space. As by default, the number of components stored by these multivariate analyses equals 5, the distance used by **HCPC** is usually based on the five first components only. This can be easily modified by changing the number of components to store, using the `ncp` argument in the different multivariate methods (*e.g.*, **PCA(experts,ncp=3)** for three components only; **PCA(experts,ncp=Inf)** for all the components).

By using the previous example, both the **PCA** (more precisely, the object `res.pca`) and the **HCPC** functions can be combined as such:

```
> res.hcpc <- HCPC(res.pca)
```

At the stage presented in Figure 2.4, the **HCPC** function requires the analyst to set a number of clusters by clicking on the hierarchical tree (also called dendrogram). **HCPC** suggests a natural cut of the hierarchical tree, based on a criterion that depends on both the ratio of the variance within-clusters and the total variance of the scatter plot. The criterion used aims at balancing the definition of clusters that are as homogenous as possible (with similar individuals), and as different as possible to each other (for which centers of gravity are different). In our example, we accept the natural cut in four groups suggested by **HCPC**. It is worth mentioning that **HCPC** will never suggest the 2-cluster solution.

Once the number of clusters is determined by the analyst, **HCPC** stores that information and plots the individuals on the factorial plane with a color-code depending on the cluster they belong to (*cf.* Figure 2.5). Such representation is obtained by adding, or representing, *supplementary information* (also referred to *illustrative information*) on a factorial plane. This feature is of utmost importance for interpreting exploratory multivariate results, and will be further developed in the next paragraph (*cf.* Section 2.3).

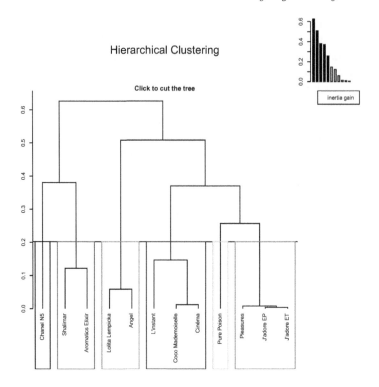

FIGURE 2.4
Representation of the dendrogram used to define the number of clusters, and
obtained with the **HCPC** function (*experts* data set).

The results provided by **HCPC** are composed of five items.

```
> names(res.hcpc)
[1] "data.clust" "desc.var"   "desc.axes"  "call"       "desc.ind"
```

The information related to the cluster each individual belongs to is stored
in the object called **res.hcpc$data.clust**. Another important output is
stored in the object **res.hcpc$desc.var**. This object presents the descrip-
tion of the clusters by the variables of the data set, by providing for each
cluster the list of significant attributes. The attributes are sorted according to
two keys: the sign of the difference between the average score for that cluster
and the overall mean for each attribute (*cf.* second and third columns), and
the significance of the *p-value* of the test that compares the mean over the
individuals of the cluster to the overall mean (*cf.* last column).

```
> res.hcpc$desc.var$quanti
$'1'
NULL
```

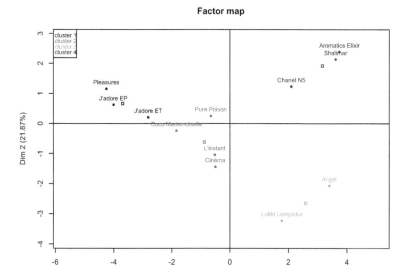

FIGURE 2.5
Representation of the clusters of perfumes on the first two dimensions resulting from a PCA, using the **HCPC** function (*experts* data set).

```
$'2'
         v.test Mean in category Overall mean sd in category Overall sd
Citrus   2.860742       1.7833333     1.0489583      0.27139898  0.4915581
Green    2.831854       2.0986111     0.8458333      0.81479614  0.8471084
Floral   2.038821       8.2763889     6.1187500      0.09206563  2.0264473
Fruity   2.034196       5.4972222     3.4868056      0.79781482  1.8924702
Vanilla -1.987935       0.5402778     3.1131944      0.25208238  2.4783308
Oriental -2.241634      1.0319444     3.6993056      0.09041656  2.2785196
Wrapping -2.715903      3.1722222     5.7368056      0.39932427  1.8081660
                p.value
Citrus   0.004226508
Green    0.004627905
Floral   0.041467867
Fruity   0.041931861
Vanilla  0.046818885
Oriental 0.024985020
Wrapping 0.006609530

$'3'
         v.test Mean in category Overall mean sd in category Overall sd
Greedy   2.996963       8.520833      2.923611       0.6333333   2.770145
Vanilla  2.703981       7.631250      3.113194       0.4479167   2.478331
Floral  -2.459626       2.758333      6.118750       0.2666667   2.026447
```

```
              p.value
Greedy   0.00272684
Vanilla  0.00685142
Floral   0.01390817
```

`$'4'`

	v.test	Mean in category	Overall mean	sd in category	Overall sd
Oriental	2.899943	7.1500000	3.699306	0.5495052	2.2785196
Spicy	2.847849	5.4000000	2.249306	1.1799296	2.1184832
Heady	2.639703	8.1361111	3.962847	0.1800956	3.0273075
Woody	2.484994	2.1694444	1.252431	0.8700012	0.7066206
Wrapping	2.008430	7.6333333	5.736806	0.2181838	1.8081660
Fruity	-2.683452	0.8347222	3.486806	0.1580529	1.8924702

```
              p.value
Oriental 0.003732307
Spicy    0.004401579
Heady    0.008297876
Woody    0.012955364
Wrapping 0.044597652
Fruity   0.007286643
```

Cluster 1 is not particularly characterized by any sensory attributes. The perfumes in cluster 1 have been perceived as some kind of average perfumes. *De facto*, these products are represented close to the center of gravity of the factorial plane. Such result can be directly linked to the notion of contrast defined previously in the *t*-test: the perfumes in cluster 1 are associated with α_i coefficients that are not significantly different from 0.

Cluster 2 is composed of products that have been perceived with a high intensity of *Citrus*, *Green* (at least higher than the average perfume), and to a lesser extent *Floral* and *Fruity*: the *p-values* associated with *Citrus* and *Green* are particularly small compared to the ones associated with *Floral* and *Fruity*. Similarly, products of cluster 2 have been perceived with a low intensity of *Wrapping*, *Oriental*, and to a lesser extent *Vanilla*.

Cluster 3 is composed of products that have been perceived with a high intensity of *Greedy*, *Vanilla*, and with a low intensity of *Floral*.

Cluster 4 is composed of products that have been perceived with a high intensity of *Oriental*, *Spicy*, *Heady*, *Woody*, to a lesser extent *Wrapping*, and with a low intensity of *Fruity*.

2.3 For experienced users: Adding *supplementary* information to the product space

This section is divided into two subsections: the first one, important in itself, introduces the notion of supplementary information and the way it can be represented on a factorial plane; the second one, important within the framework of sensory data, describes the way confidence areas, as particular supplementary information, can be represented.

2.3.1 Introduction to supplementary information

The concept of supplementary information (also called illustrative information), as well as its representation, is of utmost importance when exploring multivariate data. The idea behind the notion of supplementary elements consists in projecting additional rows and/or columns in the plane obtained from PCA performed on the "original" matrix X, in order to see how these additional elements connect to X, but without taking this additional information into account in the construction of the dimensions. Let's denote by X_+ the supplementary rows and by X^+ the supplementary columns. In practice, these two matrices X_+ and X^+ are projected on the vectors (u_s) and (v_s), respectively, after the PCA has been applied to X.

In terms of variables, supplementary information can either be continuous or categorical. As PCA only uses continuous variables in the calculation of the distances between individuals, categorical variables can only be considered as supplementary. For continuous variables, determining whether they are illustrative or not is arbitrary, and depends on the point of view adopted. Often, continuous variables are considered as supplementary if they are from a different nature (*e.g.*, the liking variable in the sensory space). Similarly, the definition of supplementary entities is arbitrary and depends on both the point of view and the sensory issue tackled.

To illustrate this feature, the data collected by C. Asselin and R. Morlat (INRA, Angers, France), who studied the effect of the soil on the quality of the wine produced in the Loire Valley, are used. These data were used by B. Escofier and J. Pagès to illustrate Multiple Factor Analysis in their paper entitled "Multiple factor analysis (AFMULT package)," published in *Computational Statistics & Data Analysis* in 1984. The data can be found either in the book website or in the SensoMineR package.

In the data set, 21 wines are described by 31 variables, amongst which are 29 continuous variables and 2 categorical variables. The 29 continuous variables are made up of 27 sensory attributes and 2 other variables of a slightly different nature, one measuring the overall quality of the wine and the other one measuring the typicality of the wine. The 2 categorical variables are related to the origin of the wine (its appellation) and the nature of the soil on which the grape was produced.

The general idea behind the study is to understand the set of wines regarding their sensory profiles, and eventually to relate these sensory profiles to the origin, the quality, and the typicality of the wines. To do so, the point of view adopted consists in considering the sensory attributes as *active* variables[1]. From this point of view, it is then possible to project the rest of the variables as *illustrative*.

After loading the FactoMineR package, import the *wines.csv* file from the book website (or load the *wine* data set from FactoMineR). Once imported, use the **colnames** function to define the position of the different variables in order to properly specify the role of each variable in the **PCA** function.

```
> library(FactoMineR)
> data(wine)
> colnames(wine)
 [1] "Label"                       "Soil"
 [3] "Odor.Intensity.before.shaking" "Aroma.quality.before.shaking"
 [5] "Fruity.before.shaking"        "Flower.before.shaking"
 [7] "Spice.before.shaking"         "Visual.intensity"
 [9] "Nuance"                       "Surface.feeling"
[11] "Odor.Intensity"               "Quality.of.odour"
[13] "Fruity"                       "Flower"
[15] "Spice"                        "Plante"
[17] "Phenolic"                     "Aroma.intensity"
[19] "Aroma.persistency"            "Aroma.quality"
[21] "Attack.intensity"             "Acidity"
[23] "Astringency"                  "Alcohol"
[25] "Balance"                      "Smooth"
[27] "Bitterness"                   "Intensity"
[29] "Harmony"                      "Overall.quality"
[31] "Typical"
```

Based on this output, it can be seen that the supplementary categorical variables are the first two variables, and the supplementary continuous variables are the two last variables. Through the `quanti.sup` and `quali.sup` arguments of the **PCA** function, we specify the appropriate role of each variable in the analysis.

```
> wine.pca <- PCA(wine,quali.sup=1:2,quanti.sup=30:31,graph=FALSE)
```

Besides the features previously shown, the **plot.PCA** function also allows representing, in a convenient way, the supplementary variables through its `col.quali`, `col.quanti`, `habillage`, and `col.hab` arguments.

In PCA, categorical supplementary information is represented on the individuals factor map (*cf.* Figure 2.6 and Figure 2.7). The representation of these categories is obtained by calculating the center of gravity of the individuals belonging to each category in question. This can be illustrated using the following code:

```
> plot.PCA(wine.pca,col.quali="gray48")
> plot.PCA(wine.pca,col.quali="gray48",habillage=2,col.hab=c("gray","azure4",
+ "black","gainsboro"))
```

---

[1]By opposition to *illustrative* variables, the *active* variables are the variables that participate in the calculation of the distance between individuals.

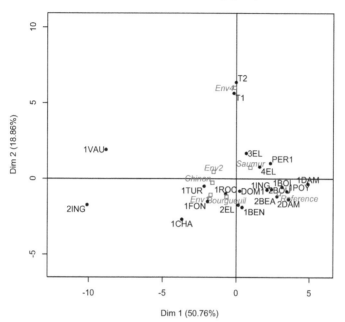

FIGURE 2.6

Representation of the categorical variables *Soil* and *Label* as supplementary variables, on the first two dimensions resulting from a PCA, using the **plot.PCA** function (*wine* data set).

The supplementary continuous variables are represented within the variables representation, using the correlation between these supplementary elements and the components obtained from the active variables (*cf.* Figure 2.8).

```
> plot.PCA(wine.pca,choix="var",col.var="gray",col.quanti="black")
```

2.3.2   The panellipse function of the **SensoMineR** package

As evoked previously, the representation of the individuals on a factorial plane can be misleading. Indeed, when individuals are represented in a subspace of lower dimensionality than the original space R^J, distances amongst individuals are necessarily reduced. This is the price to pay to visualize the individuals. Obviously, this remark is transposable to variables. Based on this observation, how is it possible to find a way to enhance the representation of the distances between individuals and between variables?

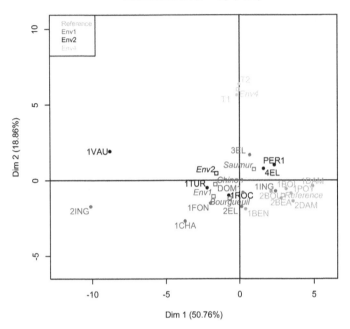

Individuals factor map (PCA)

FIGURE 2.7
Representation of the categorical variable *Soil* through the products as supplementary information, on the first two dimensions resulting from a PCA, using the **plot.PCA** function (*wine* data set).

One possible solution is to represent confidence areas around the individuals and around the variables. To do so, several strategies can be applied. The one explained in this section is implemented in the **panellipse** function of the SensoMineR package. The idea of the ellipses is based on the following question: How would the positioning of the perfumes evolve if we would slightly change the composition of the panel?

To answer this question, the original idea was to generate virtual panels from the original data (*i.e.*, the one used for the ANOVAs in Chapter 1) using simulations. To do so, new panels P_i are obtained by sampling panelists with replacement, from the original pool of panelists. For each new virtual panel P_i, a new matrix of sensory profiles, denoted X_{P_i}, is calculated. Each data set X_{P_i} (associated with the virtual panel P_i) is then combined vertically to the original data set X of sensory profiles, and is projected as a supplementary matrix of illustrative individuals in the original space obtained by PCA on X.

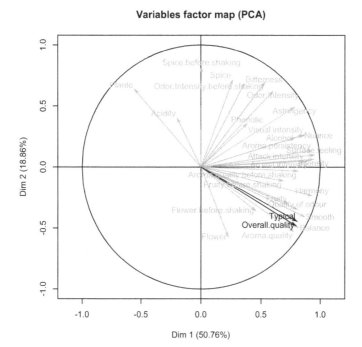

FIGURE 2.8
Representation of the continuous variables *Typical* and *Overall.quality* as supplementary variables, on the first two dimensions resulting from a PCA, using the **plot.PCA** function (*wine* data set).

Ellipses including 95% of the products associated with the virtual panels are created around each product.

This is the procedure that is automatically performed by the **panellipse** function of the SensoMineR package. To run the **panellipse** function, the position of the column related to the product information (col.p), the panelist information (col.j), and the attributes (firstvar and lastvar) should be informed. Regarding the attributes, the argument level.search.desc discards all the attributes for which the *Product* effect is associated with a *p-value* larger than the threshold defined (by default 0.2; use 1 to keep all the variables in the analysis). Let's apply this analysis on the *experts* data set used throughout this chapter.

```
> res.panellipse <- panellipse(experts,col.p=4,col.j=1,firstvar=5,
+ level.search.desc=1)
```

In Figure 2.9, the more ellipses overlap, the less distinctive (or the closer)

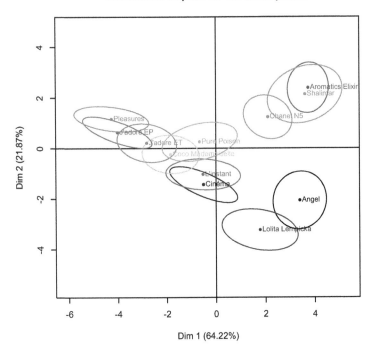

FIGURE 2.9

Representation of the perfumes on the first two dimensions resulting from a PCA, in which each product is associated with a confidence ellipse, using the **panellipse** function (*experts* data set).

the two products are. Hence, from this representation, it appears clearly that *Aromatics Elixir* and *Shalimar* are perceived as similar, whereas *Shalimar* and *Pleasures* are perceived as different.

Regarding the variables, the same idea can be applied to the representation of the sensory attributes in the correlation circle. In this case, each new data set X_{P_i} is combined vertically to the original data set X of sensory profiles. These data set X_{P_i} are then projected as a supplementary matrix of illustrative variables, in the original space obtained by PCA on X.

Such projections return the variables representation presented Figure 2.10.

Figure 2.10 is striking, as we can observe that the variability around the sensory attributes is very much dependent on their *p-value* for the *F*-test of the *Product* effect. When the *Product* effect is not significant, the positioning of the attribute appears to be very unstable. On the contrary, when the *Product* effect is significant, the positioning of the attribute appears to be very stable.

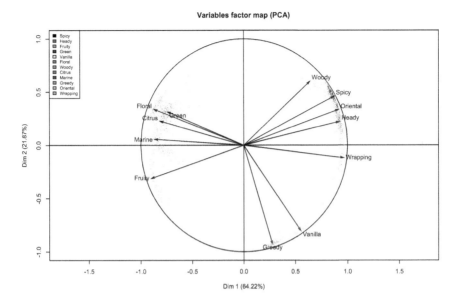

FIGURE 2.10
Representation of the sensory attributes issued from virtual panels, on the first
two dimensions resulting from a PCA, using the **panellipse** function (*experts*
data set).

This idea of representing confidence areas can be used to answer a recurrent
sensory question regarding the number of panelists in a panel: what if the
number of panelists was smaller or higher? Would they still differentiate the
products, or not? One way of answering these related questions is to generate
virtual panels by increasing or decreasing the number of panelists from the
original panel. To do so, the `nbchoix` parameter of the **panellipse** function can
be changed. By default, this parameter is set so that the virtual panel has the
same size as the original panel.

2.4 Exercises

Exercise 2.1 *Manipulating data with R*

The aim of this exercise is to increase your R skills in terms of recoding data.
Indeed, R is not only a powerful statistical software, it can also be an efficient
tool for data manipulation (transformation, or recoding).

- Import the *perfumes_qda_experts.csv* data set from the book website. Check the importation of your data.

- Change the names of the sensory attributes to *"attribute_E"* (*"_E"* for expert panel). To do so, you may use the **paste** function.

- Change the names of the modalities associated with the *Product* factor.

- Convert the *Session* and *Rank* variables into categorical variables.

Exercise 2.2 *Exploring data with R*

The aim of this exercise is to explore data with R. This will get you accustomed to some graphics in R. Don't hesitate to play with some graphical parameters to get familiar with what they are used for.

- Import the *perfumes_qda_experts.csv* data set from the book website. Check the importation of your data.

- Represent the distribution of the *Oriental* attribute, by using the **plot** function combined with the **density** function. Similarly, represent a box plot per product for the *Oriental* attribute. You may either use the **plot** function, the **boxplot** function, or the **boxprod** function of the SensoMineR package.

- After calculating the average table crossing the products in rows and the sensory attributes in columns (the **decat** function or the **averagetable** function of the SensoMineR package can be used), represent the relationship between two variables graphically (*i.e.*, a scatter plot with an attribute on the *x*-axis *versus* another attribute on the *y*-axis).

- In the R console write ?par to get a list of possible parameters to customize your graph.

Exercise 2.3 *Introducing the notion of V-test with the* decat *function*

In this exercise we introduce the notion of *V-test* (that stands for "Value-test"), introduced by a group of French statisticians, amongst them Ludovic Lebart. This indicator is very close to the notion of *z-score*. In its basic form, the *V-test* is the quantile of a standardized normal distribution (with mean equal to 0 and standard deviation equal to 1) corresponding to a given probability. It is used to transform *p-values* into scores that are more easily interpretable. The main feature of this indicator is that it can be positive, or negative, depending on the test that has been performed.

- Import the *perfumes_qda_experts.csv* data set from the book website.

- Use the **decat** function of the SensoMineR package on the imported data set, and store the results in an R object named `resdecat`. Look at the object named `resdecat$resF` [1] and compare the first and the second column using the **pnorm** and the **qnorm** functions, which give the probability and the quantile function of a standard normal distribution by default. Comment on the results.

- Look at the object named `resdecat$resT` [2]. As mentioned previously, the *V-test* is an indicator that can either be positive, or negative. In this output, the sign of this indicator depends on the positiveness of the estimation of the product coefficient α_i. Comment on the results.

Exercise 2.4 *Representing sensory profiles, with the* decat *and the* barrow *functions*

This exercise is a logical follow-up of the previous one: the notion of *V-test* is used to represent sensory profiles of products, and to understand sensory attributes in terms of products.

- Import the *perfumes_qda_experts.csv* data set from the book website.

- Use the **decat** function to get the matrix of the sensory profiles (*adjusted means*). With the **barrow** function of the SensoMineR package, represent a sensory profile per product.

- Transpose the matrix of the sensory profiles with the t[3] function. Apply the **barrow** function to the matrix once transposed. Comment on the results.

Exercise 2.5 *Representing sensory profiles with so-called spider plots*

The aim of this exercise is to show how so-called spider plots can be obtained in R.

- Import the *perfumes_qda_experts.csv* data set from the book website.

- Use the **decat** function to get the matrix of the sensory profiles. With the **stars** function from the graphics package, or the **radial.plot** function of the plotrix package, represent a spider plot per product. Then represent a spider plot for the three products *Shalimar*, *Lolita Lempicka*, and *Chanel N5*. Comment on the results.

---

[1] `resF` stands for results of the *F*-test.
[2] `resT` stands for results of the *t*-test.
[3] Given a matrix or data frame x, $t(x)$ returns the transposed of x.

The following exercises require knowledge on Multiple Factor Analysis. If you do have some knowledge about this particular multivariate technique, you can challenge yourself by trying to do the following exercises. However, if you do not know this technique or require some reminders, we advise you to read Chapter 3 first, and come back later.

Exercise 2.6 *Comparison of sensory profiles obtained from a panel of experts and a panel of consumers*

This exercise aims to become familiar with Multiple Factor Analysis. Comparing results obtained from two different points of view is not an easy task. It is then important to find a common representation in which the two different points of view can be compared.

- Import the *perfumes_qda_experts.csv* file from the book website. Use the **decat** function to get the matrix of the sensory profiles. Repeat the same procedure with the *perfumes_qda_consumers.csv* data set.

- For both the experts and consumers data, represent the two product spaces separately using PCA. To do so, you may use the **PCA** function of the FactoMineR package. Compare visually the two representations. Compare the first eigenvalue of each of the analyses. Find the proper transformation such that the first eigenvalue of each PCA equals 1.

- Once you find the transformation, apply it on each matrix of sensory profiles, and re-run a PCA on each matrix of sensory profiles separately. Comment on the results.

Exercise 2.7 *Understanding performance from a multivariate point of view with the* **panellipse.session** *function*

In the previous chapter, the performance of the panel and the panelists has been evaluated using univariate analysis (ANOVA). This following exercise highlights how multivariate analysis, combined with confidence ellipses, can help assess the reproducibility of a panel.

- Import the *perfumes_qda_experts.csv* data set from the book website.

- With the **subset** function, create two data sets, one for each session, that will be named, respectively, *perfumes_qda_session1* and *perfumes_qda_session2*. Run the **decat** function on each data set with the appropriate ANOVA model. Store the results in **res1** and **res2**.

- Run the **PCA** function on the matrix named **res1$adjmean**. Comment on the results.

- Run the **PCA** function on the matrix named **res2$adjmean**. Comment on the results.

- Merge the two matrices **res1$adjmean** and **res2$adjmean** into a new matrix named *perfumes_profiles_1_2* by using the **cbind** function.

- Run the **MFA** function of the FactoMineR package on the data set *perfumes_profiles_1_2* by considering two groups of 12 variables each, and by using the default options. Comment on the results.

- Finally, run the **panellipse.session** function on the original data set, using the adequate parameters. Compare both results.

2.5 Recommended readings

- Borgognone, M.G., Bussi, J., & Hough, G. (2001). Principal component analysis in sensory analysis: covariance or correlation matrix? *Food Quality and Preference*, 12, 323-326.

- Bro, R., Qannari, E. M., Kiers, H. A., Naes, T., & Frøst, M. B. (2008). Multi-way models for sensory profiling data. *Journal of Chemometrics*, 22, 36-45.

- Cadoret, M., & Husson, F. (2013). Construction and evaluation of confidence ellipses applied at sensory data. *Food Quality and Preference*, 28, 106-115.

- Dehlholm, C., Brockhoff, P. B., & Bredie, W. L. P. (2012). Confidence ellipses: a variation based on parametric bootstrapping applicable on multiple factor analysis results for rapid graphical evaluation. *Food Quality and Preference*, 26, 278-280.

- Dijksterhuis, G. B. (1997). Multivariate data analysis in sensory and consumer science. Trumbull: Food & Nutrition Press.

- Dijksterhuis, G. B., & Heiser, W. J. (1995). The role of permutation tests in exploratory multivariate data analysis. *Food Quality and Preference*, 6, (4), 263-270.

- Husson, F., Bocquet, V., & Pagès, J. (2004). Use of confidence ellipses in a PCA applied to sensory analysis: application to the comparison of monovarietal ciders. *Journal of Sensory Studies*, 19, 510-518.

- Husson, F., Lê, S., & Pagès, J. (2007).Variability of the representation of the variables resulting from PCA in the case of a conventional sensory profile. *Food Quality and Preference*, 18, 933-937.

- Husson, F., Le Dien, S., & Pagès, J. (2005). Confidence ellipse for the sensory profiles obtained with principal component analysis. *Food Quality and Preference*, 16, 245-250.

- Kunert, J., & Qannari, E. M. (1999). A simple alternative to generalized procrustes analysis: application to sensory profiling data. *Journal of Sensory Studies*, 14, 197-208.

- Lawless, H. T., & Heymann, H. (1998). Sensory evaluation of food: principles and practices. New York: Chapman and Hall.

- Luciano, G., & Naes, T. (2009). Interpreting sensory data by combining principal component analysis and analysis of variance. *Food Quality and Preference*, 20, 167-175.

- Meilgaard, M., Civille, G. V., & Carr, B. T. (1999). Sensory evaluation techniques (3rd ed.). Boca Raton: CRC Press.

- Meyners, M. (2003). Methods to analyse sensory profiling data, a comparison. *Food Quality and Preference*, 14, 507-514.

- Naes, T., & Risvik, E. (1996). Multivariate analysis of data in sensory science. Amsterdam: Elsevier.

- Qannari, E. M., Wakeling, I., Courcoux, P., & MacFie, H. J. H. (2000). Defining the underlying sensory dimensions. *Food Quality and Preference*, 11, 151-154.

- SSHA. 1998. Evaluation sensorielle. Manuel méthodologique (SSHA, 3e édition). Edited by Tec & Doc. Paris: Lavoisier.

- Stone, H., & Sidel, J.L. (1985). Sensory evaluation practices. Orlando: Academic Press.

- Stone, H., & Sidel, J.L. (1998). Quantitative descriptive analysis: developments, applications, and the future. *Food Technology Journal*, 52, 48-52.

- Stone, H., Sidel, J., Oliviers, S., Woolsey, A., & Singleton, R. C. (1974, November). Sensory evaluation by quantitative descriptive analysis. *Food Technology*, 24-28.

- Zook, K. L., & Pearce, J. H. (1988). Quantitative descriptive analysis. In: H. Moskowitz (Ed.), Applied sensory analysis of food, vol. 1, 43-71. Boca Raton, Florida: CRC Press.

3

When products are rated according to several lists of attributes

CONTENTS

"... or more generally, when sensory profiles stem from different sources, the use of so-called multiple data tables analyses is essential. Amongst these analyses, we can cite Generalized Canonical Analysis (GCA), Generalized Procrustes Analysis (GPA), and Multiple Factor Analysis (MFA). One of the many advantages of using MFA over other methods is that it can handle data tables of different nature: quantitative, qualitative, but also contingency tables. To perform an MFA, use the **MFA** *function of the FactoMineR package."*

3.1 Data, sensory issues, and notations

In 1984, Anthony A. Williams and Steven P. Langron proposed a truly innovative approach, an avant-gardist way of describing products, the so-called Free Choice Profiling (FCP). It was not simply about proposing a slight variation of QDA®; it was a real revolution in terms of descriptive sensory data on the one hand, and of data analysis on the other hand. One fundamental rule of the game was dramatically changing, as from now on, panelists could use their own list of sensory attributes to describe the same set of products.

This radical evolution of the way descriptive data could be collected induced many different changes. The recurrent question regarding the quality of the panel, and therefore of the data, did not arise anymore (or at least very differently, and certainly more simply). Indeed, the strategies developed to assess the performance of the panel, based on ANOVA models (*cf.* Chapter 1), could not be applied anymore. On the contrary, as the rationale behind FCP is to offer panelists the choice in the sensory attributes to describe the stimuli, we expect from the method that panelists provide reliable data, as they have used their own criteria. Intrinsically, assessing the performance of the panel may not be as crucial as it can be with classical[1] descriptive analysis methods; it may even be irrelevant, considering the idea of freedom (which does not mean that panelists cannot be trained).

Still, as a descriptive method, we expect of FCP that it provides data that would allow us to precisely describe the product space. As previously explained, each panelist j ($j = 1, \ldots, J$) describes the products using their own attributes. Each panelist then provides the sensory profiles of the products in the form of a matrix X_j of dimension $I \times K_j$, where I denotes the number of products and K_j the number of sensory attributes used by panelist j. These J matrices are usually bound together into a matrix $X = [X_1 | \ldots | X_J]$. This matrix X is then subjected to some analysis.

Before choosing and presenting the analysis of such particular data, let's present our expectations from them. The analysis of X should provide a representation of the products based on the J matrices X_j. Similarly to the evaluation of the product space in QDA (*cf.* Chapter 2), the differences between products should be understandable and interpretable. However, such product configuration should not be related to one particular panelist only (*i.e.*, the analysis should not be dominated by one matrix X_j), but should be balanced across the individual matrices. In other words, the different panelists' points of view (associated with the individual matrices X_j) should be equally balanced in the analysis. Additionally, it would be interesting to compare, within the product configuration, the individual points of view (*i.e.*, the J matrices) at different levels of interest: we would like to see what is common between the individual matrices, and what is specific to each of them. As a matrix is associated with one panelist, this representation would help us in understanding the panelists, and therefore would help us in assessing some kind of performance of the panel, at least in terms of agreement: if all the individual judgments are projected close together regarding a given product, it can be concluded that the panelists are in agreement with the description of that product.

As we will see, all these expectations are fulfilled by MFA. MFA is an exploratory multivariate method dedicated to the analysis of so-called multiple data tables, in the sense that one set of statistical individuals is described by several groups of variables. Without loss of generality, MFA is presented

---

[1] *Classical* in the sense that they are based on a single list of sensory attributes.

here within the framework of this chapter, *i.e.*, when the different groups of data contain quantitative variables. In the following chapters, other examples involving MFA, in which the groups of variables are either qualitative or contingency tables, will be presented.

Remark. The methodology presented here can also be applied to Flash Profile data. Indeed, Flash Profile is a variation of Free Choice Profile, in which panelists are asked to rank (rather than rate) the products according to their own, freely defined, sensory attributes. By considering these ranks as quantitative scores (*i.e.*, a low rank corresponds to a low intensity score), as it is often done in practice, the situation is identical to Free Choice Profile, and the same analyses do apply.

3.2 In practice

To be consistent with the previous chapters, and without loss of generality (we are more interested in the methodology rather than the results *per se*), the data set used to illustrate this methodology is directly derived from the *experts* data set presented in Chapter 2. For each panelist j ($j = 1, \dots, J$), a random selection of K_j out of 12 attributes is performed, K_j varying between 3 and 12. These K_j attributes are considered as being the K_j sensory attributes panelist j has freely selected to describe the I products. Hence, with respect to these random selections, matrices of dimension (I, K_j) are generated, forming the *FCP* data set.

These matrices are stored in the *perfumes_fcp.csv* file, which can be directly downloaded from the book website. After importing this data set in your R session, print on screen the names of the variables using the **colnames** function.

```
> FCP <- read.table(file="perfumes_fcp.csv",header=TRUE,sep=",",dec=".")
> colnames(FCP)
 [1] "Wrapping_CM"  "Green_CM"     "Floral_CM"    "Citrus_CM"    "Vanilla_CM"
 [6] "Woody_CM"     "Fruity_CR"    "Greedy_CR"    "Oriental_CR"  "Citrus_CR"
[11] "Floral_CR"    "Vanilla_CR"   "Wrapping_CR"  "Citrus_GV"    "Greedy_GV"
[16] "Woody_GV"     "Vanilla_GV"   "Fruity_GV"    "Green_GV"     "Wrapping_GV"
[21] "Oriental_GV"  "Marine_GV"    "Green_MLD"    "Marine_MLD"   "Oriental_MLD"
[26] "Woody_MLD"    "Floral_MLD"   "Vanilla_MLD"  "Wrapping_MLD" "Citrus_MLD"
[31] "Fruity_MLD"   "Greedy_MLD"   "Marine_NMA"   "Citrus_NMA"   "Woody_NMA"
[36] "Oriental_NMA" "Greedy_NMA"   "Floral_NMA"   "Fruity_NMA"   "Wrapping_NMA"
[41] "Oriental_PR"  "Wrapping_PR"  "Floral_PR"    "Woody_PR"     "Greedy_PR"
[46] "Marine_PR"    "Fruity_PR"    "Citrus_PR"    "Greedy_RL"    "Fruity_RL"
[51] "Marine_RL"    "Green_SD"     "Wrapping_SD"  "Marine_SD"    "Citrus_SD"
[56] "Oriental_SM"  "Floral_SM"    "Fruity_SM"    "Woody_SM"     "Wrapping_SM"
[61] "Marine_SM"    "Green_SM"     "Citrus_SM"    "Wrapping_SO"  "Floral_SO"
[66] "Greedy_SO"    "Vanilla_SO"   "Green_SO"     "Fruity_SO"    "Marine_SQ"
[71] "Floral_SQ"    "Wrapping_SQ"  "Greedy_SQ"    "Green_SQ"     "Vanilla_ST"
[76] "Citrus_ST"    "Wrapping_ST"  "Woody_ST"     "Fruity_ST"    "Oriental_ST"
```

A close look at the names of the variables shows that they correspond to a combination of a sensory attribute with the initials of an assessor (*e.g.*, *Citrus_NMA*). More particularly, the matrix *FCP* is composed of 12 groups of variables (or equivalently, 12 submatrices), each group (or submatrix) corresponding to the sensory profiles of the products provided by one panelist. In other words, the matrix *FCP* is a combination of 12 submatrices, forming a multiple data table. This type of data is complex, as it consists of many different connected parts, and new points of view have to be defined to fully understand it. In the following part, *FCP* is tackled using different perspectives.

3.2.1   Why can't I analyze such a table in a classical way?

Before running any analysis, it is important to understand the structure of the whole matrix by first fractionating it into its different submatrices. Such separation can either be done manually or by using the **grep** function (as we will see later). A closer look at the data shows that the number of sensory attributes per submatrix is, respectively, 6, 7, 9, 10, 8, 8, 3, 4, 8, 6, 5, and 6. Since from one submatrix to another, the number of columns is quite variable, the structure of the product space, induced by each submatrix, might be very different from one subject to another.

To evaluate the sensory profiles provided by each subject and to compare them together, a PCA is performed on each individual submatrix. To do so, the **PCA** function is applied on the *FCP* data set, restricted to the variables of one assessor at the time, as shown by the following code for the panelists *MLD* and *CM*. Since the **PCA** function belongs to the FactoMineR package, load this package first, if required.

```
> library(FactoMineR)
```

To compare the sensory profiles of these two panelists in particular, the columns associated with each of them need to be defined. Two solutions are proposed: the first one uses the **grep** function (*e.g.*, *MLD*), and the second one selects the columns manually (*e.g.*, *CM*). **PCA** is then applied on each submatrix separately.

```
> pos.mld <- grep(pattern="MLD",x=colnames(FCP))
> res.mld <- PCA(FCP[,pos.mld])
> res.cm <- PCA(FCP[,1:6])
```

The representations of the product space provided by *MLD* (*cf.* Figure 3.1) and *CM* (*cf.* Figure 3.2) seem pretty similar, at least with respect to the first dimension. Indeed, in both cases, the first dimension opposes *Angel* and *Shalimar* on the one hand, to *Pleasures* and *J'adore EP* on the other hand.

Unfortunately, this visual comparison is not straightforward, and may even be impossible in some cases. For instance, after applying **PCA** to two data sets, it may happen that the common structure to the two spaces involves different

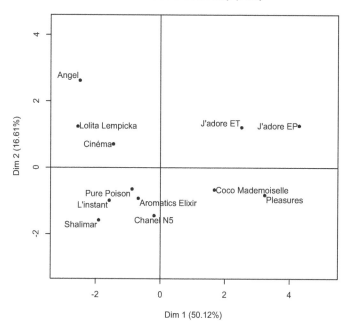

FIGURE 3.1
Representation of the perfumes as described by *MLD*, on the first two dimensions resulting from a PCA, using the **plot.PCA** function (*FCP* data set).

dimensions. In our context, it would mean that people have identified the same opposition between products, but attached to it different importance. And what is the most important for one panelist (the separation between products is observed on the first dimension, say) might be of lesser importance for another panelist (the same separation of the products is only observed on the second [or following] dimension). Such particularity can be highlighted through the example of *GV* (*cf.* Figure 3.3).

```
> res.gv <- PCA(FCP[,14:22],graph=FALSE)
> par(mfrow=c(1,2))
> plot.PCA(res.gv,axes=c(1,2))
> plot.PCA(res.gv,axes=c(2,3))
```

The first plane of the representation of the product space provided by *GV* is not so close from the one provided by *MLD* or *CM*. However, the plane defined by dimensions 2 and 3 shows higher similarities with the space provided by *MLD* or *CM*. By simply comparing representations based on the first two dimensions, it would have been quite impossible to get to the

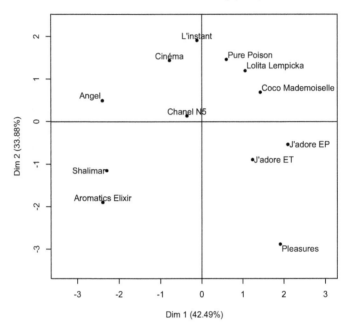

FIGURE 3.2
Representation of the perfumes as described by *CM*, on the first two dimensions resulting from a PCA, using the **plot.PCA** function (*FCP* data set).

conclusion that our three panelists share some common points of view on the products.

Based on this observation, it seems important to define an analysis that 1) searches automatically for the common structure within the different individual product configurations, and 2) summarizes the results by providing a consensual space[1]. To obtain such consensual space, it is important to give to each panelist the same importance, or weight. In other words, the panelists should be potentially contributing equally in the construction of the first dimension of the consensual space. Such procedure is not so straightforward as panelists are not providing the same number of attributes, and these attributes can be structured differently. To summarize, the panelists' product spaces are not equally multidimensional.

To evaluate the difference between individual configurations, let's have a

---

[1]The term *"consensual space"* (or consensus) refers here to the common structure shared by a majority of panelists.

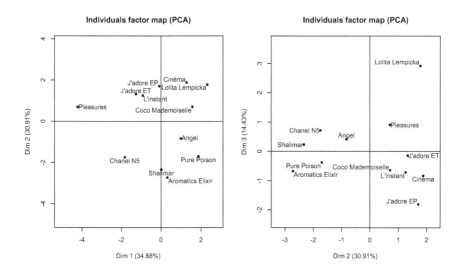

FIGURE 3.3
Representation of the perfumes as described by *GV*, on dimensions 1 and 2 (left), and dimensions 2 and 3 (right), resulting from a PCA, using the **plot.PCA** function (*FCP* data set).

look at the first eigenvalue of the separate analyses performed. The eigenvalues of each individual PCA are stored in the object $eig.

```
> res.mld$eig[1:5,]
        eigenvalue percentage of variance cumulative percentage of variance
comp 1  5.012340901           50.12340901                       50.12341
comp 2  1.661172538           16.61172538                       66.73513
comp 3  1.320380921           13.20380921                       79.93894
comp 4  0.725992634            7.25992634                       87.19887
comp 5  0.577713719            5.77713719                       92.97601

> res.cm$eig[1:5,]
       eigenvalue percentage of variance cumulative percentage of variance
comp 1  2.5493674            42.489457                        42.48946
comp 2  2.0328518            33.880864                        76.37032
comp 3  0.7135735            11.892892                        88.26321
comp 4  0.4272731             7.121218                        95.38443
comp 5  0.1650426             2.750710                        98.13514

> res.gv$eig[1:5,]
        eigenvalue percentage of variance cumulative percentage of variance
comp 1 3.138849137           34.87610153                       34.87610
comp 2 2.781578107           30.90642341                       65.78252
comp 3 1.298607466           14.42897185                       80.21150
comp 4 0.947358213           10.52620236                       90.73770
comp 5 0.362196114            4.02440127                       94.76210
```

As can be seen, *MLD* product space is dominated by its first dimension (large first eigenvalue, and relatively low eigenvalues otherwise), whereas the first two dimensions of *CM* space are almost equally important (the first two eigenvalues are almost identical): *MLD* space is defined as uni-dimensional, whereas *CM* space is bi-dimensional. If no adjustment is performed, the simultaneous analysis of these two panelists (as if they were only one) would be dominated by *MLD*, as their first dimension explains almost two times more variance than *CM* (5.01 *versus* 2.55).

It is worth reminding the reader that, in standardized PCA, the same weight is given to each variable (*i.e.*, a weight equal to 1). A direct consequence of this is that the first eigenvalue of such PCA cannot exceed the number of attributes in the data set. In practice, this upper limit is attained if all the attributes are perfectly correlated. Oppositely, the first eigenvalue cannot be smaller than 1, this lower limit being attained in a situation in which all the variables are perfectly orthogonal.

When comparing submatrices from different sizes (*e.g.*, a panelist using 3 attributes *versus* a panelist using 10 attributes), such specificity can bias the final results towards the panelist using more attributes, as they have a larger total variance than another panelist using only a few.

Through these remarks and observations, it seems clear that a classical analysis (such as PCA) is limited when considering such a particular data set. Hence, more advanced statistical methodologies that can handle multiple data tables are required here. As evoked, these analyses should balance the part of each submatrix (or equivalently of each group of variables), and should find the common structures between all the groups of variables. In other words, it should provide a representation of the product space that is based on a fair consensus.

3.2.2 How can I get a representation of the product space based on a consensus?

As mentioned previously, MFA is an analysis designed for multiple data tables. It balances the part of each submatrix X_j within the global analysis of the matrix $X = [X_1 | \ldots | X_J]$. By analogy, the procedure of balancing the submatrices in MFA is similar to the weighting of the variables in PCA when these variables are standardized. In standardized PCA, the weight of each variable equals 1, and consequently the part of each variable (when calculating distances amongst individuals) is the same from one variable to the other. Similarly in MFA, within each group j, a same weight is applied to all its variables. This weight is obtained by calculating the variance of the main dimension of variability of group j, denoted λ_1^j; and the variables of the group j are all divided by λ_1^j. Note that such weighting does not modify the structure on the individuals induced by each group of variables, as this transformation is nothing else but a homogeneous dilation of each product space associated with a group of variables, in other words a *homothety*. Still this homothety has an important effect on the global analysis of X, as the weight of the main dimension of variability of each X_j now equals 1. Consequently, such weighting of the variables guarantees that, in the simultaneous analysis of multiple groups, no group can generate the main dimension of variability all by itself. This very important feature of MFA will be detailed in Chapter 7.

When variables are continuous, which is the case here, the core of MFA is a weighted PCA, with the weights as previously defined. *De facto*, the analyst benefits from all the features of PCA: the transition formulae that link the scatter plot of the individuals and the scatter plot of the variables, the geometrical point of view that integrates the notion of contribution and of quality of representation, the notion of supplementary information (individuals and variables), *etc.* Moreover, the analyst benefits from complementary outputs that are specific to the analysis of several groups of variables (*i.e.*, several points of view).

Let's apply MFA on the *FCP* data set. To do so, the **MFA** function of the FactoMineR package is used. For this function, three main arguments need to be specified. The first argument, called `group`, defines the structure of the submatrices within the entire data set, and takes as argument the number of variables belonging to each group. The second argument, called `type`, defines the separate analysis to perform on each submatrix. Technically, this argument defines the way each group should be handled to be balanced in the global analysis. Here, the variables of each group being quantitative, the user should simply specify whether the PCA performed on each group separately should be on the standardized or unstandardized data (*i.e.*, should the PCA be based on the correlation matrix, or on the variance-covariance matrix?). To be consistent, the same point of view as in Chapter 2 is adopted, and the same importance is given to each attribute within each group. In other words, standardized PCA are considered, and the argument `type` takes an `"s"` for all the groups. Finally, the third argument, called `name.group`, facilitates the interpretation of the results, as it associates each separate group to its given name.

To run **MFA** on the *FCP* data set, the following code is used:

```
> res.mfa <- MFA(FCP,group=c(6,7,9,10,8,8,3,4,8,6,5,6),type=rep("s",12),
name.group=c("CM","CR","GV","MLD","NMA", "PR","RL","SD","SM","SO","SQ","ST"))
```

As usual, an overview of the results provided by the **MFA** function, and stored here in `res.mfa`, is given by the **names** function.

```
> names(res.mfa)
 [1] "separate.analyses" "eig"            "group"
 [4] "inertia.ratio"     "ind"            "summary.quanti"
 [7] "summary.quali"     "quanti.var"     "partial.axes"
[10] "call"              "global.pca"
```

The first element of `res.mfa`, called `res.mfa$separate.analyses`, stores the results of the separate analysis of each group. All the other objects are specific to MFA, or at least have to be interpreted by taking into account the specificities of MFA. As an example, when evaluating the eigenvalues, the analyst should bear in mind that the first eigenvalue cannot exceed the number of *active*[2] groups of variables in the analysis. This upper limit is due to the particular weighting procedure related to this method: as the main dimension of variability of each group equals 1, the upper limit is attained when the first dimensions of the separate analysis of all the groups are perfectly correlated. Oppositely, the first eigenvalue cannot be smaller than 1. This lower limit is attained if all the groups are perfectly orthogonal.

```
> round(res.mfa$eig,3)
        eigenvalue percentage of variance cumulative percentage of variance
comp 1     7.901            31.088                      31.088
comp 2     5.589            21.989                      53.076
comp 3     3.184            12.525                      65.602
comp 4     1.830             7.200                      72.802
comp 5     1.606             6.320                      79.121
comp 6     1.241             4.884                      84.006
comp 7     1.088             4.281                      88.286
comp 8     0.946             3.722                      92.008
comp 9     0.878             3.454                      95.461
comp 10    0.644             2.533                      97.995
comp 11    0.510             2.005                     100.000
```

In this example, the first eigenvalue equals 7.901 (maximum 12). This relatively high value highlights the fact that the first dimension provided by MFA is globally linked to the groups, in the sense that it represents an important direction of variability for the majority of them.

Regarding the product space thus obtained, as the variables are continuous, the representations of the products and of the attributes must be interpreted in the same way as the ones provided by PCA. Notably, Figures 3.4 and 3.5 can (and have to) be interpreted jointly.

Due to the excessive numbers of variables in the data set X, it seems tedious to graphically interpret the sensory dimensions provided by MFA (*cf.* Figure 3.5). To simplify the task, three complementary strategies can be adopted. The first one is based on graphical outputs, and on the geometrical point of view exposed in Chapter 2, in which only the most contributive, or well-represented, variables are displayed. The second one requires the use of the **dimdesc** function, as it provides a list of variables that are significantly linked to each dimension. The third one consists in performing cluster analysis on the coordinates of the variables. In this case, it is advised to perform cluster analysis by using the matrix of correlation coefficients between attributes as matrix of distance (*cf.* Exercise 3.2).

---

[2]Active, as in PCA, in the sense that the group participates in the construction of the dimensions.

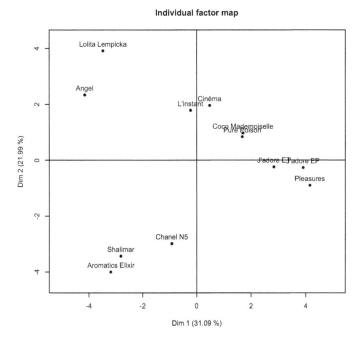

FIGURE 3.4

Representation of the perfumes on the first two dimensions resulting from an MFA on all panelists, using the **MFA** function (*FCP* data set).

For the first strategy, as exposed in Chapter 2, the `select` argument of **plot.MFA** is used:

```
> par(mfrow=c(1,2))
> plot.MFA(res.mfa,choix="var",select="cos2 0.8")
> plot.MFA(res.mfa,choix="var",select="contrib 10")
```

This code generates the simplified graphics presented in Figure 3.6. By playing with the level of `cos2` or `contrib`, more or less attributes can be shown, hence facilitating the interpretation of the results by limiting the representation to the most important variables only.

In the second strategy, the **dimdesc** function is applied to the results of MFA:

```
> dimdesc(res.mfa)
$Dim.1
$Dim.1$quanti
          correlation       p.value
Floral_SO   0.9144486 3.122001e-05
```

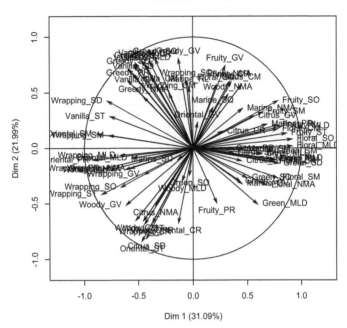

Correlation circle

FIGURE 3.5
Representation of the sensory attributes on the first two dimensions resulting
from an MFA on all panelists, using the **MFA** function (*FCP* data set).

```
Floral_MLD     0.9135088  3.291967e-05
Fruity_ST      0.8622927  3.080797e-04
Fruity_MLD     0.8218378  1.039482e-03
Green_SD       0.7905379  2.207113e-03
Fruity_RL      0.7820988  2.648216e-03
Fruity_SO      0.7808247  2.720246e-03
Floral_SM      0.7804696  2.740587e-03
Floral_CM      0.7786151  2.848695e-03
Green_SM       0.7531895  4.683229e-03
Floral_PR      0.7495261  5.007423e-03
Floral_CR      0.7483746  5.112747e-03
Green_CM       0.7282594  7.236187e-03
Marine_MLD     0.7156150  8.872644e-03
Floral_NMA     0.6794342  1.508604e-02
Marine_PR      0.6790919  1.515693e-02
Fruity_SM      0.6427840  2.416990e-02
Citrus_MLD     0.6329578  2.716306e-02
Green_MLD      0.5930702  4.209904e-02
Greedy_SO     -0.5906471  4.316157e-02
Greedy_PR     -0.5953423  4.111974e-02
```

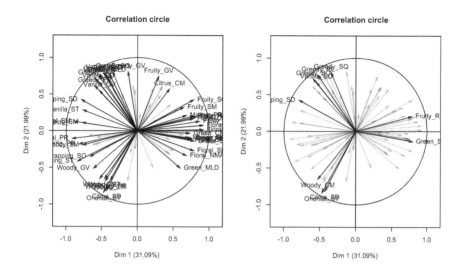

FIGURE 3.6
Representation of the sensory attributes on the first two dimensions resulting
from an MFA on all panelists, after filtering according to the cos2 (left) and
the contribution (right), using the **plot.MFA** function (*FCP* data set).

```
Woody_GV       -0.6114113 3.464998e-02
Oriental_NMA   -0.6188564 3.191422e-02
Wrapping_SO    -0.6593233 1.968315e-02
Wrapping_MLD   -0.6691841 1.731705e-02
Vanilla_ST     -0.7607128 4.067401e-03
Woody_SM       -0.7629432 3.897220e-03
Wrapping_SM    -0.7730777 3.190917e-03
Wrapping_SD    -0.7856073 2.457364e-03
Wrapping_PR    -0.8157756 1.215528e-03
Wrapping_ST    -0.8374420 6.758509e-04
Oriental_SM    -0.8819701 1.475046e-04
Oriental_PR    -0.9117125 3.637026e-05
```

The **dimdesc** function shows that the first dimension of the MFA is linked
to homologous attributes. This is, for instance, the case for *Wrapping* (negative
side), and for *Floral* and *Fruity* (positive side). Such an observation can be
facilitated using the **grep** function.

A more systematic study of the link between attributes, and more precisely
between homologous attributes, gives an insight of some kind of performance
of the panelists, at least in terms of agreement. In this case, the strong link between homologous concepts (such as *Wrapping* on the one hand, and *Floral* on
the other hand) suggests that an agreement exists between panelists on these
descriptors. Another example that emphasizes such agreement is shown in Figure 3.7 with the example of *Greedy*. Such homogeneity between homologous

attributes is expected here, since the panelists have been trained to recognize and consistently rate these attributes. However, when non-trained assessors (*e.g.*, consumers) are performing FCP or Flash Profile tests, the heterogeneity between homologous attributes does not necessarily mean "disagreement" from a product point of view, but might simply highlight a disagreement between consumers' definition of these attributes.

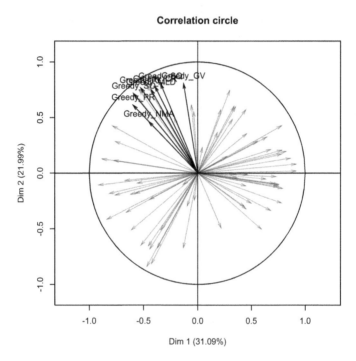

FIGURE 3.7

Assessment of the agreement amongst panelists for the attribute *Greedy* on the first two dimensions resulting from an MFA on all panelists, using the **plot.MFA** and the **grep** functions (*FCP* data set).

Additionally, this consistency between homologous attributes suggests that the MFA dimensions position products with similar sensory profiles together, and oppose products with very different profiles. To confirm this intuition, more understanding of the product space is required. Such additional information can be obtained by applying HAC on the MFA space. As already presented in Chapter 2, this can be done by using the **HCPC** function:

```
> res.hcpc <- HCPC(res.mfa)
```

The **HCPC** function proposes to separate the product space in four clusters (*cf.* Figure 3.8).

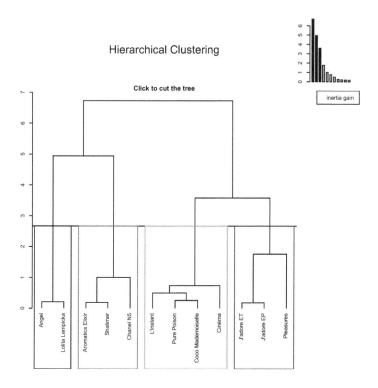

FIGURE 3.8
Representation of the dendrogram using the **HCPC** function (*FCP* data set).

The object `res.hcpc$desc.var` highlights how clusters of products are characterized by the different attributes. For instance, the following attributes are specific to cluster 1 and cluster 4:

```
> round(res.hcpc$desc.var$quanti$'1',3)
            v.test Mean/category Overall mean sd/category Overall sd p.value
Greedy_PR    3.234        9.450        2.412       0.200      3.228   0.001
Greedy_SO    3.131        9.775        2.829       0.075      3.290   0.002
Greedy_CR    2.946        8.075        2.508       0.975      2.803   0.003
Greedy_MLD   2.885        8.975        2.704       0.875      3.224   0.004
Greedy_SQ    2.834        5.625        2.550       0.125      1.609   0.005
Vanilla_GV   2.820        6.975        1.675       3.025      2.788   0.005
Greedy_RL    2.782        9.650        2.746       0.350      3.681   0.005
Vanilla_ST   2.670        9.900        3.796       0.100      3.391   0.008
Vanilla_CR   2.466        8.125        2.750       1.075      3.234   0.014
Vanilla_SO   2.357        4.650        1.925       1.650      1.715   0.018
Vanilla_MLD  2.315        9.100        3.221       0.100      3.766   0.021
Oriental_SM  2.114        9.000        3.533       0.650      3.835   0.034
```

```
Greedy_GV     2.063         7.450        3.242        2.550        3.026  0.039
Wrapping_SM   1.963         9.450        6.038        0.200        2.578  0.050
Floral_MLD   -1.993         0.850        5.663        0.750        3.581  0.046
Floral_SO    -2.159         3.775        6.842        0.425        2.107  0.031
Floral_SM    -2.213         3.050        5.996        0.750        1.974  0.027
Fruity_ST    -2.240         0.000        5.588        0.000        3.699  0.025
Floral_CR    -2.254         4.850        7.325        0.750        1.628  0.024
Floral_PR    -2.277         2.025        5.288        0.425        2.125  0.023
Floral_NMA   -2.647         2.700        7.662        2.700        2.781  0.008
```

```
> round(res.hcpc$desc.var$quanti$'4',3)
               v.test Mean/category Overall mean sd/category Overall sd p.value
Green_SD        3.003         6.433        2.125        0.703        2.748  0.003
Green_SM        2.844         4.267        1.300        1.858        1.997  0.004
Green_CM        2.779         4.650        1.579        2.133        2.116  0.005
Marine_MLD      2.685         3.150        0.904        1.844        1.602  0.007
Citrus_MLD      2.670         3.000        0.846        1.817        1.545  0.008
Green_MLD       2.528         1.150        0.517        0.248        0.480  0.011
Green_SQ        2.363         0.900        0.433        0.286        0.378  0.018
Fruity_SO       2.204         8.217        4.875        1.252        2.903  0.028
Fruity_MLD      2.196         9.167        4.263        0.826        4.277  0.028
Citrus_GV       2.176         4.400        2.425        1.736        1.738  0.030
Floral_MLD      2.105         9.600        5.663        0.356        3.581  0.035
Fruity_PR       2.097         3.733        2.154        1.281        1.442  0.036
Fruity_CR       2.071         8.150        4.546        0.495        3.333  0.038
Marine_GV       1.989         1.200        0.425        1.071        0.746  0.047
Oriental_PR    -2.094         0.650        3.417        0.122        2.530  0.036
Wrapping_MLD   -2.472         1.050        6.054        0.648        3.877  0.013
Wrapping_PR    -2.648         2.917        6.354        0.170        2.486  0.008
Wrapping_SD    -2.675         2.117        6.108        0.649        2.858  0.007
```

As expected, these two clusters of products are described with homologous sensory attributes. Indeed, cluster 1 is defined by products, which were perceived by most panelists as intense for *Greedy* and *Vanilla*, and not intense for *Floral*. The cluster 4 is defined by products, which are perceived by most panelists as intense for *Green*, *Fruity*, *Marine*, and *Citrus*, and not as intense for *Wrapping*.

Tips: Selecting "homologous" attributes using the grep function

In this tip, we will show you how "homologous" attributes can be searched automatically using the **grep** function, and how such a tool can be applied to generate useful graphics for the interpretation of FCP data (*cf.* Figure 3.7 for an example).

In practice, the **grep** function searches for a particular `pattern` within a vector of elements. We are particularly interested here in searching a given attribute, within the list of attributes. For instance, the attribute *Floral* (using `pattern="Floral"`) can be automatically extracted from the entire set of attributes (using x=**colnames**(FCP)) by using the following code:

```
> Floral <- grep(pattern="Floral",x=colnames(FCP),value=TRUE)
> Floral
```

```
[1] "Floral_CM"  "Floral_CR"  "Floral_MLD" "Floral_NMA" "Floral_PR"
[6] "Floral_SM"  "Floral_SO"  "Floral_SQ"
```

In this example, since the argument `value=TRUE` is used, the object `Floral` contains the values of the elements that present the particular pattern. However, if `value` is set to `FALSE` (by default), the object `Floral` contains the position within the vector of elements where the pattern is found.

Based on this automatic search of pattern, lists of homologous attributes can easily be created. These lists can then be used in the **plot.MFA** function in order to represent simultaneously homologous variables only (*cf.* Figure 3.9).

```
> Wrapping <- grep(pattern="Wrapping",x=colnames(FCP),value=TRUE)
> Greedy <- grep(pattern="Greedy",x=colnames(FCP),value=TRUE)
> par(mfrow=c(1,2))
> plot.MFA(res.mfa,choix="var",select=Wrapping)
> plot.MFA(res.mfa,choix="var",select=Floral)
```

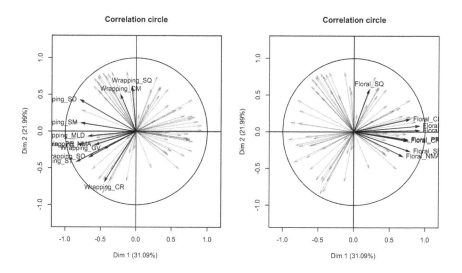

FIGURE 3.9

Assessment of the agreement amongst panelists for the sensory attributes *Wrapping* and *Floral* on the first two dimensions resulting from an MFA on all panelists, using the **plot.MFA** and the **grep** functions (*FCP* data set).

A similar procedure can be applied to panelists. To do so, the `pattern` to search should be related to the panelist of interest, and not to the attributes. Such selection then generates a graphic, which highlights the sensory attributes of that panelist. To represent the attributes defined by *CR* (*cf.* Figure 3.10), the following code is used:

```
> CR <- grep(pattern="CR",x=colnames(FCP),value=TRUE)
> CR
[1] "Fruity_CR"    "Greedy_CR"    "Oriental_CR" "Citrus_CR"    "Floral_CR"
[6] "Vanilla_CR"   "Wrapping_CR"
> plot.MFA(res.mfa,choix="var",select=CR)
```

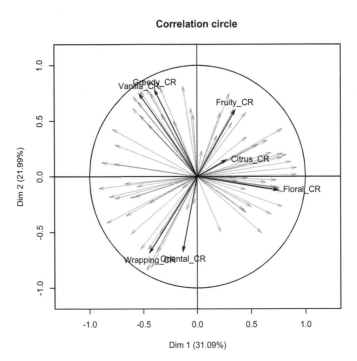

FIGURE 3.10
Representation of the sensory attributes of *CR* on the first two dimensions
resulting from an MFA on all panelists, using the **plot.MFA** and the **grep**
functions (*FCP* data set).

3.2.3 How can I integrate the group structure in my inter-pretation?

In fine, when integrating different groups of variables in the analysis, beyond
getting a global representation of the products, it is important to evaluate
how much common information is shared between the groups. In other words,
do we have some common factors amongst our groups of variables? If so, how
are the common factors amongst our groups of variables structured?

These main questions were the ones raised by Hotelling in 1936, when he first introduced Canonical Analysis for assessing relations between two groups of variables. His approach has then been generalized by Carroll (1968) to the situation in which more than two groups of variables are involved. Without loss of generality, when two groups of attributes only are considered, the idea of Canonical Analysis is to find two linear combinations of variables, one for each group, that are most highly correlated together. These two linear combinations (one for each group) are the so-called canonical variables. In practice, they are qualified as "canonical" as they are representative of something, in our case a group of sensory attributes.

In the case of Generalized Canonical Analysis (GCA), canonical variables are not defined directly by comparing sets of variables two by two, but indirectly by first defining a variable that would be representative of all the groups of variables. In other words, the idea is to find a sequence of z_s such that the following criterion is maximized:

$$\sum_j R^2(z_s, X_j).$$

Here R^2 denotes the coefficient of determination, and we have:

$$\text{Var}(z_s) = 1 \text{ and } \text{Cov}(z_s, z_{s'}) = 0, \forall s \neq s'.$$

The z_s are the common factors to the group of variables. To obtain the canonical variables, z is simply regressed on the variables of X_j. A drawback of GCA is related to the use of the coefficient of determination, and *de facto* the use of multiple linear regression. Unfortunately, this method is very sensitive to multi-collinearity, a phenomenon that is very often observed with sensory data.

The way MFA extracts its dimensions of variability from the groups of variables can be seen as a variant of GCA. In MFA, the coefficient of determination is replaced by another measure of relationship between a variable and a group of variables: the so-called Lg measure. By definition, the $L_g(z, X_j)$ coefficient, measured between a variable z and X_j, equals the inertia of all the variables of X_j orthogonally projected onto z. When the variables v_k of X_j are continuous (which is our case), with weights equal to m_k, the coefficient $L_g(z, X_j)$ is defined by:

$$L_g(z, X_j) = \sum_k m_k r^2(z, v_k).$$

By analogy, this measure is precisely the one used in PCA when extracting its main dimensions of variability. But of course, in the context of PCA, it does not suffer from multi-collinearity problems. As mentioned previously, in MFA, the variables of a group j are all weighed by $1/\lambda_1^j$. In that particular case, the $L_g(z, X_j)$ coefficient measured between a variable z and X_j can be written as follows:

$$L_g(z, X_j) = \frac{1}{\lambda_1^j} \sum_k r^2(z, \nu_k).$$

This measure equals 1 if and only if z corresponds to the first dimension of the PCA performed on the variables of X_j.

The canonical variables are then obtained by performing a partial least squares regression (PLS regression) of z on the variables of X_j. In fact, it can be shown that a weighted PCA using the MFA weights, and a variant of GCA using the L_g measure, lead to the same solution, *i.e.*, a representation of the individuals based on orthogonal dimensions. Indeed, in both cases, the sequence of z_s is nothing else than the sequence of F_s, the coordinates of the individuals provided by PCA (*cf.* Chapter 2).

This representation can be enhanced using the canonical variables that correspond to the vectors of coordinates of the individuals for a given group of variables and a given dimension. Hence, the following notation F_s^j is used to denote the vector of coordinates of the individuals regarding group j on dimension s.

Geometrically, within the so-called "global" representation of the stimuli (or consensual space) obtained from all the variables (the part of each group being balanced), the representation of the stimuli defined by each group of variables separately can be added. This superimposed representation (also called "partial" points representation) is of utmost importance, as it shows how close the different points of view can be, within each product. This representation is simply obtained by projecting the point of view induced by each X_j onto the dimensions issued from MFA. One main feature of this representation is the fact that the positioning of the partial representations can be interpreted in a similar way: for a given stimuli, the point of view provided by group j can be very extreme compared to points of view provided by other groups, regarding the first dimension for instance. In the FCP example, the partial point representation corresponds to Figure 3.11. By default, the **MFA** function provides this graphic, but only shows the partial points for the two products with the smaller within inertia (*i.e.*, the partial points are close to mean point), and the two products with the larger within inertia (*i.e.*, the partial points are distant from the mean point). To represent all the partial points, the following code is required:

```
> plot.MFA(res.mfa,choix="ind",partial="all",habillage="group")
```

The products *Lolita Lempicka* and *Pure Poison* are the products associated with the largest within inertia. For these two products, the points of view diverge more than for any other products. Oppositely, *J'adore EP* and *Chanel N5* are the products associated with the smaller within inertia. For these two products, the individual points of view converge more.

Remark. With FCP (and Flash Profile), panelists are free to evaluate the

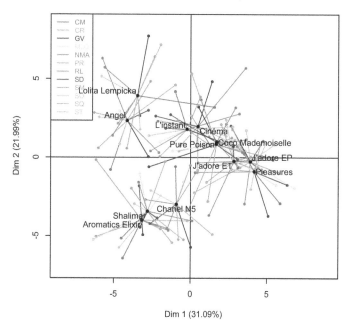

FIGURE 3.11
Representation of all the partial individuals on the first two dimensions of the product space, using the **plot.MFA** function (*FCP* data set).

products using their own criteria. Hence, divergence of opinion between panelists cannot be interpreted as a strict disagreement in terms of product perception, but as a disagreement in the importance given to the differences perceived between products. In other words, in a particular situation of FCP (and Flash Profile), the best common space summarizes the dominating differences existing between products as highlighted by the panelists.

To evaluate the relation between each panelist point of view in the final product space, the correlation between the MFA dimensions (or common factors) and the canonical variables is investigated. This correlation coefficient is stored in **res.mfa$group$correlation**. In this matrix, the intersection of row j and column s corresponds to the correlation coefficient $r(F_s^j, F_s)$ between the canonical variable of group j and each dimension of the MFA. This correlation coefficient shows to which extent the dimensions can be considered as a structure of group j. Note that this coefficient is always positive.

```
> res.mfa$group$correlation
```

	Dim.1	Dim.2	Dim.3	Dim.4	Dim.5
CM	0.8420411	0.8076817	0.7147114	0.5142463	0.2279936
CR	0.9386000	0.8238809	0.6996418	0.5302432	0.4402471
GV	0.8191236	0.8752828	0.8207007	0.4298697	0.7613124
MLD	0.9041807	0.8610971	0.5506487	0.3983491	0.7618902
NMA	0.7205395	0.8164912	0.6208808	0.4069858	0.7666184
PR	0.9676199	0.8536618	0.6201820	0.5699056	0.7013625
RL	0.8302574	0.8705330	0.3269824	0.7330304	0.2283961
SD	0.8707191	0.9005601	0.5925741	0.4393523	0.2383805
SM	0.9358547	0.6504157	0.7123539	0.4556074	0.6241345
SO	0.9430347	0.9545336	0.5949590	0.6398699	0.5407440
SQ	0.6718014	0.8696481	0.7288451	0.6976033	0.4543576
ST	0.9147933	0.8202435	0.8463279	0.5811781	0.5139779

In this example, the first dimension of the MFA is strongly correlated with the canonical variable of *PR*, *SO*, *SM*, and *CR*. These panelists separated the products in a similar way as the one of the MFA consensual space. *SQ*, and *NMA*, on the other hand, separated the products differently: the canonical variables associated with their group are less correlated with the first dimension of the MFA.

3.3 For experienced users: Comparing different panels with hierarchical multiple factor analysis (HMFA)

In the previous section, the analysis of multiple data tables has been presented within the context of FCP (or Flash Profile). For such particular data, MFA appears to be a good solution, as it balances the different groups in the analysis, and the consensual space defined corresponds to the common information shared by a majority of panelists. The natural extension of such methodology consists in considering any situations, in which the same stimuli are described by different groups of variables, and the configurations related to these different groups of variables need to be compared.

In practice, two different situations involving multiple data tables exist. In the first situation, the different groups of variables come from different sources or are of a different nature and the analysis aims at measuring the part in common across them. This corresponds to situations such as FCP data, Napping® data (*cf.* Chapter 7), or assessment of the relationship between data of a different nature (*e.g.*, sensory and instrumental).

In the second situation, the different groups of variables are of a similar nature, and the analysis aims to challenge the consistency of the individual configurations. This is, for instance, the case when panelists provide the sensory profiles of the same products in duplicate or triplicate, and the analyst wants to ensure the similarity of the repeated assessments. Such situation can be extended to cases in which the sensory profiles of the same stimuli are obtained from different panels (as in cross-cultural studies, for instance).

Although the two situations (and the analysis involved) show similarities, the sensory objectives, and hence the interpretations of the results, are clearly different. Indeed, in the first case, the analysis aims at finding the common part shared by a majority of individual configurations. It is therefore assumed that the individual configurations can be different, as they come from different sources. Hence, the goal of MFA is to measure the strength of the link between the different configurations, and to highlight, in decreasing order, which dimensions are in common within the different groups. In the second case, the different groups are expected to be identical. Therefore, the goal of the analysis is not the evaluation of the similarities between the different configurations, but in the evaluation of the differences (if any). Ideally, the different configurations are perfectly identical.

To illustrate the second case, let's put ourselves in a situation where the same samples were evaluated by two different panels (experts *versus* consumers), using different settings (QDA *versus* Ideal Profile Method[1]), different number of attributes (with some overlap), and different number of sessions (in duplicate for the experts). With this setting, the purpose of this analysis is to evaluate the differences between the sensory profiles provided from these two different panels. More precisely, through this comparison of sensory profiles, an element of response to the following "burning" question can be answered: Can consumers profile products?

The data sets used for illustration are the ones presented in Chapter 2 and in Chapter 10. One the one hand, 12 experts rated 12 perfumes in duplicate on 12 sensory attributes. This data set is stored in the file *perfumes_qda_experts.csv*. On the other hand, the same 12 perfumes were rated by 103 consumers on 21 attributes. This data set is stored in the file *perfumes_consumers.csv*. Both data sets can be downloaded from the book website.

After importing the two data sets in R, let's transform the adequate variables into factors (use the **summary** function to check).

```
> experts <- read.table("perfumes_qda_experts.csv",header=TRUE,sep=",",dec=".",
+ quote="\"")
> experts$Session <- as.factor(experts$Session)
> experts$Rank <- as.factor(experts$Rank)
> consumers <- read.table("perfumes_consumers.csv",header=TRUE,sep=",",dec=".",
+ quote="\"")
> consumers$consumer <- as.factor(consumers$consumer)
```

To compare the sensory profiles of the two panels, the following three-step procedure can be considered:

1. For each panel, generate the matrix of sensory profiles.

2. After setting the names of the products and attributes correctly, combine the two matrices of sensory profiles.

---

[1] For more information concerning the Ideal Profile Method (IPM), refer to Chapter 10.

3. Perform MFA on the resulting multiple data table.

Although this procedure answers the sensory issue raised previously, a slightly different methodology is considered here.

Let's consider that, besides comparing the experts' profiles to the consumers' profiles, we also would like to distinguish between the different profiles provided by the experts (*i.e.*, Session 1 and Session 2).

By taking this new information into account, a first solution consists in following the same procedure as before, except that 1) the average sensory profiles of the experts is now separated into two blocks of sensory profiles (*i.e.*, one for each session), and 2) the final MFA is not performed on two (*i.e.*, experts and consumers), but on three (*i.e.*, expert session 1, expert session 2, and consumers) groups. Intuitively, this solution has drawbacks, as the MFA space is probably more influenced by the experts than the consumers. Indeed, our intuition (legitimately) states that the experts' configurations across sessions are more similar than the configuration of the consumers. In the case this intuition is verified, the MFA space is mainly driven by the relation between the different expert sessions, rather than the common part between experts and consumers.

To integrate this new information properly, the variables need to be structured according to a hierarchy leading to groups and subgroups of variables. In this example, the 12 perfumes were profiled by 12 sensory experts in two sessions on the one hand, and by a consumer panel on the other hand. Analyzing such data implies balancing the part of the consumer panel with the part of expert panels on the one hand, but also the part of the experts across sessions on the other hand. To do so, it seems necessary to consider some hierarchy on the data (as represented in Figure 3.12). As explained previously, the usual methods (MFA, Generalized Procrustes Analysis [GPA], *etc.*) do not suit this type of problem, since they lead to outputs where a point of view of a group of variables may be preponderant in comparison to the point of view of other groups.

The approach to consider for such structure on the variables in a global analysis involves balancing the groups of variables within every node of the hierarchy. It is also necessary to have appropriate outputs: in particular, we would like to have a graphical display of each perfume globally, as profiled by the consumer panel, as profiled by the expert panel, and as profiled by the expert panel within each session. Such graphical displays make it possible to investigate the repeatability of the experts and to compare, product by product, the assessment of the consumer panel to the expert panel, which is the aim of this study.

HMFA is an extension of MFA to the case where variables are structured according to a hierarchy. In MFA, we consider a partition on the variables: observations are described by variables partitioned into groups. One of the main features of MFA is that it balances the importance of each group before performing PCA on the merged groups. The particular weights applied to the different groups provide interesting properties: in particular, it allows gen-

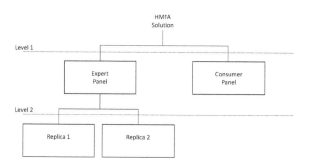

FIGURE 3.12
Representation of the hierarchical structure of the data considered for the comparison of the consumers (*consumers* data set) and the experts (*experts* data set) sensory profiles using HMFA.

erating different graphical displays which highlight the similarities amongst individuals through different points of view. In HMFA, a succession of MFA is applied to each node of the hierarchy, in order to balance the groups of variables within every node, by going through the hierarchical tree from the bottom up. In the example outlined above, first an MFA is applied to the perfumes profiled by the 12 expert panelists, in which each group of variables corresponds to one session. Second, an MFA is applied to the groups associated with the consumer panel on the one hand, and with the expert panel considered as one group of variables on the other hand, the variables of this latter group being balanced according to the first step.

Based on this information, the procedure used to analyze such data involves the following steps:

1. For each panel, generate the adequate matrix of sensory profiles (one for the consumer panel, one per session for the expert panels).

2. After setting the names of the products and attributes correctly, combine the matrices of sensory profiles.

3. (Either) Perform MFA on the two experts' sessions only, extract the scores on the entire set of dimensions, combine it to consumer profiles, and run a second MFA.

4. (Or) Perform HMFA directly on the resulting multiple data table.

The solution involving HMFA is presented here, whereas the succession of MFA is proposed as an exercise (*cf.* Exercise 3.3).

In order to follow this procedure step-by-step, let's first generate the sensory profiles for the consumer and for the experts. These profiles can be obtained using the **averagetable** function of the SensoMineR package.

```
> profile.conso <- averagetable(consumers,formul="~product+consumer",firstvar=3)
> profile.expert1 <- averagetable(experts[experts$Session=="1",],
+ formul="~Product+Panelist",firstvar=5)
> profile.expert2 <- averagetable(experts[experts$Session=="2",],
+ formul="~Product+Panelist",firstvar=5)
```

To facilitate the differentiation between the different profiles, the attribute names are changed according to the sensory profile they belong to. To do so, the **paste** function is applied to the **colnames** function, as follows:

```
> colnames(profile.conso) <- paste(colnames(profile.conso),"_C",sep="")
> colnames(profile.expert1) <- paste(colnames(profile.expert1),"_E1",sep="")
> colnames(profile.expert2) <- paste(colnames(profile.expert2),"_E2",sep="")
```

Then merge the three tables together. Here, it is of utmost importance to make sure that the products are ordered in the same way within the different tables to avoid any mistakes (*e.g.*, combining the data of two different products). As the names are exactly identical between data sets, the following code is used:

```
> data.hmfa <- cbind(profile.expert1,
+ profile.expert2[rownames(profile.expert1),],
+ profile.conso[rownames(profile.expert1),])
```

Finally, HMFA is performed on *data.hmfa* using the **HMFA** function of the FactoMineR package. This function takes as inputs three major arguments. The first argument called H defines the hierarchy between groups. This hierarchy is defined here as a list with one vector for each hierarchical level; in each vector the number of variables or the number of groups constituting the group is defined. In this example, at the first level, three groups of variables are present: expert session 1, expert session 2, and consumer. In the second level of the hierarchy, the first two groups (*i.e.*, the experts) are grouped together, whereas the consumer panel constitutes a group by itself. The second important argument, called **type**, defines the type of variables of each group, within the first partition. The same point of view as usual is adopted, and standardized PCA is considered for each group. The third argument, called **name.group**, follows the same structure as the argument H, and takes one name per table and per group of tables.

To perform **HMFA** on the *data.hmfa* data set, the following code is used:

```
> hierar <- list(c(12,12,21),c(2,1))
> name.groups <- list(c("Expert S1","Expert S2","Consumer"),
+ c("Experts","Consumer"))
> res.hmfa <- HMFA(data.hmfa,H=hierar,type=rep("s",3),name.group=name.groups)
```

```
> names(res.hmfa)
[1] "eig"        "group"        "ind"          "partial"     "quanti.var"
[6] "call"
```

The consensual space obtained by HMFA is shown in Figure 3.13. The interpretation rules that apply here are similar to the one of PCA. Since the interpretation of this space is of less interest here, it is not done.

Individuals factor map

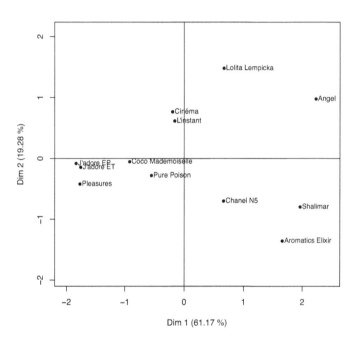

FIGURE 3.13
Representation of the common product space on the first two dimensions, using the **HMFA** function (*consumers* and *experts* data set).

Since it is of our first interest to compare the configurations provided from the experts and consumers, the partial points related to the highest level of the hierarchy are projected onto this product space. This combined representation is given Figure 3.14.

As the partial points are all relatively close to their respective mean point, the configurations provided from the experts and consumers are similar. This is particularly true for the products *J'adore EP* and *J'adore ET*. Oppositely, the products showing larger differences between the two panels are *Angel*, *Shalimar*, and *Lolita Lempicka*. A closer look to the partial points shows that the expert panel separated more the products on the second dimension than the consumer one.

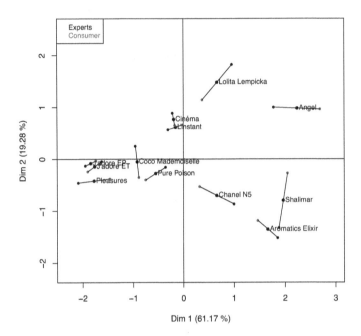

FIGURE 3.14
Representation of the products and their partial representations at the first level of the hierarchy (experts *versus* consumers), on the first two dimensions, using the **HMFA** function (*consumers* and *experts* data set).

In multivariate analysis, the notion of close and/or large distance is quite subjective. To evaluate more objectively the closeness of the configurations provided from the two panels, a statistical criterion is required. As in MFA, the first eigenvalue can play the role of such indicator here. Let's remember that the first eigenvalue is comprised between 1 (*i.e.*, the consumer and expert configurations are perfectly orthogonal) and 2 (*i.e.*, the first dimension of both configurations are perfectly correlated). To obtain the eigenvalue of the HMFA, the following code is used:

```
> res.hmfa$eig
        eigenvalue percentage of variance cumulative percentage of variance
comp 1  1.94345389            61.1729019                         61.17290
comp 2  0.61250531            19.2794528                         80.45235
comp 3  0.14849063             4.6739481                         85.12630
```

```
comp 4   0.11204465            3.5267605              88.65306
comp 5   0.09726771            3.0616359              91.71470
comp 6   0.08081816            2.5438634              94.25856
comp 7   0.07151431            2.2510119              96.50957
comp 8   0.05312035            1.6720365              98.18161
comp 9   0.03026155            0.9525243              99.13414
comp 10  0.01427355            0.4492798              99.58342
comp 11  0.01323483            0.4165847             100.00000
```

The first eigenvalue being close to 2 (2 being the upper limit), it can be concluded here that the two configurations show strong similarities. Although the two configurations are not perfectly identical, the differences observed here can be considered as random noise.

Finally, the last comparison that we are interested in concerns the assessment of the experts between replicas. To investigate this, the partial points related to the second level of the hierarchy are additionally projected on the product space. This corresponds to splitting the expert partial point into two partial points, one for each session. The corresponding representation is provided in Figure 3.15.

The additional split of the experts by replica shows that for *Angel*, for instance, the expert panel is very consistent across session. Indeed, the partial points are very closely projected to the mean expert point. However, for other products such as *Lolita Lempicka* and *Pleasures*, for instance, the split shows some larger disagreement, as the partial points are more distant from the expert mean point.

Based on the results presented here, it seems fair to conclude that the configurations provided from experts and consumers are similar. Indeed, the differences observed between configurations seem to be rather random noise than clear differences. To fully appreciate the similarity between the consumer and expert configurations, the two spaces should not only be visually similar, but should also be interpreted as such.

To interpret the product space, the variable representation is used (*cf.* Figure 3.16). Similarly to MFA, this representation appears to be tedious to interpret, as a large number of attributes are involved. To facilitate the interpretation of this representation, the same three strategies presented previously can be used. Here, we propose to use the **dimdesc** function of the FactoMineR package on res.hmfa.

```
> res.dimdesc <- dimdesc(res.hmfa)
> res.dimdesc$Dim.1$quanti
             correlation      p.value
musk_C         0.9817951   1.527452e-08
incense_C      0.9527733   1.708810e-06
leather_C      0.9319083   1.027512e-05
Wrapping_E1    0.9312953   1.073473e-05
woody_C        0.9248572   1.661522e-05
Heady_E2       0.9152533   2.982038e-05
Spicy_E2       0.9139044   3.219550e-05
nutty_C        0.9045027   5.318614e-05
Heady_E1       0.9038050   5.509099e-05
```

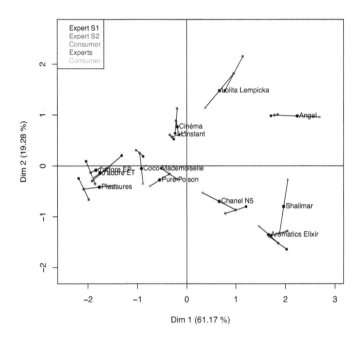

FIGURE 3.15

Representation of the products and their partial representations at the second
level of the hierarchy (replica 1 *versus* replica 2 for the experts), on the first
two dimensions, using the **HMFA** function (*consumers* and *experts* data set).

```
Oriental_E2     0.8920477 9.607862e-05
Wrapping_E2     0.8874647 1.173407e-04
animal_C        0.8821313 1.465415e-04
Oriental_E1     0.8801411 1.587857e-04
earthy_C        0.8780683 1.723735e-04
spicy_C         0.8762217 1.852290e-04
Spicy_E1        0.7995216 1.801699e-03
intensity_C     0.7350580 6.455990e-03
anis_C          0.6427558 2.417812e-02
Woody_E1        0.6242097 3.004544e-02
caramel_C       0.6080048 3.595606e-02
Vanilla_E1      0.6035530 3.771537e-02
Woody_E2        0.5979171 4.002961e-02
Green_E2       -0.6007532 3.885278e-02
Citrus_E2      -0.6027217 3.805060e-02
Marine_E1      -0.7371594 6.228149e-03
Green_E1       -0.7748437 3.078553e-03
rose_C         -0.7992477 1.813146e-03
Floral_E2      -0.8154015 1.227096e-03
```

Correlation circle

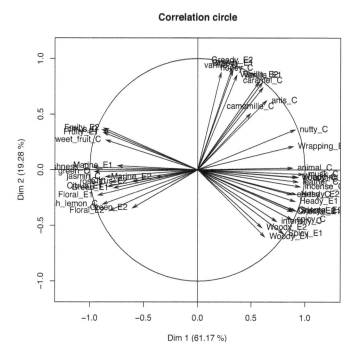

FIGURE 3.16
Representation of sensory attributes on the first two dimensions using the
HMFA function (*consumers* and *experts* data set).

```
Citrus_E1        -0.8292097  8.526424e-04
sweet_fruit_C    -0.8454013  5.332940e-04
jasmin_C         -0.8503061  4.578517e-04
Fruity_E2        -0.8638968  2.913702e-04
Fruity_E1        -0.8701717  2.326657e-04
fresh_lemon_C    -0.8781417  1.718774e-04
citrus_C         -0.8798709  1.605074e-04
Floral_E1        -0.9207748  2.149516e-05
green_C          -0.9531522  1.642406e-06
freshness_C      -0.9621273  5.757976e-07
```

The results of the **dimdesc** function shows that the first dimension of the
HMFA opposes attributes related to oriental notes on its positive side (*e.g.*,
incense, spicy, musk, leather, etc.), to attributes related to fruity and flowery
notes on its negative side (*e.g., freshness, green, floral, citrus, etc.*). Such opposition is true for both panels (and between sessions for the expert panel), hence
confirming the reliability of the consumer panels when profiling products.

Besides providing an argument in favor of the use of consumers for profiling products (for many years, the possible role of consumers in sensory tests was a controversial discussion), this example also provides you more knowledge on the use of MFA and its derivatives. In particular, and although the **MFA** function is used the same way in practice, the objectives, and hence the interpretation of the results, should be adapted to each situation.

3.4   Exercises

Exercise 3.1 *Introducing the notion of illustrative (or supplementary) groups of variables in Multiple Factor Analysis*

The data used in this exercise were collected by C. Asselin and R. Morlat (INRA, Angers, France), who studied the effect of the soil on the quality of the wine produced in the Loire Valley. In particular, these data were used by B. Escofier and J. Pagès to illustrate MFA in their paper entitled "Multiple factor analysis (AFMULT package)" (*Computational Statistics & Data Analysis*, 1994). In this data set, 21 wines are described by 31 variables, amongst which 29 sensory attributes and 2 categorical variables are related to the origin of the wine (its appellation, and the nature of the soil on which the grape was produced). As explained in their paper, the sensory attributes can be naturally gathered into five groups, each of them corresponding to a particular phase when tasting a wine: first the olfaction of the wine at rest, then its visual perception, followed by its olfaction after being agitated, its gustation, and its overall assessment.

In this study, we want to describe the wines regarding the different phases, each of them being balanced within a global analysis, and we certainly want to see whether the structure on the wines induced by this multivariate (and multiple[1]) description can be related to the origin of the wines and/or to the way they are globally perceived.

- To answer these issues, first import the *wines.csv* file from the book website.

- Perform an MFA on the data set by justifying carefully the constitution of the groups of variables and their status, *i.e.*, whether you consider them as active or illustrative. Comment on the results.

Exercise 3.2 *Introduction to the notion of clustering attributes*

In this chapter, the issue of interpreting the variable representation has been raised in the context of multiple data tables analysis. Indeed, such a task can become tedious, as often a large number of variables are involved. However,

---

[1]By definition, involving several parts.

when possible, it is of utmost importance to evaluate whether homologous attributes are used similarly throughout the groups of variables. Such an investigation plays the role of assessing the performance of the panelists (at least in terms of agreement).

- Import the file *perfumes_fcp.csv* file from the book website.

- Run the analysis on this data set through MFA.

- Calculate the correlation matrix between attributes using the **cor** function.

- The matrix of correlations is a similarity[2] matrix. Transform it into a matrix of dissimilarity (or matrix of distance).

- Perform cluster analysis using the **agnes** function of the cluster package and the method of your choice. Plot the results to represent the dendrogram. Use the **cutree** function to cut the tree at the level of interest, and combine the words/attributes with their classification information. Conclude regarding the agreement between homologous attributes.

Exercise 3.3 *Introduction to Hierarchical Multiple Factor Analysis as a succession of Multiple Factor Analysis*

HMFA is a particular application of MFA to multiple data tables structured in groups and subgroups. More precisely, and depending on the complexity of the hierarchy, it consists in performing successive MFA on the adequate tables. This exercise has been designed for you to get a deeper understanding of how these successive MFA are performed. To do so, you will be asked to apply a similar analysis as in Section 3.3, *For experienced users*, except that you will be asked to perform it both manually and automatically.

- Import the *biscuits_france.csv* and the *biscuits_pakistan.csv* files from the book website.

- Generate the matrix of sensory profiles for each session for the French panel and for the Pakistani panel.

- Perform MFA between the two sessions (French panel only) by setting the option `ncp=Inf` (this option is used so that the coordinates of the products are retrieved on all the possible dimensions).

- After extracting the coordinates of the products (stored in `$ind$coord`), combine them to the matrix of the sensory profile provided by the Pakistani panel.

---

[2]A similarity matrix is a matrix in which the value x_{AB} between two elements A and B is inversely proportional to the distance between A and B: the higher x_{AB}, the closer A and B.

- Run a second MFA on these two groups (coordinates on one side, Pakistani on the other side), by setting the analysis appropriately (justify the settings used).

- Once the results are obtained, run the same analysis using the **HMFA** function and compare the results.

The two following exercises are peculiar, as, rather than teaching you R and/or developing your skills in answering sensory issues, they are meant to satisfy your curiosity by providing unconventional examples of application of MFA. Read attentively these different documents: surely, they will inspire you just as much as they inspired us.

Exercise 3.4 *Some considerations about Gaëtan Lorho's dissertation for the degree of Doctor of Science in Technology: "Perceived Quality Evaluation: An Application to Sound Reproduction over Headphones"*

Gaëtan Lorho's dissertation is notably a source of inspiration in terms of data structure. You can find it and download it at the following address[3]: `http://lib.tkk.fi/Diss/2010/isbn9789526031965/`.

- While reading his dissertation, we recommend you to think carefully about the different structures on the variables (in other words, ways of grouping) he has considered in order to answer his different research questions.

Exercise 3.5 *A small journey in Tapio Lokki's work*

The aim of this exercise is to develop your critical sense by "reviewing" one of Tapio Lokki's papers. The idea is to enter into a completely new world as Lokki is studying a very unique type of product: concert halls.

The paper can be downloaded from the following link: `https://mediatech.aalto.fi/~ktlokki/Publs/JAS000835_lokki2011.pdf`.

3.5   Recommended readings

- Arnold, G. M. (1986). A generalised Procrustes macro for sensory analysis. *Genstat Newsletter*, 18, 61-80.

- Arnold, G. M., & Williams, A. A. (1985). Use of generalised Procrustes techniques in sensory analysis. In: J. R. Piggott (Ed.), Statistical procedures in food research, 233-253.

---

[3]It is also available on the book website.

- Blancher, G., Lê, S., Siefferman, J.-M., & Chollet, S. (2008). Comparison of visual appearance and texture profiles of jellies in France and Vietnam and validation of attribute transfer between the two countries. *Food Quality and Preference*, 19, (2), 185-196.

- Carroll, J. D. (1968). Generalisation of canonical analysis to three or more sets of variables. *Proceedings of the 76th Convention of the American Psychological Association*, 3, 227-228.

- Clapperton, J. F., & Piggott, J. R. (1979). Flavor characterization by trained and untrained assessors. *Journal of the Institute of Brewing*, 85, (5), 275-277.

- Delarue, J., & Sieffermann, J.-M. (2004). Sensory mapping using Flash profile. Comparison with a conventional descriptive method for the evaluation of the flavour of fruit dairy products. *Food Quality and Preference*, 15, (4), 383-392.

- Deliza, R., MacFie, H., & Hedderley, D. (2005). The consumer sensory perception of passion-fruit juice using free-choice profiling. *Journal of Sensory Studies*, 20, (1), 17-27.

- Dijksterhuis, G. B., & Gower, J. C. (1991). The interpretation of generalized procrustes analysis and allied methods. *Food Quality and Preference*, 3, (2), 67-87.

- Dijksterhuis, G. B., & Punter, P. (1990). Interpreting generalized procrutes analysis "analysis of variance" tables. *Food Quality and Preference*, 2, (4), 255-265.

- Escofier, B., & Pagès, J. (1994). Multiple factor analysis (AFMULT package). *Computational Statistics & Data Analysis*, 18, (1), 121-140.

- Escofier, B. & Pagès, J. (1998). Analyses factorielles simples et multiples; objectifs, méthodes et interpretation, 3e édition. Paris: Dunod (284 p).

- Follet, C., Lê, S., McEwan, J. A., & Pagès, J. (2006). Comparaison d'evaluations sensorielles descriptives fournies par trois panels de deux pays différents. Agro-Industrie et Méthodes Statistiques - 9èmes Journées Européennes, Janvier 2006, Montpellier, France, pp. 51-59.

- Forde, C. G. (2006). Cross-cultural sensory analysis in the Asia-Pacific region. *Food Quality and Preference*, 17, 646-649.

- Gower, J. C. (1975). Generalized procrustes analysis. *Psychometrika*, 40, (1), 33-51.

- Hanafi, M., & Kiers, H. (2006). Analysis of K sets of data, with differential emphasis on agreement between and within sets. *Computational Statistics and Data Analysis*, 51, (3), 1491-1508.

- Hirst, D., Muir, D. D., & Naes, T. (1994). Definition of the sensory properties of hard cheese: a collaborative study between Scottish and Norwegian panels. *International Dairy Journal*, 4, (8), 743-761.

- Hotelling, H. (1936). Relations between two sets of variates. *Biometrika*, 28, (3-4), 321-377.

- Husson, F., Le Dien, S., & Pagès, J. (2001). Which value can be granted to sensory profiles given by consumers? Methodology and results. *Food Quality and Preference*, 12, (5), 291-296.

- Husson, F., & Pagès, J. (2003). Comparison of sensory profiles done by trained and untrained juries: methodology and results. *Journal of Sensory Studies*, 18, (6), 453-464.

- Lawless, H. T. (1984). Flavor description of white wine by "expert" and non-expert wine consumers. *Journal of Food Sciences*, 49, (1), 120-123.

- Lê, S., Husson, F., & Pagès, J. (2006). Confidence ellipses applied to the comparison of sensory profiles. *Journal of Sensory Studies*, 21, (3), 241-248.

- Lê, S., Pagès, J., & Husson, F. (2008). Methodology for the comparison of sensory profiles provided by several panels: application to a cross-cultural study. *Food Quality and Preference*, 19, (2), 179-184.

- Le Dien, S., & Pagès, J. (2003a). Analyse factorielle multiple hiérarchique. *Revue de Statistique Appliquée*, 51, (2), 47-73.

- Le Dien, S., & Pagès, J. (2003). Hierarchical multiple factor analysis: application to the comparison of sensory profiles. *Food Quality and Preference*, 14, (5), 397-403.

- Le Mée, D. (2006). Using a descriptive sensory panel across different countries and cultures. *Food Quality and Preference*, 17, (7-8), 647-648.

- Lokki, T., Pätynen, J., Kuusinen, A., & Tervo, S. (2012). Disentangling preference ratings of concert hall acoustics using subjective sensory profiles. *The Journal of the Acoustical Society of America*, 132, (5), 3148-3161.

- Lokki, T., Pätynen, J., Kuusinen, A., Vertanen, H., & Tervo, S. (2011). Concert hall acoustics assessment with individually elicited attributes. *The Journal of the Acoustical Society of America*, 130, (2), 835-849.

- Lorho, G. (2010). Perceived quality evaluation: an application to sound reproduction over headphones. ISBN 978-952-60-3196-5.

- Lorho, G., Vase Legarth, S., & Zacharov, N. (2010). Perceptual validation of binaural recordings for mobile multimedia loudspeaker evaluations. AES 38*th* International Conference, Piteå, Sweden, 2010, June 13-15.

- MacFie, H. J. H., & Hedderley, D. (1993). Current practice in relating sensory perception to instrumental measurements. *Food Quality and Preference*, 4, (1), 41-49.

- McEwan, J. A., Colwill, J. S., & Thomson, D. M. H. (1989). The application of two free-choice profile methods to investigate the sensory characteristics of chocolate. *Journal of Sensory Studies*, 3, (4), 271-286.

- Moskowitz, H.R. (1996). Experts versus consumers: a comparison. *Journal of Sensory Studies*, 11, (1), 19-37.

- Pagès, J. (1995). Eléments de comparaison entre l'analyse factorielle multiple et la méthode Statis. *Comptes rendus des XXVII Journées ASUm Jouy en Josas*.

- Pagès, J. (2002). Comparaison entre évaluations sensorielles descriptives fournies par des jurys entrainés et un jury initié : méthodologie et résultats. *Sciences des Aliments*, 22, 557-575.

- Pagès, J. (2005). Analyse factorielle multiple et analyse procustéenne. *Revue de Statistique Appliquée*, 53, (4), 61-86.

- Pagès, J., Bertrand, C., Ali, R., Husson, F., & Lê, S. (2007). Sensory analysis comparison of eight biscuits by French and Pakistani panels. *Journal of Sensory Studies*, 22, (6), 665-686.

- Pagès, J., & Husson, F. (2001). Inter-laboratory comparison of sensory profiles. Methodology and results. *Food Quality and Preference*, 12, (5-7), 297-309.

- Pagès, J., & Husson, F. (2005). Multiple factors analysis with confidence ellipses: a methodology to study the relationships between sensory and instrumental data. *Journal of Chemometrics*, 19, (3), 138-144.

- Qannari, E. M., Courcoux, P., Lejeune, M., & Maystre, O. (1997). Comparaison de trois Stratégies de Détermination d'un Compromis en Evaluation Sensorielle. *Revue de Statistique Appliquée*, 45, (1), 61-74.

- Risvik, R., Colwill, J. S., McEwan, J. A., & Lyon, D. H. (1992). Multivariate analysis of conventional profiling data: a comparison of a British and a Norwegian trained panel. *Journal of Sensory Studies*, 7, (2), 97-118.

- Siefferemann, J. M. (2002). Flash profiling. A new method of sensory descriptive analysis. In AIFST 35*th* convention, July 21-24, Sidney, Australia.

- Van der Burg, E., & Dijksterhuis, G. (1996). Generalized canonical analysis of individual sensory profiles and instrumental data. In T. Naes & E. Risvik (Eds.), Multivariate analysis of data in sensory science. Elsevier.

- Williams, A. A., & Langron, S. P. (1984). The use of free-choice profiling for the evaluation of commercial ports. *Journal of the Science of Food and Agriculture*, 35, (5), 558-568.

- Worch, T., Lê, S., & Punter, P. (2010). How reliable are the consumers? Comparison of sensory profiles from consumers and experts. *Food Quality and Preference*, 21, (3), 309-318.

Part II

Qualitative descriptive approaches

Part II

Qualitative descriptive approaches

What could justify to bring together, in a same part, free text comments, discrimination test data, and specific approaches such as sorting task and Napping®? One possible answer is that these data are by nature *qualitative*, in the etymological meaning of the word[1]. Another possible answer is that all these approaches can be considered as *holistic*[2], although in sensory science, "holistic" mainly refers to sorting task and Napping®. Indeed, the word "holistic" describes methodologies which focus their attention on the entire product, and not on its decomposition into its different sensory characteristics only.

By going back to the different sensory tasks described in this part, it seems obvious that free text comments, which consist in depicting products using one's own words, are by essence qualitative. With regard to the data provided by discrimination tests, sorting task, or Napping®, it may not be that straightforward. To justify this, let's go back to the purpose of all these tasks, and the information that can be extracted from them. From a practical point of view, the main purpose of considering products holistically is to evaluate the degree of similarity and/or difference between products. For discrimination tests, the name of the task speaks for itself. Still, let us remind you the purpose of such a task: given a configuration of two different products, typically it consists in indicating which of the products are the same or different. Such tasks seem to have existed forever and are taught in every curriculum in sensory evaluation. On the other hand, sorting task and Napping have experienced a renewed interest lately, in the last couple of decades, and both tasks consist in aggregating or separating products based on their perceived similarities or differences. This justifies the regroupment of these sensory tasks together in this part.

In terms of structure, this second part is organized from the least "constrained" to the most "constrained" methodology, the constraint relying on the number of products involved during one assessment. Free text comments are usually provided when assessing products in monadic sequence. Consequently, as one assessment consists in depicting one product only, the subject

---

[1] Qualitative is derived from the Latin word *qualitas* ("way of being," "kind," "character," "nature"), a word formed from the word *qualis* meaning "as such."

[2] Holistic comes from the Greek word *holos* meaning "whole." Holism is the idea that all the properties of a given system cannot be determined or explained by its component parts alone. Instead, the system as a whole determines in an important way how the parts behave (source: http://en.wikipedia.org/wiki/Holism).

doesn't feel any constraint whatsoever. In discrimination tests, the number of different products to be assessed during one trial is most commonly equal to two. In that case, the subject has to remember two products during each assessment. For sorting and Napping®, however, the number of products is usually higher than two. In the case of Napping®, the subjects have to stand back from the whole product space to provide their own plane representation. From this point of view, the subjects have to remember all the products, so to speak. Such constraint is reduced in a sorting task, as the subjects usually focus on a subset of products to ensure its homogeneity. For these reasons, we decided to present sorting before Napping®. Starting this second part with the word association task is also coherent with the structure of the book, as this methodology is the natural extension of free choice profile (presented at first part of Chapter 3). As we will see, such information is also of utmost importance for the interpretation of sorting and Napping data.

Finally, the order of presenting these chapters within this part is also consistent, from a statistical point of view. Indeed, Chapter 4 introduces the notion of contingency table and presents a very important method through its application to textual data: Correspondence Analysis (CA). Additionally, a very common situation that can only be handled by an extension of Multiple Factor Analysis (MFA) to contingency tables, is also presented in this chapter. Chapter 5 introduces models, such as the Bradley-Terry and the Thurstonian models, which are conceptually more complicated to apprehend than the ANOVA models presented in Chapter 1. Chapter 6 presents the Multiple Correspondence Analysis (MCA), the reference method for exploring multivariate qualitative data. This method can be seen as a natural extension of Correspondence Analysis (CA), and as some sort of Principal Component Analysis (PCA) for qualitative data, but with its own interpretation rules (due to the nature of the data). In particular, the distance between individuals in MCA is more complicated to understand than the one in PCA. Chapter 7 finally presents a special case of MFA, in which groups of variables are composed of two variables only and are not standardized: this case study is a very good complement to Chapter 3.

4

When products are depicted by comments

CONTENTS

"... don't lose hope in statistics: it's still possible to analyze systematically such data using, amongst other methods, Correspondence Analysis (CA) based strategies. Due to the intrinsic heterogeneity of the data, a long and tedious recoding phase is necessary. Once done, a description of the products by the words and a representation of the product space can be obtained using respectively the **textual** *function, the* **descfreq** *function, and the* **CA** *function of the FactoMineR package."*

4.1 Data, sensory issues, and notations

"It's too salty," "it reminds me of my childhood," "it smells like gasoline." Although their value may be questionable, comments are indispensable to understand what a consumer thinks, or feels, when assessing a product[1]. Moreover, these spontaneous data are of utmost importance as consumers usually stress the sensory properties of the product that are the most significant to them. Unfortunately, this information is also very dense, and exploiting it

---

[1]In particular, in the framework of holistic approaches, as will be explained in the next chapters.

seems less straightforward than with numerical data. Still, these qualitative data are more and more studied.

The analysis of textual data, from their collection to their statistical analysis and their understanding, is a scientific discipline in itself, often referred to as *textual analysis* or *text mining*. Textual data are, for instance, extracts from books, articles, emails, or simply answers to open questions in surveys. The large domain of application of textual analysis includes corpus analysis, authorship, SPAM filtering, *etc.*

Most of the time, consumers evaluate products in monadic sequence, and for each product, they describe the product assessed using their own words. As in Chapter 1, the statistical unit of interest is one assessment. As such, this statistical unit is described by (at least) three variables: a first one that corresponds to the consumer who provided the comment, a second one that corresponds to the product on which the comment has been given, and a third one which is a qualitative measure consisting of the words that constitute the comment. Such data set allows to obtain a description of the products using the consumers' own words.

Due to its particularity, it is fair asking ourselves, what can we expect from the statistical analysis of such data? As we are essentially interested in understanding the products, we want to get a description of each of them using all the consumers' comments (the equivalent of a sensory profile through the words). Additionally, and based on their different "profiles," we also want to position each product on a map, such as two products being close if they are described in a similar way, *i.e.*, with the same words. Eventually, we can be interested in understanding the way consumers express themselves (from a sociological, anthropological, or even marketing perspective), in comparing the vocabularies from one panel of consumers to another, or even from one culture to another (hence, one language to another).

To obtain such results, similarly to Chapter 2 the "raw data set" needs to be "summarized" into a contracted form, in which the products are placed in rows and the different words used to describe them in columns. Such a data set can be obtained by submitting the "raw data set" to a transformation that leads to a very important type of data table: the so-called *contingency table*. Within this contingency table, the intersection of row i and column j highlights the number of times word j has been associated with product i to describe it. This contingency table is then submitted to different statistical methods, such as the *Chi-square test* and Correspondence Analysis (CA), for instance.

In this chapter, functions of the FactoMineR package that are related to textual analysis (creation of the contingency table, CA) and to the description of a categorical variable by other categorical variables are presented.

Note that, in this chapter, although we limit ourselves to the very restricted framework of free text comments, the methodologies provided encompass a lot of practical situations. For instance, Check All That Apply data (CATA)[2] can be seen as a simplified case of free text comments, in which consumers choose within a predetermined list of items rather than using their own words. *In fine*, the data we are dealing with are "bags of words" provided by consumers to describe products (or more generally stimuli).

Tips: Introducing the notion of lemmatization

Practically, the analysis of free comments often requires pre-processing steps. These pre-processing steps include:

- deleting "linking" words such as "and," "or," "you," "I" *etc.* to facilitate readability;

- combining words (or concepts) to avoid misinterpretation of the data: "not sweet" might be replaced by "not_sweet" so that it counts as one entity and does not lead to misinterpretation;

- regrouping words with similar meanings (including corrections of typing mistakes): "sweetness" might be grouped with "sweet."

The last step is called lemmatization[3]. This last step is a critical point since it is usually done manually and a part of subjectivity is *de facto* involved in this process. Roughly, the scientific community is opposing people who consider that lemmatization is necessary, and people who consider that it should never be done. As a rule of thumb, the literature suggests that if you want to perform lemmatization, only words you are sure they have the same meaning should be grouped together. In other words, if you doubt whether two words should be grouped together, you should keep them separated. It is also advised to perform the lemmatization with colleagues and only regroup the words that everybody agrees on.

4.2   In practice

---

[2]In a broader sense, *i.e.*, not limited to the sole words but extended to pictures, sound logos, *etc.*

[3]Lemmatization in linguistics, is the process of grouping together the different inflected forms of a word so they can be analyzed as a single item. (...) In many languages, words appear in several inflected forms. For example, in English, the verb "to walk" may appear as "walk", "walked", "walks", "walking". The base form, "walk", that one might look up in a dictionary, is called the lemma for the word. (...) Lemmatization is closely related to stemming. The difference is that a stemmer operates on a single word without knowledge of the context, and therefore cannot discriminate between words which have different meanings depending on part of speech.(http://en.wikipedia.org/wiki/Lemmatisation)

4.2.1   How can I approach *textual data*?

Studying comments made on perfumes is particularly challenging as those products usually evoke a lot of different things to consumers (olfactory memory). Despite the variability of the consumers in terms of words used to describe perfumes, and also in terms of consumer experience, one of our main questions of interest was on the feasibility of using and interpreting such data. We will see that, in our case, we managed to get a description and a representation of the products that are consistent.

To get a better understanding of the data we are going to analyze, let's first import the file *perfumes_comments.csv* (which can be downloaded from the book website) in your R session, and let's have a look at the structure of this data set[1]. To do so, the **read.table** function and the **summary** function are used.

```
> comments <- read.table("perfumes_comments.csv",header=TRUE,sep="\t",dec=".",
+ quote="\"")
```

```
> summary(comments)
           Stimulus         Subject        Comment
 Pleasures        : 81    1      : 12    flowery: 45
 Lolita Lempicka : 79    2      : 12    fruity : 35
 Shalimar        : 78    3      : 12    sweet  : 26
 J'adore ET      : 76    4      : 12    light  : 22
 Aromatics Elixir: 76    6      : 12    strong : 21
 Cinéma          : 76    10     : 12    soft   : 20
 (Other)         :434    (Other):828    (Other):731
```

As previously explained, the *comments* data set is structured according to three qualitative variables. Moreover, the **summary** function highlights the fact that 81 free comments were provided for *Pleasures*, 79 free comments for *Lolita Lempicka*, etc. (*cf.* first column). In other words, consumers did not provide a comment on all 12 products.

In more detail, let's visualize five random rows of the data set. To do so, a vector x which has as dimension the number of assessments (to obtain this information we use the **nrow** function) is created. By using the **sample** function, a sample of the specified size from the elements of x without replacement is taken.

```
> x <- sample(nrow(comments),5,replace=FALSE)
> comments[x,]
            Stimulus Subject                   Comment
486         L'instant       4                  odorless
820         Pleasures      23    talcum powder for baby
181         J'adore ET     95                   flowery
456 Aromatics Elixir       91                  medicine
55          J'adore ET     11 bubble;shampoo;soap;mossy
```

---

[1]These data were also collected by Mélanie Cousin, Maëlle Penven, Mathilde Philippe, and Marie Toularhoat as part of their master's degree project.

Let's first notice that, in the column *Comment*, "words" are separated by semi-colon. Hence, in the data set, "talcum powder for baby" should be considered as one unique "word", in the sense of one concept. We can also notice that consumer 95 used one word (*i.e.*, "flowery") to describe *J'adore ET*, whereas consumer 11 used four words (*i.e.*, "bubble," "shampoo," "soap," and "mossy") to describe the same perfume: the way a product is described by two different consumers can be different with respect to the words that have been used (from "flowery" to "shampoo") and their number (from one to four, in this case).

As mentioned in the previous section, such data set is not directly usable, as our statistical unit of interest is a product (and not a combination *Stimulus × Subject*), and we would like to understand the products with respect to the words used to describe them. Hence, the *comments* data set needs to be transformed into a more functional data set, $X = (n_{ij})$, where rows correspond to products and columns correspond to words: at the intersection of row i and column j, the number of occurrences word j has been used to describe product i is given. Such data set is called a *contingency table*.

Typically, a contingency table is calculated when one wants to study the associations between the modalities of two categorical variables. Usually, when analyzing textual data, raw data are transformed into a contingency table that crosses the categories of a categorical variable of interest and the words: again, at the intersection of row i (category i) and column j (word j), the number of times word j has been associated with category i is given. The categorical variable of interest depends on the question of interest. If the analyst is interested in understanding products, the categorical variable of interest is the one related to the stimuli, in our case the first one. If the analyst is interested in understanding consumers, the categorical variable of interest is the one related to the subjects, in our case the second one.

To generate the contingency table, in which the rows correspond to the products and the columns to the words used to describe the products, the **textual** function of the FactoMineR package is applied to the raw data set. In this function, it is important to specify the position of the textual variable within the data set (in our case, the variable *Comment*, `num.text=3`), as well as the position of the categorical variable of interest (in our case, the variable *Stimulus*, `contingence.by=1`). Last but not least, it is also very important to specify the way words are separated (`sep.word=";"`) in the data set.

```
> library(FactoMineR)
> res.textual <- textual(comments,num.text=3,contingence.by=1,sep.word=";")
> names(res.textual)
[1] "cont.table" "nb.words"
```

As shown by the **names** function, the **textual** function provides two outputs. The first output called `"cont.table"` contains the contingency table we are interested in; the second output called `"nb.words"` lists the words that appear in the textual variable, and describes them by their total number of occurrences and the number of comments in which they appear.

Let's have a look at the first seven columns of the contingency table and let's study the sum over the rows of the occurrences n_{ij} for each column j. To do so, the **apply** function is used. This function returns a vector with the sum (FUN=sum) computed on each column (MARGIN=2) of the contingency table. This very important vector, denoted $(n_{.j})$, is called the *row margin*, as the sum is computed over the rows. Similarly, the *column margin* is the vector $(n_{i.})$ of the sums computed over the columns (MARGIN=1). Let $n = \sum_i n_{i.} = \sum_j n_{.j}$ be the total number of occurrences of words used to describe all the products.

```
> res.textual$cont.table[,1:7]
                   a bit strong acidic adult aggressive airduster airy alcohol
Angel                  0     0      0     5                  0    0       2
Aromatics Elixir       0     0      0     4                  0    0       5
Chanel N5              0     1      0     2                  0    0       2
Cinéma                 1     1      0     1                  0    0       0
Coco Mademoiselle     0     1      1     0                  0    0       0
J'adore EP            0     1      0     0                  0    1       0
J'adore ET            0     0      1     0                  0    1       0
L'instant             1     0      1     0                  0    0       0
Lolita Lempicka       0     1      0     0                  0    0       0
Pleasures             0     1      1     0                  0    1       0
Pure Poison           0     2      0     0                  0    1       1
Shalimar              0     0      0     5                  1    0       4
```

```
> apply(res.textual$cont.table[,1:7],MARGIN=2,FUN=sum)
a bit strong          acidic        adult   aggressive      airduster          airy
          2               8            4           17              1             4
    alcohol
         14
```

As expected, the contingency table crosses the products on the one hand, and the words that have been used by the consumers on the other hand. As shown by the output, the number of occurrences may be very different from one word to the other (from 1 for *airy* to 17 for *aggressive*). It seems intuitive that words that are rarely used will influence exaggeratedly the statistical analysis of such table (this refers to the notion of *outliers* in statistics) as they are very specific to some products, and would drive the results of the analysis. For that reason, the analyst tends to get rid of these low-frequent words.

To remove these outliers, let's select only the words that have been used at least five times. To do so, the following code can be used:

```
> words_selection <- res.textual$cont.table[,
+ apply(res.textual$cont.table,2,sum)>=5]
> word_selection[,1:7]
                   acidic aggressive alcohol amber black pepper candy caramel
Angel                  0        5       2     0          4     1       0
Aromatics Elixir       0        4       5     2          2     0       0
Chanel N5              1        2       2     1          1     0       0
Cinéma                 1        1       0     0          1     0       2
Coco Mademoiselle     1        0       0     1          1     1       0
J'adore EP            1        0       0     0          0     0       0
J'adore ET            0        0       0     0          0     0       1
```

L'instant	0	0	0	0	1	3	0
Lolita Lempicka	1	0	0	0	0	4	1
Pleasures	1	0	0	0	0	0	0
Pure Poison	2	0	1	2	0	0	1
Shalimar	0	5	4	1	3	1	0

As we can see, R got rid of the columns *a bit strong, adult, airduster, airy*, as their respective number of occurrences is lower than five. In other words, *word_selection* is a subset of `res.textual$cont.table` which contains the words whose total frequency is higher (or equal) than five.

Now that the data set has been "cleaned," what can we do with it? Let's recall that, at the intersection of one row i and one column j, n_{ij} represents the number of times word j has been used to describe product i. This number can be used in different ways, each of them corresponding to a given point of view, to a specific statistical analysis, and to different types of results.

A first idea consists in interpreting n_{ij} as some kind of absolute quantity, *i.e.*, as some kind of intensity. Let's take the example of an attribute: a product is perceived as all the more intense for this attribute, as it has been cited a great number of times. With respect to this point of view, two products may be considered very differently - as very intense for one of them, and not intense at all for the other one - although, these two products may have been described similarly, when conditioned to the number of occurrences of words used to describe each product (*i.e.*, conditionally to the $n_{i.}$). In other words, for two products i and i', n_{ij} and $n_{i'j}$ may be very different, but nevertheless the proportions $n_{ij}/n_{i.}$ and $n_{i'j}/n_{i'.}$ may be very similar.

A second idea consists in taking into account the number of occurrences of words used to describe a product. In other words it consists in considering the so-called *row-profile*. This row-profile is obtained by dividing the rows of the contingency table by the elements of its column margin ($n_{i.}$).

To illustrate the differences between points of view, let's consider the data set entitled *perfumes_comments_sim*. This contingency table is a simulated table that was built based on *comments* in which a selection of products and a small selection of words have been done. After importing this data set in R, let's have a quick look at it using the **summary** function.

```
> comments_sim <- read.table("perfumes_comments_sim.csv",header=TRUE,sep=",",
+ row.names=1,quote="\"")
> comments_sim
```

	flowery	fruity	strong	sweet	fresh	soap
J'adore ET	118	113	21	21	56	15
L'instant	154	123	62	139	62	15
Shalimar	8	8	123	16	16	23
Lolita Lempicka	92	277	77	339	62	31
Pleasures	108	42	19	4	35	19
Chanel N5	276	45	924	45	45	555
Angel	15	154	246	123	15	15
Cinéma	200	231	46	200	77	15

To understand the first point of view and the notion of intensity, let's count the number of occurrences of words used to describe each perfume (*i.e.*, the column-margin). As previously, the **apply** function is used. This function returns, in our case, a vector presenting the sum (FUN=sum) computed on the rows (MARGIN=1) of our data set.

```
> row.sum <- apply(comments_sim,MARGIN=1,FUN=sum)
> row.sum
      J'adore ET          L'instant       Shalimar Lolita Lempicka
             344                555            194              878
       Pleasures          Chanel N5          Angel           Cinéma
             227               1890            568              769
```

As we can see, for some reasons, comments on *Chanel N5* are particularly numerous (1890), compared to comments on *Shalimar*, for instance (194). Does it mean that *Chanel N5* has been globally perceived as ten times more intense than *Shalimar*, and that these two products should be considered as completely different?

To understand the second point of view, let's divide our contingency table by its column margin, and let's highlight these proportions as percentage (we multiply the resulting proportions by 100).

```
> round(comments_sim/row.sum*100,2)
                flowery fruity strong sweet fresh  soap
J'adore ET        34.30  32.85   6.10  6.10 16.28  4.36
L'instant         27.75  22.16  11.17 25.05 11.17  2.70
Shalimar           4.12   4.12  63.40  8.25  8.25 11.86
Lolita Lempicka   10.48  31.55   8.77 38.61  7.06  3.53
Pleasures         47.58  18.50   8.37  1.76 15.42  8.37
Chanel N5         14.60   2.38  48.89  2.38  2.38 29.37
Angel              2.64  27.11  43.31 21.65  2.64  2.64
Cinéma            26.01  30.04   5.98 26.01 10.01  1.95
```

With respect to this data set of row-profiles, *Shalimar* and *Chanel N5* appear to be perceived very similarly.

This is precisely this second point of view that we will adopt in this chapter, in order to get an individual description of each product on the one hand, and a representation of all the products on the other hand.

Tips: On a different use of the textual function

In the previous example, the **textual** function of the FactoMineR package generated a contingency table from free comments by creating a matrix crossing the products in rows and the words used to describe these products in columns. At the intersection of row i and column j, the number of times the word j has been used to describe the product i is given. If this function is specific to one type of data, its use can be extended to other purposes.

For example, let's substitute the column containing the free comments by the column related to the consumer information.

```
> variant.textual <- textual(comments,num.text=2,contingence.by=1)
> head(variant.textual$cont.table[,1:5])
                        1 10 100 11 12
Angel                   1  1   1  1  1
Aromatics Elixir        1  1   1  1  1
Chanel N5               1  1   0  1  1
Cinéma                  1  1   0  1  1
Coco Mademoiselle       1  1   1  1  1
J'adore EP              1  1   0  1  1
```

As can be seen, the resulting contingency table is a binary table which highlights whether a consumer described a product with words or not. In this example, consumer 1 described all the products shown here while consumer 100 did not describe *Chanel N5*, *Cinéma*, and *J'adore EP*.

4.2.2 How can I get an individual description of each product?

To get an individual description of each product that is consistent with the second point of view adopted (*i.e.*, the one considering the notion of row-profile), the proportions $n_{ij}/n_{i.}$ and $n_{.j}/n$ are compared for each pair composed of product i and word j. Let's recall that n denotes the total number of occurrences of words used to describe all the products, $n_{.j}$ denotes the number of occurrences of the word j used to describe all the products, $n_{i.}$ denotes the number of occurrences of words used to describe product i, and n_{ij} denotes the number of occurrences of the word j used to describe the product i.

This comparison is legitimate, as when the two proportions $n_{ij}/n_{i.}$ and $n_{.j}/n$ are approximately equal, it can be concluded that the product i was not specifically associated with word j. On the contrary, when $n_{ij}/n_{i.}$ and $n_{.j}/n$ are really different, it can be concluded that the product i was specifically associated with word j ("positively" or "negatively", in a sense that will be specified latter).

Let's define the null hypothesis, H_0, associated with this test. By construction, the null hypothesis represents the *statu quo*. This hypothesis states that the product i is not specifically characterized by the word j. In other words, under H_0, these two proportions should be equal. This hypothesis test is based on a *hypergeometric distribution*.

By definition, the hypergeometric distribution is a discrete probability distribution that describes the probability of N_{ij} successes in $N_{i.}$ draws without replacement from a finite population of size N containing exactly $N_{.j}$ successes. Under the null hypothesis, the random variable N_{ij} follows the hypergeometric distribution $\mathcal{H}(N, N_{.j}, N_{i.})$. The probability of having a more extreme value than the observed value can therefore be calculated.

Let's focus on *Aromatics Elixir* and the word *strong*, for instance. To compare $n_{ij}/n_{i.}$ and $n_{.j}/n$, we first need to determine the position of the word of interest. To do so, the **grep** function can be used:

```
> grep("strong",colnames(words_selection))
> 57
```

In our case, the word *strong* is at the 57th position, *i.e.*, it corresponds to the 57th column of *words_selection*.

As in the previous section, $n_{.j}$ and n_{ij} are calculated using the **apply** function. This time, the row-margin (MARGIN=2) is calculated, and the column associated with the word of interest is visualized.

```
> col.sum <- apply(words_selection,MARGIN=2,FUN=sum)
> col.sum[57]
strong
  102
> words_selection[,57]
           Angel  Aromatics Elixir        Chanel N5          Cinéma
              15                27               19               2
Coco Mademoiselle        J'adore EP       J'adore ET        L'instant
               5                 2                3               3
  Lolita Lempicka         Pleasures      Pure Poison         Shalimar
               4                 4                3              15
```

The word *strong* has been used 102 times to describe all the perfumes and 27 times for the sole perfume *Aromatics Elixir*. To determine the number of occurrences of words for all the products (n) and for the sole perfume *Aromatics Elixir* ($n_{i.}$), the following code can be used:

```
> n <- sum(words_selection)
> n
[1] 1274
> row.sum <- apply(words_selection,MARGIN=1,FUN=sum)
> row.sum
           Angel  Aromatics Elixir        Chanel N5          Cinéma
              92               104               86             117
Coco Mademoiselle        J'adore EP       J'adore ET        L'instant
             104               118              118              95
  Lolita Lempicka         Pleasures      Pure Poison         Shalimar
             118               124              105              93
```

To finally accept (or reject) the null hypothesis (of independence), the two proportions $N_{ij}/N_{i.}$ and $N_{.j}/N$ are compared. In this case, since $27/104$ and $102/1274$ are not equal, the null hypothesis states that the two proportions $N_{ij}/N_{i.}$ and $N_{.j}/N$ are dependent. Such result can be obtained using the **phyper** function. However, performing such test for all products and all words can become rapidly tedious.

To overcome this problem, the **descfreq** function of the FactoMineR package can be used. Its main parameters are the name of the contingency table to analyze (**donnee**) and the significance threshold (**proba**) which is set to 0.05 by default.

```
> res.descfreq <- descfreq(words_selection)
> res.descfreq
$Angel
```

	Intern %	glob %	Intern freq	Glob freq	p.value	v.test
strong	16.304348	8.006279	15	102	0.0091164679	2.607655
aggressive	5.434783	1.334380	5	17	0.0108769061	2.546629
black.pepper	4.347826	1.020408	4	13	0.0219782582	2.290743
light	1.086957	6.043956	1	77	0.0382855820	-2.071784
fresh	0.000000	4.395604	0	56	0.0272887525	-2.207362
flowery	0.000000	9.419152	0	120	0.0001556205	-3.781925

```
$'Aromatics Elixir'
        Intern %    glob % Intern freq Glob freq      p.value    v.test
strong  25.961538 8.0062794          27       102 8.530222e-09  5.757632
cologne  4.807692 0.7064364           5         9 6.385959e-04  3.414669
alcohol  4.807692 1.0989011           5        14 7.271575e-03  2.684144
male     2.884615 0.5494505           3         7 2.903622e-02  2.182994
old      2.884615 0.6279435           3         8 4.373025e-02  2.016667
fresh    0.000000 4.3956044           0        56 1.519919e-02 -2.427598
sweet    0.000000 5.4160126           0        69 4.734761e-03 -2.824546
flowery  1.923077 9.4191523           2       120 3.511306e-03 -2.919023
fruity   1.923077 9.7331240           2       124 2.572938e-03 -3.014629
light    0.000000 6.0439560           0        77 2.294789e-03 -3.049164

$'Chanel N5'
       Intern %   glob % Intern freq Glob freq      p.value    v.test
soap   12.79070 2.276295          11        29 1.790556e-06  4.775731
strong 22.09302 8.006279          19       102 3.636406e-05  4.129445
fresh   0.00000 4.395604           0        56 3.648177e-02 -2.091517
sweet   0.00000 5.416013           0        69 1.401318e-02 -2.456925
fruity  0.00000 9.733124           0       124 2.164889e-04 -3.698955

$Cinéma
        Intern %   glob % Intern freq Glob freq      p.value    v.test
vanilla 5.982906 2.197802           7        28 0.021301776  2.302593
sweet  10.256410 5.416013          12        69 0.038735935  2.066981
strong  1.709402 8.006279           2       102 0.005251509 -2.791187

$'Coco Mademoiselle'
       Intern %    glob % Intern freq Glob freq      p.value   v.test
subtle 2.884615 0.3924647           3         5 0.009370041 2.598249
warm   3.846154 1.0989011           4        14 0.044265037 2.011572

$'J'adore EP'
            Intern %    glob % Intern freq Glob freq      p.value    v.test
fruity     18.644068 9.7331240          22       124 0.002557131  3.016498
shower.gel  3.389831 0.6279435           4         8 0.007307944  2.682476
shampoo     3.389831 0.6279435           4         8 0.007307944  2.682476
fresh       9.322034 4.3956044          11        56 0.021776627  2.294242
strong      1.694915 8.0062794           2       102 0.004872130 -2.815369

$'J'adore ET'
        Intern %   glob % Intern freq Glob freq      p.value    v.test
flowery 18.644068 9.419152          22       120 0.001598620  3.156158
fruity  17.796610 9.733124          21       124 0.006182667  2.737933
light   11.016949 6.043956          13        77 0.040950439  2.044031
strong   2.542373 8.006279           3       102 0.019794365 -2.330223

$'L'instant'
NULL

$'Lolita Lempicka'
        Intern %   glob % Intern freq Glob freq      p.value    v.test
sweet   17.796610 5.4160126          21        69 4.570183e-07  5.043531
vanilla  7.627119 2.1978022           9        28 1.128484e-03  3.256366
candy    3.389831 0.7849294           4        10 1.892356e-02  2.347033
flowery  4.237288 9.4191523           5       120 4.767475e-02 -1.980256
```

```
$Pleasures
         Intern %   glob % Intern freq Glob freq       p.value    v.test
flowery 21.774194 9.419152          27       120 1.786898e-05   4.289979
strong   3.225806 8.006279           4       102 4.202183e-02  -2.033304
sweet    0.000000 5.416013           0        69 1.389492e-03  -3.196825


$'Pure Poison'
        Intern %   glob % Intern freq Glob freq      p.value    v.test
strong 2.857143 8.006279           3       102 0.04602473  -1.995166


$Shalimar
             Intern %    glob % Intern freq Glob freq        p.value     v.test
medicine    5.376344 0.7849294           5        10 0.0007014223   3.389023
male        4.301075 0.5494505           4         7 0.0015739600   3.160690
intense     5.376344 1.0204082           5        13 0.0030002417   2.967713
strong     16.129032 8.0062794          15       102 0.0101465733   2.570794
spicy       8.602151 3.0612245           8        39 0.0111118334   2.539162
aggressive  5.376344 1.3343799           5        17 0.0114009208   2.530164
toiletry    3.225806 0.5494505           3         7 0.0212612581   2.303313
alcohol     4.301075 1.0989011           4        14 0.0302037281   2.167408
soft        0.000000 7.1428571           0        91 0.0015477362  -3.165581
flowery     0.000000 9.4191523           0       120 0.0001398214  -3.808484
fruity      0.000000 9.7331240           0       124 0.0000998651  -3.890919
```

As already mentioned in Chapter 2, the concept of *V-test* (*cf.* last column) is very close to the notion of *z-score*. In its basic form, this index is the quantile of a standardized normal distribution (with mean equal to 0 and standard deviation equal to 1) corresponding to a given probability. It is used to transform *p-values* into scores that are more easily interpretable. The main feature of this indicator is that it can be positive, or negative, depending on the test that has been performed.

According to the output, the word *strong* was specifically associated with *Aromatics Elixir*, in the sense that consumers used that word more often to characterize that product (27/104*100=25.962%) than for all the products (102/1274*100=8.006%). In this situation the *V-test* is positive. On the contrary, the word *flowery* was specifically associated with *Aromatics Elixir*, in the sense that consumers used that word less often to characterize that product (2/104*100=1.923%) than for all the products (120/1274*100=9.419%). In this situation the *V-test* is negative. In this output the *v.test* column is sorted by descending order.

Similar results are observed for *Shalimar* regarding the words *strong* and *flowery*, whereas the inverse is observed for *J'adore ET*. Indeed, for *J'adore ET*, the *V-test* is positive for *flowery* and negative for *strong*. This result suggests that *Aromatics Elixir* and *Shalimar* have similar profiles, which are different from the one of *J'adore ET*.

4.2.3 How can I graphically represent the product space?

In order to obtain a representation of the product space that would summarize
the information in the data set, the first idea that may come to mind is to
use Principal Component Analysis (PCA), as the nature of the information is
quantitative. Let's assess the results obtained by PCA on our simulated data
set *comments_sim*.

```
> par(mfrow=c(1,2))
> res.pca <- PCA(comments_sim,graph=FALSE)
> plot.PCA(res.pca,choix="ind",title="PCA (data: comments_sim)")
> plot.PCA(res.pca,choix="var",title="PCA (data: comments_sim"))
```

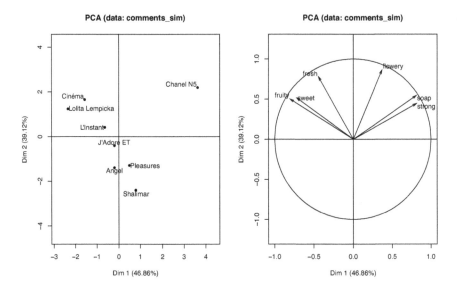

FIGURE 4.1
Representation of the perfumes and the words on the first two dimensions
resulting from a PCA on the raw data, using the **plot.PCA** function (*comments_sim* data set).

According to Figure 4.1, the representations of the PCA are not consistent
with our second point of view, as we expected *Chanel N5* and *Shalimar* to be
close on the plane. An alternative could be to consider the row-profile data
set and to see what happens if we *naively* apply a PCA on such data set.

```
> row.sum <- apply(comments_sim,MARGIN=1,FUN=sum)
> par(mfrow=c(1,2))
> res.pca <- PCA(comments_sim/row.sum,graph=FALSE)
> plot.PCA(res.pca,choix="ind",title="PCA (data: row-profile of comments_sim)")
> plot.PCA(res.pca,choix="var",title="PCA (data: row-profile of comments_sim)")
```

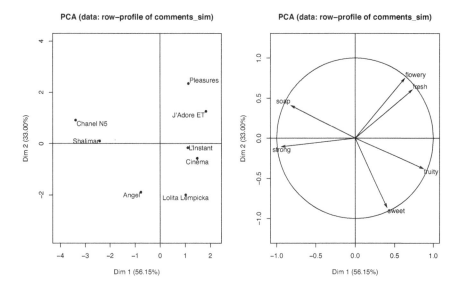

FIGURE 4.2

Representation of the perfumes and the words on the first two dimensions resulting from a PCA on the row-profiles, using the **plot.PCA** function (*comments_sim* data set).

As shown in Figure 4.2, these representations seem to be closer to our second point of view: the expected proximity between *Chanel N5* and *Shalimar* is better represented.

Still, this transformation is not sufficient, and a PCA on the row-profile data set is not "enough." In PCA, the analysis of the individuals on the one hand, and of the variables on the other hand, lie on two different distances. In the case of a contingency table, rows and columns seem to play an identical part.

The method we are going to use is called Correspondence Analysis (CA). This method is a reference method for studying the link between two categorical variables, and *de facto* for analyzing contingency tables. It is somehow a graphical and exploratory variant of the Chi-square test.

Indeed, in CA, the distance between two rows i and i' is defined the following way:

$$d^2_{\chi^2}(i, i') = \sum_j \frac{n}{n_{.j}} \left(\frac{n_{ij}}{n_{i.}} - \frac{n_{i'j}}{n_{i'.}} \right)^2.$$

As we can see, this distance, called rightfully the Chi-square distance, depends on the profile of the rows, in other words, on the ratios $n_{ij}/n_{i.}$ and $n_{i'j}/n_{i'.}$.

As in most *principal component methods*, CA seeks for the dimensions that best represent distances amongst rows, as previously defined. In its principle, CA is very similar to PCA, but still CA has its own specificities in terms of interpretation.

Those specificities are due to the intrinsic nature of the data to be analyzed: a contingency table in which rows and columns are of the same nature. Consequently the distance used to differentiate between rows on the one hand, and columns on the other hand, is the same. Similarly, the distance between two columns j and j' is defined the following way:

$$d_{\chi^2}(j, j') = \sum_i \frac{n}{n_{i.}} \left(\frac{n_{ij}}{n_{.j}} - \frac{n_{ij'}}{n_{.j'}} \right)^2.$$

Due to that particular distance, within our framework, the plane that best represents distances amongst the products is the one for which the products are the most dependent in terms of words that have been used to describe them.

Before running a CA on our data set, let's first run a Chi-square test. To do so, the **chisq.test** function is used:

```
> res.chisq <- chisq.test(words_selection)
> res.chisq

        Pearson's Chi-squared test

data:  word_selection
X-squared = 1293.054, df = 737, p-value < 2.2e-16
```

With such a small *p-value*, the null hypothesis of independence between the perfumes and the way they are associated with words is rejected. In other words, some particular associations between products and words is expected in the graphical outputs provided by CA.

Let's run the **CA** function of the FactoMineR package on the contingency table *words_selection*, and let's have a look at the eigenvalues.

```
> res.ca <- CA(words_selection,graph=FALSE)
> names(res.ca)
[1] "eig"  "call" "row"  "col"  "svd"
> round(res.ca$eig,4)
```

	eigenvalue	percentage of variance	cumulative percentage of variance
dim 1	0.4628	45.5958	45.5958
dim 2	0.1798	17.7132	63.3090
dim 3	0.0835	8.2298	71.5388
dim 4	0.0581	5.7222	77.2610
dim 5	0.0518	5.1043	82.3653
dim 6	0.0447	4.4027	86.7680
dim 7	0.0431	4.2486	91.0166
dim 8	0.0327	3.2250	94.2415
dim 9	0.0236	2.3277	96.5693
dim 10	0.0193	1.9058	98.4751
dim 11	0.0155	1.5249	100.0000

Technically, in CA, the eigenvalues λ_s lie between 0 and 1. These two extreme values correspond to two typical situations that may occur in the study of two categorical variables.

When $\lambda_s = 1$, rows and columns are exclusively associated. In other words, the rows and the columns of the contingency table can be organized as a block diagonal matrix, with non null matrices on the diagonal and blocks of zero matrices on the off-diagonal. This situation corresponds to a very strong association between the categories of one variable and the categories of the other variable (an exclusive association actually!). On the contrary, when $\lambda_s = 0$, no association at all is observed between the categories of one variable and the categories of the other variable.

In our example, the first eigenvalue ($\lambda_1 = 0.4628$) is particularly important, which highlights a strong association between products and words on the first dimension. With a value of 0.1798, the second eigenvalue is not negligible. It's worth considering the second dimension in the interpretation as it will also highlight some association between products and words.

Remark. Studying the eigenvalues is particularly important in CA. As in PCA, the inertia of the scatter plot of the rows N_I, and the inertia of the scatter plot of the columns N_J are equal. Moreover, in CA, this inertia equals the square of the *Phi* coefficient between the two categorical variables:

$$I(N_I) = I(N_J) = \phi^2 = \frac{\chi^2}{n}.$$

As shown below, if we sum the eigenvalues we obtain $I(N_I)$ (or equivalently $I(N_J)$), and if we multiply this value by the total number of occurrences of words used to describe the products, we obtain the Chi-square distance between the rows and the columns of our contingency table (*cf.* res.chisq, the output of our Chi-square test).

```
> sum(res.ca$eig[,1])*sum(words_selection)
[1] 1293.054
```

Since the rows and columns relie on the same distance, the CA represents in the same graphs the rows and columns. Let's now successively represent the rows and the columns within a same graph, before representing the rows and the columns separately.

```
> plot.CA(res.ca,title="Perfumes: free comments (CA)",col.row="grey30",
+ col.col="grey70")
> plot.CA(res.ca,invisible="col",title="Perfumes: free comments (CA)",
+ col.row="grey30")
> plot.CA(res.ca,invisible="row",title="Perfumes: free comments (CA)",
+ col.col="grey70")
```

This superimposed representation of the rows and the columns is legitimate in CA, as both objects (rows and columns) play a symmetric part in the

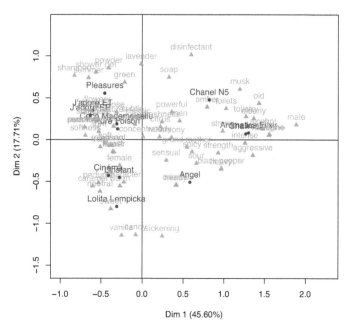

FIGURE 4.3
Representation of the perfumes and the words on the first two dimensions resulting from a CA, using the **plot.CA** function (*words_selection* data set).

analysis. Still this representation can be hard to interpret (due to the amount of information), as in the case of Figure 4.3. Hence, for practical reasons, the rows and columns are often represented separately (*cf.* Figure 4.4 and Figure 4.5).

The joint interpretation of Figure 4.4 and Figure 4.5 has to be done cautiously row by row (*resp.* column by column), by considering the position of a given row (*resp.* column) with respect to the columns (*resp.* rows).

Tips: On lemmatization and *distributional equivalence*

One of the most important properties of CA is what is called the *distributional equivalence*, which states that if two profiles are equal (rows or columns), merging them doesn't change the final result of the analysis. In a textual analysis framework, it means that it is possible to combine words into a same concept if they have been used similarly to describe rows (here, products).

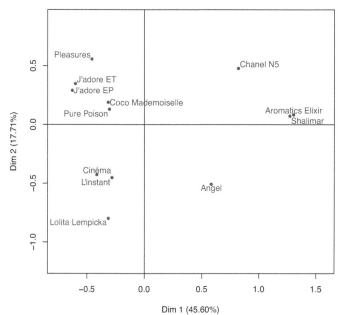

FIGURE 4.4
Representation of the sole perfumes on the first two dimensions resulting from a CA, using the **plot.CA** function (*words_selection* data set).

The different issues raised by the lemmatization are somehow resolved by this very important property.

To illustrate this property, let's have a look at the profiles associated with the two words *heady* and *oriental*.

```
> grep(colnames(words_selection),pattern="heady")
[1] 25
> grep(colnames(words_selection),pattern="oriental")
[1] 38
> words_selection[,c(25,38)]
                  heady oriental
Angel                 3        2
Aromatics Elixir      4        2
Chanel N5             2        1
Cinéma                4        1
Coco Mademoiselle     2        0
J'adore EP            0        0
J'adore ET            0        0
L'instant             4        2
Lolita Lempicka       4        2
```

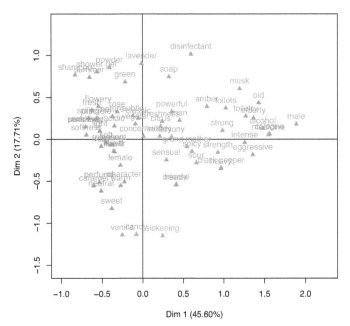

Perfumes: free comments (CA)

FIGURE 4.5

Representation of the sole words on the first two dimensions resulting from a CA, using the **plot.CA** function (*words_selection* data set).

Pleasures	0	0
Pure Poison	1	2
Shalimar	3	1

To get a better vision, let's transform these raw values into row-profiles and represent the resulting values as percentage.

```
> col.sum <- apply(words_selection,MARGIN=2,FUN=sum)
> col.pro <- 100*sweep(words_selection,MARGIN=2,col.sum,FUN="/")
> round(col.pro[,c(25,38)],3)
                   heady oriental
Angel             11.111   15.385
Aromatics Elixir  14.815   15.385
Chanel N5          7.407    7.692
Cinéma            14.815    7.692
Coco Mademoiselle  7.407    0.000
J'adore EP         0.000    0.000
J'adore ET         0.000    0.000
L'instant         14.815   15.385
Lolita Lempicka   14.815   15.385
```

```
Pleasures          0.000    0.000
Pure Poison        3.704   15.385
Shalimar          11.111    7.692
```

As we can see, these two profiles are close. This impression is reinforced by Figure 4.5 in which we can see that the two words *oriental* and *heady* are almost superimposed. If we sum up these two columns into one unique variable we arbitrarily call *heady_oriental*, and perform CA on this new table, the results obtained are very similar to the one obtained by CA on the original table.

```
> words_selection_new <- as.data.frame(words_selection[,-c(25,38)])
> words_selection$heady_oriental <- words_selection[,25]+words_selection[,38]
> res.ca.new <- CA(words_selection_new,graph=FALSE)
> plot.CA(res.ca.new,invisible="row",title="Perfumes new: free comments (CA)",
+ col.col="grey70")
```

As with PCA and MFA, a better understanding of Figure 4.5 can be obtained by applying the **dimdesc** function on `res.ca`. This function describes automatically the dimensions provided by any *principal component methods* implemented in the FactoMineR package.

```
> res.dimdesc <- dimdesc(res.ca)
> res.dimdesc$'Dim 1'$col
                  coord
shampoo       -0.825792661
perfumed      -0.699857363
sensitivity   -0.699857363
woman         -0.699857363
softness      -0.690957376
spring        -0.664139219
summer        -0.650917662
fresh         -0.621454589
caramel       -0.599584950
shower gel    -0.564283193
discrete      -0.561973246
flowery       -0.543739055
perfume       -0.530748117
light         -0.526736871
neutral       -0.511935637
...
strength       0.929549481
heavy          0.966038533
black pepper   0.968080357
strong         0.984103038
toilets        1.027368730
musk           1.196715201
intense        1.260405850
toiletry       1.272258277
aggressive     1.357779570
elderly        1.367114574
old            1.430956082
```

```
alcohol        1.480292512
medicine       1.549030276
cologne        1.562132652
male           1.887375312
```

In our example, the **dimdesc** function characterizes each dimension on its positive side and on its negative side, by ordering the words based on their importance in each side. The words that have been significantly used to describe the negative side of the first dimension are the words *shampoo, perfumed, sensitivity, woman,* while the words that have been significantly used to describe the positive side of the first dimension are *male, cologne, medicine, alcohol, old.* Similarly, the products that are related to the negative side of the first dimension are *J'adore EP* and *J'adore ET,* whereas the products associated with the positive side of the first dimension are *Aromatics Elixir* and *Shalimar.* The correspondence between words and products shows that *J'adore EP* and *J'adore ET* are described as *shampoo, perfumed, etc.*

The **dimdesc** function can also be used to evaluate the link between the words used often and the words used rarely. To do so, we propose to split the contingency tables into two tables, one including the words often used, and one including the words rarely used. Hence, the first step consists in creating a data set of "active" words, *words_active,* by selecting the words that appear at least five times (five being an arbitrary threshold). Similarly, we create a second data set of "supplementary" words, *words_supp,* by selecting the words that appear less than five times. These two data sets are created in order to facilitate the call of **CA**, and notably the selection of the supplementary columns in the analysis: the data set *words,* being the combination of *words_active* and *words_supp,* is *de facto* constituted of a block of active columns and another block of supplementary columns.

```
> words_active <- res.textual$cont.table
+ [,apply(res.textual$cont.table,2,sum)>=5]
> words_supp <- res.textual$cont.table[,apply(res.textual$cont.table,2,sum)<5]
> words <- cbind(words_active,words_supp)
> res.ca <- CA(words,col.sup=(ncol(words_active)+1):ncol(words_supp),
+ ncp=Inf,graph=FALSE)
```

Once the function **CA** is called, the dimensions are automatically described by all the words with the **dimdesc** function. As, by definition, *active* words are participating in the construction of the dimensions, it seems more judicious to understand these dimensions with respect to *supplementary* words only. In our case, the **dimdesc** function provides a list of meaningful words for each dimension (*e.g.*, dimdesc(res.ca)$'Dim 1'$col for the first dimension). To select the *supplementary* words only, a vector of their position in each list is created (in our example, select.dimdesc for the first dimension).

```
> res.dimdesc <- dimdesc(res.ca)
> select.supp <- which(rownames(res.dimdesc$'Dim 1'$col)
+ %in% colnames(words_supp))
> res.dimdesc.select <- as.matrix(res.dimdesc$'Dim 1'$col[select.supp,])
> rownames(res.dimdesc.select) <- rownames(res.dimdesc$'Dim 1'$col)
+ [select.supp]
> colnames(res.dimdesc.select) <- "coord"
> head(res.dimdesc.select)
                              coord
businessman        -0.9259757
flower in toilets  -0.9259757
jasmin             -0.9259757
skin cream         -0.9259757
sea water          -0.9060149
cottony            -0.8860541
```

The negative side of the first dimension is also associated with *business-man*, *flower in toilets*, *etc*. Thanks to this methodology and the distributional equivalence in CA, it is possible to merge those low frequent words with frequent words into groups (lemmatization) in order to facilitate and enrich the interpretation of the results.

Tips: Extensive application of textual analysis, the **twitteR** package

In 2009, Jeff Gentry shared to the R community a package called twitteR by uploading it on the CRAN. The aim of this package is to connect directly your RGUI to twitter, and to allow searching directly tweets within R via the **searchTwitter** function. These tweets are then imported automatically into the R environment and are ready to be analyzed. This function is particularly powerful since it allows searching tweets only in a certain language, and/or provided from a certain location, at a certain time.

Due to restrictions imposed by twitter, the access to twitter was limited and only a maximum of 1500 tweets could be searched within each query. This policy has changed with the new security system, and no (at least less) limitations in the number of tweets searched are imposed. But to access twitter database, the user needs to register, and authenticate. If you go through all these steps, nothing stops you to search all the words you want, and analyze tweets regarding your favorite brand, for example.

```
> searchTwitter("#dataanalysis",n=3,lang="eng")
[[1]]
[1] "tmmdata: Saving time is huge deal working in data. TED talk on ideas to
+ save time to solve problems-big deal in #dataanalysis"

[[2]]
[1] "isatools: Looking to learn more advanced R? Have a look here
+ http://t.co/70LWEhJTOa #r #bioconductor #dataanalysis"

[[3]]
[1] "jmmoraleslopez: "The greatest value of a picture is when it forces us to
+ notice what we never expected to see" #DataAnalysis"
```

As a good complement to the twitteR package, we recommend you to also look at the tm package (Feinerer, 2013), which is a package dedicated to text mining, and which proposes very useful functions to make your textual analysis easier.

Additionally to textual analysis, a common practice consists in representing a "bag of words" using a "word cloud." Word clouds consist in representing words together in a cloud, in which the size of each word is proportional to the frequency of use of that word. Hence, this graphical representation summarizes the information about a product or concept by highlighting the most important words used to describe it. Due to its simplicity, word clouds are very popular nowadays. It is hence not surprising to know that R can generate such representations. To generate such graphics, the wordcloud package and its **wordcloud** function are used. This function takes as inputs the list of words (`words`) and their frequencies (`freq`). In the previous example related to the perfume, the word cloud for products *Coco Mademoiselle* and *Shalimar* (*cf.* Figure 4.6) are generated using the following code:

```
> library(wordcloud)
> wordcloud(colnames(res.textual$cont.table),freq=res.textual$cont.table[5,],
+ min.freq=3,max.words=max(res.textual$cont.table))
> wordcloud(colnames(res.textual$cont.table),freq=res.textual$cont.table[12,],
+ min.freq=3,max.words=max(res.textual$cont.table))
```

FIGURE 4.6
Representation of *Coco Mademoiselle* (left) and *Shalimar* (right) based on comments, using the **wordcloud** function (*comments* data set).

4.2.4 How can I *summarize* the comments?

Obviously, contingency tables are intrinsically symmetrical as they cross the information issued from two categorical variables: studying such tables or their transposition through CA is exactly the same.

Still from an application point of view, as rows and columns are either products or words, it can make sense to study words through clusters of homogeneous words.

The purpose of this paragraph is to present how words issued from free comments can be clustered. As shown previously, exploratory multivariate methods and unsupervised classification (or clustering) are often complementary. The **HCPC** function of the FactoMineR package is conceived to cluster statistical units with respect to their coordinates on the dimensions issued from any *principal component methods* implemented in the FactoMineR package.

To cluster the words based on the results of CA, the **HCPC** function is applied to the *res.ca* objects:

```
> res.hcpc <- HCPC(res.ca,proba=1,cluster.CA="columns")
```

The 3-cluster solution proposed by the **HCPC** function is considered here (*cf.* Figure 4.7). To see how each of the clusters is characterized, the object res.ca$data.clust is used. More precisely, the words associated with each cluster are printed on screen by using the following code:

```
> rownames(res.hcpc$data.clust[res.hcpc$data.clust$clust=="1",])
 [1] "man"        "grand mother" "spicy"        "sour"        "amber"
 [6] "strength"   "heavy"        "black pepper" "strong"      "toilets"
[11] "musk"       "intense"      "toiletry"     "aggressive"  "elderly"
[16] "old"        "alcohol"      "medicine"     "cologne"     "male"
> rownames(res.hcpc$data.clust[res.hcpc$data.clust$clust=="2",])
 [1] "caramel"    "perfume"   "neutral"   "sweet"       "female"   "warm"
 [7] "vanilla"    "character" "candy"     "sickening" "heady"     "oriental"
> rownames(res.hcpc$data.clust[res.hcpc$data.clust$clust=="3",])
 [1] "shampoo"    "perfumed"  "sensitivity"  "woman"     "softness"
 [6] "spring"     "summer"    "fresh"        "shower gel" "discrete"
[11] "flowery"    "light"     "fruity"       "soft"      "powder"
[16] "pleasant"   "nature"    "young"        "forest"    "acidic"
[21] "flower"     "rose"      "green"        "subtle"    "vegetal"
[26] "classic"    "lavender"  "concentrated" "woody"     "citrus"
[31] "freshness"  "sensual"   "soap"         "lemony"    "powerful"
[36] "disinfectant"
```

For instance, the first cluster contains words that can be qualified as rather "masculine" (*man, male, strong, heavy, etc.*).

Another way to describe the clusters is by considering particular entities (here, words) in the cluster. These particular entities are divided into two groups: the paragons (the entities that are the closest to the center of the class), and the more typical entities (the entities that are the farthest from the center of the class). These particular entities are stored in the object res.ca$desc.ind.

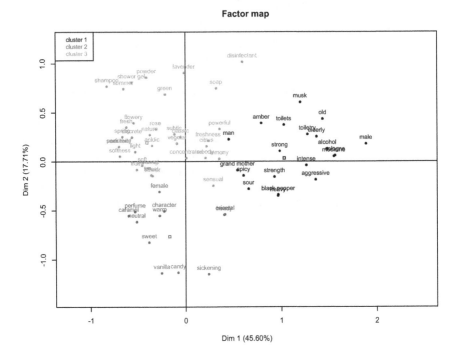

FIGURE 4.7
Representation of the words clustered into three groups on the first two dimensions resulting from a CA, using the **HCPC** function (*comments* data set).

```
> res.hcpc$desc.ind
$para
cluster: 1
  alcohol    strong   elderly     spicy     heavy
0.4186589 0.4712412 0.5128171 0.6673431 0.7209848
-------------------------------------------------------
cluster: 2
    sweet   vanilla     heady   oriental    female
0.5184493 0.6026299 0.6600909 0.7228054 0.7994711
-------------------------------------------------------
cluster: 3
    light   flowery      soft  discrete     fresh
0.4221930 0.4252818 0.4455259 0.4800539 0.5131613

$dist
cluster: 1
    male   cologne  medicine       old   alcohol
2.620707 2.215031 2.188499 1.878003 1.803803
-------------------------------------------------------
```

```
cluster: 2
    candy sickening   caramel    vanilla   neutral
 1.751303  1.624069  1.556570  1.488896  1.348107
-------------------------------------------------------------
cluster: 3
    subtle      shampoo    shower gel disinfectant     summer
  2.306724     2.135957     2.066596    1.924413      1.834059
```

These two particular entities can be used to better understand the different clusters.

4.3 For experienced users: Comparing free comments from different panels, the Rorschach test revisited

To understand cultural differences, it can be interesting to ask consumers from different countries to describe a same set of stimuli. The comparison of the results obtained from the different countries is of first interest since it highlights the similarities and differences between cultures. However, this comparison may not be straightforward, in particular when the description of the stimuli is based on free comments. In that case, the exercise consists in comparing contingency tables resulting from the different panels, with common rows (the stimuli) but with different columns (the words used to describe the stimuli).

To compare two contingency tables (involving the same rows), a first solution consists in performing CA on one of the two tables, and project the second table as supplementary. However, since the principal components are only related to the active table, the comparison is not complete, in the sense that both points of view are not taken into account.

To compare multiple contingency tables, a variant of MFA is required. Like any other MFA, this variant performs a separate analysis on each group (here, CA is performed on each contingency table). After weighting appropriately the rows and columns of the different subtables, a meta analysis (PCA) is performed. This variant of the MFA is known as Multiple Factor Analysis for Contingency Tables (MFACT).

To illustrate this methodology, the 10 cards of the Rorschach test were presented to two panels, one French and one Asian (*cf.* Figure 4.8). Each panel was asked to describe with their own words the 10 different cards. To facilitate the comparison, the free comments elicited were translated in English. The data were collected by Marianne Buche, Sophie Birot, and Lisa Defeyter during their master's degree studies.

The data from the two panels are presented in the files *rorschach France.csv* and *rorschach Asia.csv*. After downloading the two data sets, import them in R.

```
> Rorschach_Fr <- read.table("rorschach France.csv",sep=",",header=TRUE)
```

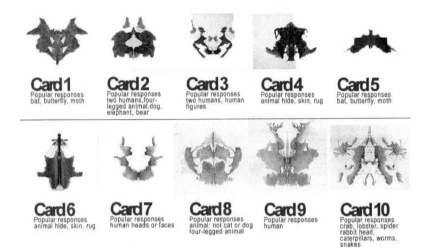

FIGURE 4.8
The ten cards of the Rorschach test.

```
> Rorschach_As <- read.table("rorschach Asia.csv",sep=",",header=TRUE)
```

For each panel, the contingency table crossing the cards (in rows) and the words that were used to describe the cards (in columns) are generated using the **textual** function (FactoMineR).

```
> textualFr <- textual(Rorschach_Fr,num.text=3,contingence.by=1,sep.word="--")
> contingencyFr <- textualFr$cont.table
> ncol(contingencyFr)
[1] 93
> textualAs <- textual(Rorschach_As,num.text=3,contingence.by=1,sep.word="--")
> contingencyAs <- textualAs$cont.table
> ncol(contingencyAs)
[1] 91
```

Both the French and the Asian panels used sensibly the same amount of words to describe the 10 cards (93 *versus* 91 words). In order to understand the way the different panels described the cards, a separate CA is performed on each contingency table.

```
> CA.Fr <- CA(contingencyFr,graph=FALSE)
> plot.CA(CA.Fr,title="CA: Rorschach (France)",new.plot=TRUE)
> round(CA.Fr$row$contrib,3)
         Dim 1  Dim 2  Dim 3  Dim 4  Dim 5
plate1   0.001  0.004  2.329 55.090 15.667
plate10  7.353 33.100  0.941  7.724 33.488
plate2   5.390  9.135  0.200  0.457  3.270
plate3   6.090 10.147  2.709  0.004  0.109
plate4  12.960  0.004 12.160  0.454  1.090
```

```
plate5  35.369  0.165 55.000  0.027  0.025
plate6  17.466  1.138 24.952  2.398  3.234
plate7   9.238 22.897  1.473  4.864  2.153
plate8   3.105  5.881  0.006  4.290 39.646
plate9   3.029 17.529  0.231 24.693  1.319


> CA.As <- CA(contingencyAs)
> plot.CA(CA.As,title="CA: Rorschach (Asia)",new.plot=TRUE)
> round(CA.As$row$contrib,3)
          Dim 1  Dim 2  Dim 3   Dim 4   Dim 5
plate1   10.741  0.654  1.008   0.272  15.446
plate10   0.444 17.490  0.001  17.176   5.671
plate2    8.302  4.892 21.656   0.244   1.121
plate3   26.596 10.908  8.003   1.690   0.875
plate4   15.060  2.601  0.127   0.012   0.109
plate5   10.835  2.744  0.611   0.200  15.908
plate6   14.797  1.594  0.036   3.904  58.616
plate7   10.617  8.803 66.217   0.676   0.030
plate8    1.160 35.218  1.365  49.023   2.203
plate9    1.448 15.095  0.976  26.803   0.019
```

The results for the French panel are presented in Figure 4.9. The first dimension opposes *plate5*, *plate4*, and *plate6* to the rest of the plates. Together, these three plates contribute to 65% of the construction of dimension 1. On the second dimension, *plate10*, *plate9*, and *plate8* are opposed to *plate7*, *plate3*, and *plate2*.

For the Asian panel, three groups of plates are observed (*cf.* Figure 4.10). The first dimension opposes *plate1*, *plate4*, *plate5*, and *plate6* to *plate2*, *plate3*, and *plate7*. The second dimension opposes *plate8*, *plate9*, and *plate10* to the rest of the plates.

The comparison of the two configurations is done through MFACT. This analysis is part of the **MFA** function of the FactoMineR package. To perform MFACT, the type of the group (**type** argument) should be set to "f" for contingency tables (or frequency). After combining the two data sets together, let's run the MFACT on the total data set.

```
> Rorschach_All <- cbind(contingencyFr,contingencyAs[rownames(ContingencyFr),])
> res.mfa <- MFA(Rorschach_All,group=c(ncol(contingencyFr),ncol(contingencyAs)),
+ type=c("f","f"),name.group=c("France","Asia"),graph=FALSE)
> plot.MFA(res.mfa,choix="ind",habillage="none")
```

The representation of the product space thus obtained (*cf.* Figure 4.11) highlights three groups of plates: the dimension 1 opposes *plate4*, *plate5*, and plate6 to the rest of the plates, whereas dimension 2 opposes *plate8*, *plate9*, and *plate10* to *plate2*, *plate3*, and *plate7*.

To assess the strength of the relationship between the two configurations, a numerical indicator is required. Here, we consider the *RV* coefficient (*cf.* Chapter 7). This coefficient takes a value between 1 (*i.e.*, the two configurations are homothetic) and 0 (*i.e.*, the two configurations are orthogonal).

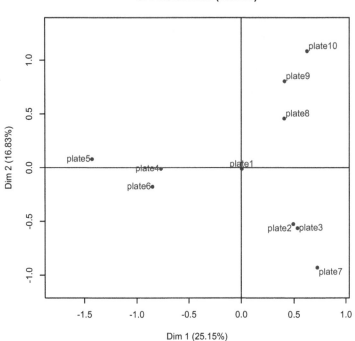

FIGURE 4.9
Representation of the plates of the Rorschach test for the French panel on the first two dimensions resulting from a CA, using the **plot.CA** function (*Rorschach_Fr* data set).

```
> round(res.mfa$group$RV,3)
       France  Asia   MFA
France 1.000  0.920 0.976
Asia   0.920  1.000 0.983
MFA    0.976  0.983 1.000
```

The *RV* coefficient between the two configurations being 0.92, the two configurations are almost identical.

A closer look at the similarities and differences per plate is observed by projecting the configuration from each country within the representation of the product space.

```
plot.MFA(res.mfa,choix="ind",partial="all",habillage="group",new.plot=FALSE)
```

These projections, also known as partial points representations (*cf.* Figure 4.12), show that *plate4* and *plate6* are the ones with the highest degree

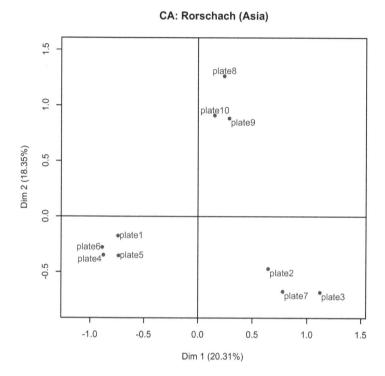

FIGURE 4.10
Representation of the plates of the Rorschach test for the Asian panel on the first two dimensions resulting from a CA, using the **plot.CA** function (*Rorschach_As* data set).

of agreement (*i.e.*, the partial points representations are close to the average point), while *plate1* and *plate9* have the largest disagreement.

The group representation (*cf.* Figure 4.13) finally highlights the fact that the Asian representation is slightly more multidimensional than the French group, at least on the second dimension. This can be explained by the fact that the Asian panel defined *plate9* as more typical on that dimension than the French panel did.

```
plot.MFA(res.mfa,choix="group",habillage="none",new.plot=FALSE)
```

This difference of multi-dimensionality is shown by the difference in the L_g coefficient, which is a particular indicator of MFA. Indeed, the Asian configuration (L_g=3.291) is slightly more multi-dimensional than the French configuration (L_g=2.356).

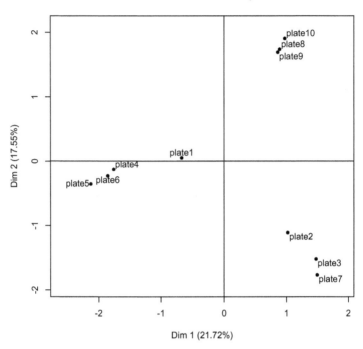

FIGURE 4.11
Representation of the plates of the Rorschach test for both panels on the
first two dimensions resulting from an MFA on contingency tables, using the
plot.MFA function (*Rorschach_All* data set).

```
> round(res.mfa$group$Lg,3)
         France  Asia   MFA
France   2.356  2.562 2.527
Asia     2.562  3.291 3.007
MFA      2.527  3.007 2.844
```

4.4   Exercises

Exercise 4.1 *Using the IRaMuTeQ interface*

IRaMuTeQ is an R interface dedicated to the multivariate analysis of textual

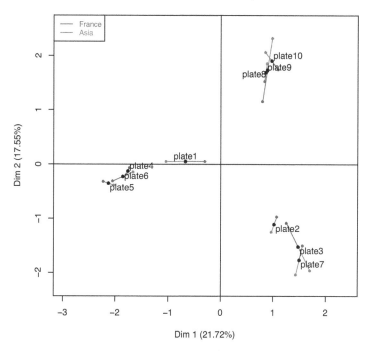

FIGURE 4.12
Representation of the plates and their partial representations for both panels on the first two dimensions resulting from MFA, using the **plot.MFA** function (*Rorschach_All* data set).

data and questionnaires. This interface contains a lot of very useful features especially designed for studying corpus. The main drawback of this interface is the format of the data to be analyzed.

- Transform the perfumes data into the right format. You can use the file IRaMuTeQ *Example.txt* as example of format required by the software.

- Describe the perfumes by using IRaMuTeQ.

Exercise 4.2 *Integrating supplementary words: the perfumes data*

This exercise is particularly important as we introduce here the concept of supplementary elements in Correspondence Analysis. These elements can be either rows or columns.

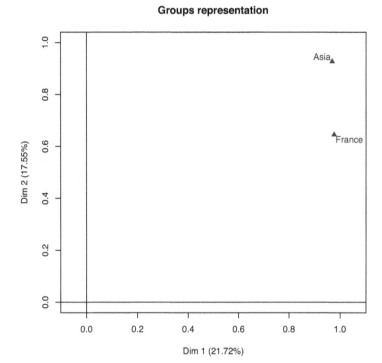

FIGURE 4.13
Representation of the French panel and the Asian panel on the first two dimensions resulting from an MFA, using the **plot.MFA** function (*Rorschach_All* data set).

- Import the file *perfumes_comments.csv* from the book website.

- By using the **textual** function of the FactoMineR package, create the contingency table crossing the products in rows, and the words in columns. Then, separate into two data sets words whose occurrences are lower than five from words whose occurrences are higher than five. Merge the two data sets by using the **cbind** function.

- Run the **CA** function of the FactoMineR package by specifying that the supplementary elements correspond to the words whose occurrences are lower than five.

- Plot the supplementary elements only. Describe the first two dimensions by using supplementary elements only.

Exercise 4.3 *Analyzing CATA data: the Sennheiser case study (1)*

"Sound branding (also known as audio branding, music branding, sonic branding, acoustic branding or sonic mnemonics) is the use of sound to reinforce brand identity. Sound branding is increasingly becoming a vehicle for conveying a memorable message to targeted consumers, taking advantage of the powerful memory sense of sound." (`http://en.wikipedia.org/wiki/Sound_trademark`)

In this experience, 27 variants of the Sennheiser sound logo were generated according to three factors with three modalities. Those factors were the pitch (three different octaves), the tempo (three different speeds: 80, 120, 160), and the instrument (accordion, piano, violin). Consumers had to listen to 12 out of the 27 variants, and then had to answer the following question: according to you what are the values that the company wants to highlight with that variant? To answer that question, consumers had to check amongst the following list of 18 words: Innovation, Traditional, Premium, Popular, Technology, Artisanal, International, Proximity, Sobriety, Passion, Fun, Serious, Simplicity, Complexity, Freedom, Conviviality, Performance, Human.

- Import the *Sennheiser_CATA.csv* file from the book website.

- Using the **aggregate** function combined with the **sum** function, generate the contingency table crossing the products in rows and the CATA terms in columns.

- Import the design of experiment used to generate the 27 sounds, and which can be found in the *Sennheiser_design.csv* file from the book website. Combine this table to the previous one using the **merge** function.

- Represent the sound space by performing CA on the contingency table. Project the modalities as supplementary elements on that space.

- Apply the **descfreq** function of the FactoMineR package to get a description of the variants and of the instruments function of the CATA terms.

Exercise 4.4 *Analyzing JAR data: the Sennheiser case study (2)*

In the previous exercise, we showed you one way to analyse CATA questions when the data are stored as 0 and 1. In this large study, JAR questions have also been asked. Consumers had to evaluate the variants of the logo on a JAR scale, with respect to a list of 8 emotions or emotional states (Serious, Surprising, Exciting, Sensual, Joyful, Energetic, Sad): according to you, do you think this variant is Too sensual...Not sensual enough with respect to Sennheiser's values? We will show here an original way to analyse the data.

- Import the file *Sennheiser_JAR.csv* from the book website.

- Combine to your data set the description of the 27 different sounds. Such information is available in the *Sennheiser_design.csv* file, also available on the book website. To combine the two files, use the **merge** function.

- Using the **textual** function of the FactoMineR package, get a description of the different sounds in function of the different JAR attributes.

- Similarly, get a description of the different instruments in function of the different JAR attributes. To do so, we recommend you to use the by.quali argument of the **textual** function.

- Using the **descfreq** function of the FactoMineR package, describe each sound using the Just About Right information. You can also describe each instrument using the same procedure.

- Represent the different variants of the sounds and the different levels of JAR for each emotion on the same graphical output. Project on that space the composition of the different sounds.

Exercise 4.5 *Analyzing CATA data with the Cochran's Q test and the McNemar test*

Eight orange juices have been evaluated by 76 consumers and characterized using 15 CATA terms: Color, Pulp, Natural Appearance, Intensity, Orange, Sweet, Acidic, Bitter, Off-Taste, Natural Taste, Astringent, Irritant, Thick, Refreshing and Aftertaste.

- Import the data *orange_CATA.csv* from the book website.

- Using the **aggregate** function together with the **sum** function, generate the contingency table crossing the products in rows and the CATA terms in columns. Use the **descfreq** function from the FactoMineR package to get a description of the products function of the CATA terms.

- Using the **cochran.qtest** function from the RVAideMemoire package, run the Cochran's Q test for each of these CATA terms using the product and subject information. For which attribute is the Cochran's Q test significant?

- Then, for a significant attribute, compute for each pair of product a 2×2 matrix defining whether the attribute has been ticked for the two products (N_{11}), for none of the products (N_{00}), or for one but not the other product (N_{10} and N_{01}).

- Use the anti-diagonal of this matrix (N_{10} and N_{01}, *i.e.*, the number of times only one of the two products has been characterized by that CATA term considered) to compute the McNemar exact test. Such test is obtained by comparing N_{10} (or equivalently N_{01}) to ($N_{10} + N_{01}$) in a binomial test with the probability of chance of 0.5.

Exercise 4.6 *Sensitivity of the McNemar test*

The McNemar test depends on the anti-diagonal of this 2×2 contingency table. Hence, 2 products with the same overall frequency can or cannot be significantly different. An example is shown here.

Let's consider two products A and B evaluated by 100 consumers for one specific attribute. Let's consider that product A received 50 ticks whereas product B received 40 ticks.

Let's consider 5 different situations, and let's have a look at the evolution of the *p-value* of the McNemar test in each situation:

1. all the consumers who ticked Yes for B ticked Yes for A

2. 5 consumers who ticked Yes for B ticked No for A

3. 10 consumers who ticked Yes for B ticked No for A

4. 15 consumers who ticked Yes for B ticked No for A

5. 20 consumers who ticked Yes for B ticked No for A

- Construct the five 2×2 matrices.

- In each case, compute the McNemar exact test (*cf.* Exercise 4.5). Conclude.

4.5 Recommended readings

- Bécue-Bertaut, M., Álvarez-Esteban, R., & Pagès, J. (2008). Rating of products through scores and free-text assertions: Comparing and combining both. *Food Quality and Preference*, 19, (1), 122-134.

- Bécue-Bertaut, M., & Lê, S. (2011). Analysis of multilingual labeled sorting tasks: application to a cross-cultural study in wine industry. *Journal of Sensory Studies*, 26, (5), 299-310.

- Bécue-Bertaut, M., & Pagès, J. (2004). A principal axes method for comparing contingency tables: MFACT. *Computational Statistics & Data Analysis*, 45, (3), 481-503.

- Bécue-Bertaut, M., & Pagès, J. (2008). Multiple factor analysis and clustering of a mixture of quantitative, categorical and frequency data. *Computational Statistics & Data Analysis*, 52, (6), 3255-3268.

- Cousin, M., Penven, M., Toularhoat, M., Philippe, M., Cadoret, M., & Lê, S. (2008). Construction of a products space from consumers' data (Napping and ultra flash profile). Application to the sensory evaluation of twelve luxury fragrances. 10th European Symposium on Statistical Methods for the Food Industry, Louvain-la-Neuve, Belgium, January 2008.

- Cousin, M., Penven, M., Toularhoat, M., Philippe, M., Cadoret, M., & Lê, S. (2008). Construction of a products space based on consumers' words from the Napping method. Application to the sensory evaluation of twelve luxury fragrances. 10th European Symposium on Statistical Methods for the Food Industry, Louvain-la-Neuve, Belgium, January 2008.

- Feinerer, I. (2012). tm: Text Mining Package. *R package* version 0.5-7.1.

- Feinerer, I. (2014). Introduction to the tm Package. Text Mining in R. Retrieved from: `http://cran.r-project.org/web/packages/tm/vignettes/tm.pdf`.

- Gentry, J. (2014). Twitter client for R. Retrieved from: `http://geoffjentry.hexdump.org/twitteR.pdf`.

- Husson, F., Lê, S., & Pagès, J. (2010). *Exploratory multivariate analysis by example using R*. Chapman & Hall/CRC Computer Science & Data Analysis.

- Kostov, B., Bécue-Bertaut, M., & Husson, F. (2013). Multiple factor analysis for contingency tables in the FactoMineR package. *The R Journal*, 5.

- Kostov, B., Bécue-Bertaut, M., & Husson, F. (2014). An original methodology for the analysis and interpretation of word-count based methods: Multiple factor analysis for contingency tables complemented by consensual words. *Food Quality and Preference*, 32, 35-40.

- Lebart, L., Salem, A., & Berry, L. (1998). *Exploring textual data*. Kluwer Academic Publishers.

- Meyners, M., Castura, J., C., & Carr, B. T. (2013). Existing and new approaches for the analysis of CATA data. *Food Quality and Preference*, 30, (2), 309-319.

- Sauvageot, F., Urdapilleta, I., & Peyron, D. (2006). Within and between variations of texts elicited from nine wine experts. *Food Quality and Preference*, 17, (6), 429-444.

- Taddy, M. (2013). Measuring political sentiment on Twitter: factor optimal design for multinomial inverse regression. *Technometrics*, 55, (4), 415-425.

- Ten Kleij, F., & Musters, P. A. D. (2003). Text analysis of open-ended survey responses: a complementary method to preference mapping. *Food Quality and Preference*, 14, (1), 43-52.

5

When two different products are compared in various situations

CONTENTS

"... specific models such as the Thurstonian model, and the Bradley-Terry model have to be considered. These models allow evaluating if the products are perceived similarly or not, or if one is better than the other. In practice, three very nice packages are dedicated to such models, the sensR package for discrimination tests, the BradleyTerry2 and the prefmod packages for paired comparisons."

5.1   Data, sensory issues, and notations

It seems that discrimination tests have existed forever and are taught in every *curriculum* in sensory evaluation. Their objective is to evaluate whether products can be considered as different, or similar. These tests are often used in situations where a company is facing an ingredient or recipe change and wants to evaluate whether the changes will be perceived or not compared to the original recipe: possible contexts requiring their use are compliance with health initiatives (*e.g.*, salt or sugar reduction), cost reductions, quality control, *etc.* In practice, difference is for instance tested if the reformulation is supposed to improve the current formulation. Oppositely, similarity is usually tested if the change in the new formulation should not be detected by the consumers.

The similarities (or differences) between products can be evaluated overall or according to one particular characteristic (sweetness, for instance). In the first situation, the test can be considered as holistic (*cf.* Chapters 6 and 7); in the second situation, the approach is said to be *oriented*. Still, both approaches require similar analyses and are both described here.

A large variety of discrimination tests and protocols exist. These protocols vary in particular in their accuracy, number of products, and number of samples to assess. Although more than two products may be involved in the experiment, in most cases products are compared two by two, in different experimental contexts, *i.e.*, regarding different combinations of samples. For the triangle test for instance, amongst the three samples that are presented, two samples are associated with one of the two products involved, the other sample being associated with the other product. However, the statistical unit is the same for all these tasks, as it corresponds to one assessment, *i.e.*, one subject testing a pair of products, according to a given protocol.

For most discrimination protocols, the information reported is whether the subject answered correctly or not. For example, in the triangle test, did the panelist detect correctly the odd sample from the 2 similar ones? Finally, we are interested in two single values: the number of correct answers and the total number of subjects performing the test. In other protocols such as the paired comparison, the information reports which of the two products won its match, from an attribute point of view or from a hedonic point of view. Finally, an assessment is described by four qualitative variables: a first one indicating the subject, two others for the two products involved in the match, and a fourth one indicating the result of the match.

Since discrimination tests (whether they are oriented or not, and whether they involve two or more products) aim at evaluating the degrees of similarities/differences between each pair of products, we expect from the statistical analysis of such data that it returns the distance between each pair of products. Such distance (whether it is a sensory or hedonic distance, depending on the task performed) should reflect the reality perceived by the subjects: the larger the distance, the easier to detect the difference between the two products. When more than two products are evaluated, a map based on these distances can be generated (*cf.* Chapter 5, Exercise 5.5, for a similar example).

The estimation of these distances between pairs of products is obtained through different models. The model used depends on the task, and more especially on the number of products involved. When the entire set of products is limited to two products only, the guessing model is used. In this case, the distance between pairs of products used is defined by the proportion of discriminators p_d. When the product space involves more than two products and products are compared by pair, the Bradley-Terry model is used. In this case, the distance between product is defined by the abilities. Finally, another model that can be applied in both situations is presented: the Thurstonian model. In this case, the distance (noted δ) between each pair of product is given by its estimated measure noted d'.

It is naturally in this order that Section 5.2 (*In practice*) is organized. First, the guessing model applied to the discrimination tests is developed (using the SensoMineR and the sensR packages). Then, the Bradley-Terry model is applied to multiple paired comparison data (using the BradleyTerry2 package). In Section 5.3 (*For experienced users*), an alternative model that can be applied to both discrimination tests and multiple paired comparison tests is presented: the Thurstonian model (using the sensR package).

5.2 In practice

5.2.1 How can I measure the distance between two products?

Although the purpose of discrimination tests is fairly simple and straightforward, many different protocols (implying different number of products, different instructions, different levels of difficulties for the subjects *etc.*) exist. An exhaustive list of protocols and associated instructions, involving two products, is given in Table 5.1.

TABLE 5.1
List of discrimination tests including the protocol and the probability of guessing.

Holistic Test	Products	Protocol	Prob.
Duo-trio	Ref. *versus* A B	Which product is the same as the reference?	1/2
Triangle	A A B	Which product is different?	1/3
Tetrad	A A B B	Group the products into groups of 2 identical products	1/3

Oriented Test	Products	Protocol	Prob.
2-AFC	A B	Which product is more ...?	1/2
3-AFC	A A B	Which product of the 3 is more ...?	1/3
m-AFC	A A A ... A B	Which product of the m is more ...?	1/m
Specified Tetrad	A A B B	Which 2 products are the most ...?	1/3
Same/Different	A A or A B	Are they the same or different? How sure are you?	

Although these tests follow different protocols, they are all designed in a way that the answer of each subject is either right or wrong. Mathematically, this means that the responses to these tests follow a binomial distribution. In practice, these responses can be analyzed in two different ways, depending on the point of view adopted. If the focus is on the subjects, and more precisely on their capacity to detect differences between the products, the guessing model should be adopted. If the focus is on the products and how they are perceived, the Thurstonian approach taken from signal detection theory should be considered.

In this section, we focus our attention on the guessing model only as it is historically the most commonly used. However, the Thurstonian model is described in more detail in Section 5.3.

In all these protocols, the subjects are forced to answer the question, whether they do or do not perceive a difference between products. This implies that the subjects have a non-null probability to get the right answer by picking the odd product simply by chance. This probability is denoted as probability p_g of guess: in a situation where no differences between the products are perceptible, the probability of correct responses is defined by the probability of guessing right p_g. This probability p_g depends directly on the protocol. For example, the probability of guess p_g equals $1/2$ for the duo trio and 2-AFC tests and $1/3$ for the triangle, tetrad, and 3-AFC tests.

The core of the guessing model relies on this probability of guess. More precisely, the guessing model assumes that the population can be divided into two groups of subjects: "ignorant people" who are always guessing the answer and "detectors" who are always correctly discriminating the products (and hence providing the correct answer). In this case, the second group of subjects (*i.e.*, detectors or discriminators) are of utmost importance since the sensory distance between products is measured through the proportion of subjects (denoted p_d for *proportion of discriminators*) belonging to that category. To estimate the proportion of discriminators p_d, both the proportion of correct answers, p_c (defined by $\frac{n_c}{n_{total}}$), and the probability of a correct guess p_g are needed.

To illustrate this principle, let's consider the duo-trio protocol. In a duo trio, each subject is presented with an identified reference product and with two coded products, one of them matching the reference. The subject is asked to determine which of the two coded products is most similar to the reference. By chance, the subject has a probability of $1/2$ of guessing correctly. Let's consider that four subjects perform the duo trio and three answer correctly. One of the four answers being wrong, it is an incorrect guess. Based on strict probability, since $p_g = 1/2$, one wrong guess implies one correct guess. Hence, amongst the three correct answers, one is considered as a guess, and two (out of four) are provided from true discriminators (*i.e.*, $p_d = 2/4 = 50\%$).

By generalizing the results, the calculation of the proportion of discriminators is defined according to the following equation:

$$p_c = p_d + p_g(1 - p_d).$$

Hence,

$$p_d = \frac{p_c - p_g}{1 - p_g}.$$

To analyze data from the discrimination test, the SensoMineR and sensR packages are used. To load these packages, the **library** function is required:

```
> library(sensR)
> library(SensoMineR)
```

In order to extend your product portfolio, your company decided to create a new variant of cookie. Your task is to make sure that the new prototype is different enough from the current variant. For that, you decided to perform a triangle test involving the current version and the new prototype.

Before evaluating the products, it is of utmost importance to correctly set up this test. This setup includes defining four main parameters:

- the α-error of the test, *i.e.*, the probability of considering two products as different when they are not;

- the power of the test through the β-error, β being the probability of considering two products as different if they are similar. The power is defined as $1 - \beta$;

- the minimum proportion of discriminators p_d that you would consider as sufficient to consider the products as different;

- the number of assessors to use in the test.

These four parameters are interrelated: once three of them are defined, the fourth one is deducted. For example, by strictly defining the α-error, the power, and the p_d threshold value, the number of assessors needed under these conditions is deduced. Let's consider a triangle test with $\alpha = 0.05$, $\beta = 0.10$ (*i.e.*, power of 0.90), and $p_d = 0.20$ (at least 20% of the population should detect the difference). By using the **discrimSS** function of the sensR package, we estimate the minimum number of subjects required:

```
> n <- discrimSS(pdA=0.20,target.power=0.90,alpha=0.05,pGuess=1/3,
+ test="difference",statistic="exact")
> n
[1] 117
```

In this condition, 117 subjects are required. In your panel, you only have 80 potential subjects available. Since recruiting 37 additional subjects is not possible, you would like to know the potential loss in power of the test if we decrease the sample size from 117 to 80. To evaluate the power of such test, the **discrimPwr** function is used:

```
> pwr <- discrimPwr(pdA=0.20,sample.size=80,alpha=0.05,pGuess=1/3,
+ test="difference",statistic="exact")
> pwr
[1] 0.7366518
```

Being happy with this power of 0.74, you decide to continue with this setup.

The next step consists in designing the order of presentation for a triangle test involving 80 subjects. Such design can be generated using the **triangle.design** function from the SensoMineR package.

```
> design_triangle <-  triangle.design(nbprod=2,nbpanelist=80)
```

To visualize a subset of the design (here the six first presentations), the **head** function is used on the object design_triangle:

```
> head(design_triangle)
```

```
                   Product X Product Y Product Z
Panelist1.Test1        1         2         1
Panelist2.Test1        1         2         2
Panelist3.Test1        1         1         2
Panelist4.Test1        2         1         1
Panelist5.Test1        2         1         2
Panelist6.Test1        1         2         1
```

In order to define how many correct answers are required to conclude for significant differences between the two products, the **findcr** function from the sensR package is used.

```
> correct <- findcr(sample.size=80,alpha=0.05,p0=1/3,test="difference")
> correct
[1] 35
```

With these settings, 35 correct answers out of 80 subjects are required to conclude for significant difference between the two products.

Let's consider you performed the test and 41 out of 80 subjects provided correct answers. To evaluate the probability of this test, we use the **discrim** function of the sensR package.

```
> results <- discrim(correct=41,total=80,method="triangle",statistic="exact",
+ test="difference")
> results
```

```
Estimates for the triangle discrimination protocol with 41 correct
answers in 80 trials. One-sided p-value and 95 % two-sided confidence
intervals are based on the 'exact' binomial test.

        Estimate Std. Error   Lower  Upper
pc        0.5125    0.05588 0.39810 0.6259
pd        0.2687    0.08383 0.09716 0.4389
d-prime   1.5314    0.28926 0.86572 2.1081

Result of difference test:
'exact' binomial test:  p-value = 0.0007063
Alternative hypothesis: d-prime is greater than 0
```

As expected, the test is significant. Hence, it can be concluded that the products are different. From these results, it can also be seen that the actual proportion of discriminators is estimated at $p_d = 27\%$ (larger than the threshold set at 20%). As a conclusion, the new prototype can be added to the market as it is different enough from the current version.

In this fictive example, only the triangle test has been presented. For all the other protocols, the same R code fits except for only a few parameters that might need to be readjusted: the probability of guess through the `pGuess` parameter, the threshold value for p_d through the `p0` parameter (*cf.* Section 5.4), and the nature of the protocol through the `method` parameter. Similarly, when a test of similarity is performed rather than a test of difference, the `test` parameter should be changed to `test="similarity"` (*cf.* Section 5.4).

5.2.2 How can I measure the inter-distance between products when compared in pairs?

Although discrimination testing often involves the comparison of two products, situations in which more than two products are compared exist. In such situations, multiple paired comparisons are often used. In the multiple paired comparison protocol, each subject is asked to compare pairs of products one by one. For each pair, the subject is asked to evaluate which of the two products is the more intense or more liked. To some extent, ranking tests, in which products are ordered increasingly according to one particular characteristic (may it be liking), can be seen as a natural extension to the multiple paired comparison task.

To illustrate this methodology, a real case study involving 10 different mascaras is used. The objective of this study is to define which mascaras are the more/less appreciated. In order to answer this very common problem in sensory science, some practical limitations regarding the products should be noticed: since human beings have only two eyes, it is difficult to test more than two products simultaneously. For that reason, the choice has been made to compare them in pairs using multiple paired comparison, which seemed to be the appropriate task. During the task, subjects were placed in a situation where they had to compare two eyes on which two different mascaras had been applied. In practice, pairs of pictures were provided to the subjects. Since 10 products were compared in pairs, the subjects had to compare multiple pairs, and assess for each pair which of the two products they preferred. The data of this study were collected by Sophie Birot, Célia Pontet, and Lisa Defeyter during their master's degree studies. Since this task implies the concept of match between each pair of mascaras, the analysis of the data should reflect this concept. For that reason, the model of Thurstone and the model of Bradley-Terry seem the most appropriate. Here, the data are analyzed using the model of Bradley-Terry. The model of Bradley-Terry deals with situations in which p samples are compared to one another in paired contests. The odds that sample i beats j are defined by α_i/α_j where α_i and α_j are positive parameters which could be interpreted as "abilities." This model is expressed in the log linear form:

$$\log(P(i \text{ beats } j)) = \lambda_i - \lambda_j$$

where $\lambda_i = \log(\alpha_i)$.

By assuming independence between each comparison of pairs, the parameters of this model can be estimated by maximum likelihood. For this analysis, the package BradleyTerry2 is required (Turner & Firth, 2012). After installing this package, load it with the **library** function.

```
> library(BradleyTerry2)
> options(contrasts=c("contr.treatment","contr.poly"))
```

The data set of this study can be found in the *mascaras.csv* file. After downloading the data set from the book website, import it in your R session under the name *mascaras* using the **read.table** function.

```
> mascaras <- read.table("mascaras.csv",header=TRUE,sep=",")
> summary(mascaras)
      Judge          Preferred          Other
 Min.   :  1.00   Le 2    :225    Ex.vol  :201
 1st Qu.: 37.00   Sexy    :203    Oscil.  :186
 Median : 75.00   Iconic  :175    Everlong:164
 Mean   : 74.88   HD      :162    1001    :156
 3rd Qu.:114.00   Extreme :119    Extreme :156
 Max.   :150.00   Pulse   :109    Pulse   :151
                  (Other) :379    (Other) :358
```

As can be seen from the summary table, the first column related to the subject is considered as numeric while it should be considered as a factor. To change the subject column into a factor, the **as.factor** function is used.

```
> mascaras[,1] <- as.factor(mascaras[,1])
> summary(mascaras)
     Judge           Preferred          Other
 1      :  10   Le 2    :225    Ex.vol  :201
 2      :  10   Sexy    :203    Oscil.  :186
 3      :  10   Iconic  :175    Everlong:164
 4      :  10   HD      :162    1001    :156
 6      :  10   Extreme :119    Extreme :156
 8      :  10   Pulse   :109    Pulse   :151
 (Other):1312   (Other) :379    (Other) :358
```

Now that the first column is properly set as a factor, it is important to notice that the data set is not in the right format for the analysis. Indeed, the data set is structured in three columns: one for the subject (or "Judge"), one with the preferred product ("Preferred"), and one with the other product of the pair ("Other"). We would like to restructure the data in a way to have two columns related to the product information, and two columns with the win/lose information.

In order to transform the data set, let's first generate a contingency table crossing the products in rows and in columns, and which presents the results of the comparison. In this contingency table, the cell c_{ij} corresponds to the number of time product i won against product j when these two products were compared. Such table can be generated directly using the **xtabs** function.

```
> Contingency <- xtabs(~Preferred+Other,data=mascaras)
> Contingency
          Other
```

Preferred	1001	Everlong	Ex.vol	Extreme	HD	Iconic	Le 2	Oscil.	Pulse	Sexy
1001	0	11	18	15	13	7	7	16	12	8
Everlong	13	0	21	9	9	5	4	15	13	6
Ex.vol	13	13	0	10	6	6	5	9	11	8
Extreme	9	13	24	0	11	14	5	18	14	11
HD	21	13	23	23	0	16	12	25	19	10
Iconic	26	22	22	19	16	0	13	25	23	9
Le 2	26	31	30	30	22	18	0	33	18	17
Oscil.	10	17	19	12	8	9	2	0	16	3
Pulse	14	17	21	16	6	10	6	15	0	4
Sexy	24	27	23	22	22	14	16	30	25	0

In order to convert this contingency table of wins to a four-column data set containing the number of wins and losses for each pair of products, the **countsToBinomial** function from the BradleyTerry2 package is used.

```
> mascaras.BT <- countsToBinomial(Contingency)
```

After renaming the columns of this newly created data set (called here mascaras.BT) using the **colnames** function, it is interesting to look at the summary of this table to ensure that the structure of the data is as expected.

```
> colnames(mascaras.BT) <- c("ProductA","ProductB","WinA","WinB")
> summary(mascaras.BT)
    ProductA       ProductB         WinA             WinB
 1001    :9   Sexy    :9    Min.   : 3.00    Min.   : 2.00
 Everlong:8   Pulse   :8    1st Qu.: 8.00    1st Qu.:13.00
 Ex.vol  :7   Oscil.  :7    Median :11.00    Median :18.00
 Extreme :6   Le 2    :6    Mean   :12.44    Mean   :18.04
 HD      :5   Iconic  :5    3rd Qu.:16.00    3rd Qu.:23.00
 Iconic  :4   HD      :4    Max.   :33.00    Max.   :31.00
 (Other) :6   (Other) :6
```

Since the table is now in the required format, we can apply the model of Bradley-Terry to mascaras.BT. To do so, the **BTm** function of the BradleyTerry2 package is used. This function requires as parameters the names of the columns to analyze (**outcome**), as well as the names related to product 1 (**player1**) and to product 2 (**player2**). Finally, the name of the data set from where to read the results should be mentioned (**data**). The results of this analysis are stored in the object called Model.

```
> Model <- BTm(outcome=cbind(WinA,WinB),player1=ProductA,
+ player2=ProductB,data=mascaras.BT)
> Model
Bradley Terry model fit by glm.fit

Call:  BTm(outcome = cbind(WinA, WinB), player1 = ProductA, player2 = ProductB,
    data = paires.sf)
```

```
Coefficients:
..Everlong    ..Ex.vol    ..Extreme       ..HD    ..Iconic      ..Le 2
 -0.21698     -0.56611     0.09500     0.71138     0.85676     1.44526

..Oscil.     ..Pulse      ..Sexy
-0.29380     -0.03671     1.26574

Degrees of Freedom: 45 Total (i.e. Null);  36 Residual
Null Deviance:      302.8
Residual Deviance: 28.99          AIC: 210.1
```

Another way to look at these results is by printing the **summary** of this model.

```
> summary(Model)

Call:
BTm(outcome = cbind(WinA, WinB), player1 = ProductA, player2 = ProductB,
    data = paires.sf)

Deviance Residuals:
    Min       1Q    Median        3Q       Max
-1.3583   -0.6223   -0.1288    0.4398    1.7648

Coefficients:
            Estimate Std. Error z value Pr(>|z|)
..Everlong -0.21698    0.18059  -1.202  0.22955
..Ex.vol   -0.56611    0.17909  -3.161  0.00157 **
..Extreme   0.09500    0.17606   0.540  0.58950
..HD        0.71138    0.17480   4.070 4.71e-05 ***
..Iconic    0.85676    0.17701   4.840 1.30e-06 ***
..Le 2      1.44526    0.18566   7.784 7.01e-15 ***
..Oscil.   -0.29380    0.17718  -1.658  0.09727 .
..Pulse    -0.03671    0.17751  -0.207  0.83618
..Sexy      1.26574    0.18420   6.871 6.36e-12 ***
---
Signif. codes:  0 '***' 0.001 '**' 0.01 '*' 0.05 '.' 0.1 ' ' 1

(Dispersion parameter for binomial family taken to be 1)

    Null deviance: 302.764  on 45  degrees of freedom
Residual deviance:  28.987  on 36  degrees of freedom
AIC: 210.14

Number of Fisher Scoring iterations: 4
```

In this model, the contrats on the coefficients are such that the coefficient related to the first product is set to 0, the other products being rated according to that product of reference. In our situation, the product "1001" is set automatically as the reference product since it is the first product alphabetically. To interpret the results, a coefficient is positive if the corresponding product has been preferred over "1001." Inversely, a coefficient is negative if "1001" has been preferred over that corresponding product. The *p-values* highlighting the significance of these abilities are also provided.

Tips: Changing the product of reference

The product of reference can be easily changed using the **update** function. To do so, the `refcat` parameter should be used to reassign the product of reference within the model. By considering "Iconic" as the new reference, the new model updates all the abilities according to that new reference. To do so, the following code is used:

```
> Model.update <- update(Model,refcat="Iconic")
> Model.update
Bradley Terry model fit by glm.fit

Call:  BTm(outcome = cbind(WinA, WinB), player1 = ProductA,
player2 = ProductB, refcat = "Iconic", data = mascaras.BT)

Coefficients:
    ..1001   ..Everlong    ..Ex.vol    ..Extreme       ..HD     ..Le 2
   -0.8568      -1.0737     -1.4229      -0.7618     -0.1454     0.5885
   ..Oscil.     ..Pulse     ..Sexy
   -1.1506      -0.8935      0.4090

Degrees of Freedom: 45 Total (i.e. Null);  36 Residual
Null Deviance:      302.8
Residual Deviance: 28.99        AIC: 210.1
```

Note that in this case, the global results of the model are unchanged, and only the abilities are updated.

By reordering the abilities, it is possible to assess which products were preferred and how distant products are from each other. To do so, we extract the abilities from the model by using the **BTabilities** function from the BradleyTerry2 package.

```
> abilities <- BTabilities(Model)
> abilities
              ability       s.e.
1001       0.00000000 0.0000000
Everlong  -0.21697772 0.1805864
Ex.vol    -0.56610851 0.1790869
Extreme    0.09499577 0.1760601
HD         0.71138033 0.1747969
Iconic     0.85676190 0.1770125
Le 2       1.44526012 0.1856642
Oscil.    -0.29380100 0.1771773
Pulse     -0.03670704 0.1775105
Sexy       1.26573753 0.1842042
```

Rather than comparing the abilities numerically, we propose to compare them graphically. The first possibility consists in representing these abilities by also including the standard error associated with their estimates (*cf.* Figure 5.1). To do so, the **qvcalc** function from the qvcalc package (for Quasi Variances for Model Coefficients) is used.

```
> library("qvcalc")
> abilities.qv <- qvcalc(abilities)
> plot(abilities.qv,levelNames=rownames(abilities),cex.axis=0.7)
```

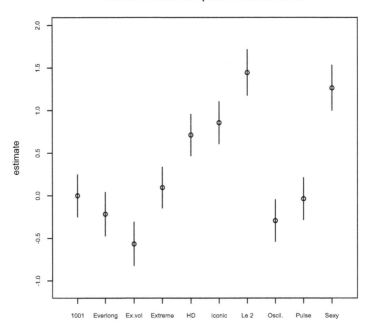

FIGURE 5.1
Representation of the mascaras with confidence intervals (*mascaras.BT* data set).

The abilities can also be represented linearly, on a *preference line*, using the **plot** function (*cf.* Figure 5.2).

```
> plot(x=abilities[,1],y=rep(0,10),type="p",xlim=c(-1,2),ylim=c(-0.2,0.2),
+ axes=FALSE,xlab=" ",ylab=" ",main="Preference Line",pch=20)
> abline(h=0)
> text(x=abilities[,1]-0.1,y=rep(-0.025,10),labels=rownames(abilities)[-1],
+ srt=45)
> points(x=abilities[1,1],y=0,pch=20,col="red")
> text(x=abilities[1,1]+0.1,y=0.025,labels=rownames(abilities)[1],srt=45,
+ col="red")
```

In Figure 5.2, the product of reference (*i.e.*, "1001") is written on top. Here, it appears clearly that the most liked products are "Le 2" and "Sexy," while the least liked product is "Ex.vol."

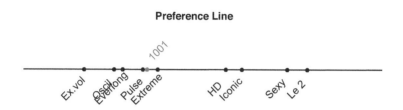

FIGURE 5.2
Representation of the mascaras on a *preference line* (*mascaras.BT* data set).

5.3 For experienced users: The Thurstonian approach

In the previous section, data obtained from discrimination tests were analyzed using the guessing model. This model focuses its attention on the subjects by separating them into two categories, the ones who are guessing right and the ones who are detecting differences between the products. This second category of subjects (*i.e.*, the discriminators) is of main interest since the proportion of discriminators helps conclude whether two products could be considered as different or not.

If this model of analyzing the data has been widely used, mainly thanks to its simplicity, it presents some drawbacks. The proportion of discriminators p_d is linearly linked to the probability of guess p_g. Hence, protocols (such as 3-AFC, triangle, and tetrad) that are associated with the same p_g (here 1/3) provide the same p_d regardless the intrinsic difficulty of the task (in this case, it has been proven that the 3-AFC yields more correct answers than the tetrad, which also yields more correct answers than the triangle test). A direct consequence of that is that the different tests performed on the same products can produce different conclusions if the same threshold for p_d is used. For this reason, the guessing model has been qualified as "method-specific."

In order to correct for this undesired effect, another more stable model has been considered. This model originally comes from the signal detection theory, and is know as the Thurstonian model. As opposed to the guessing model, the Thurstonian approach is oriented towards the product by associating each product with a perceptual distribution. The Thurstonian approach assumes that a variability exists around each product, and even two products obtained from the same batch are not perfectly identical (hence not perceived identically). This variation of the perception of the (supposedly) same product is estimated by a certain distribution, called the perceptual distribution. This perceptual distribution is approximated by a normal distribution.

When two products are compared, the two perceptual distributions are graphically represented together: for two clearly different products, the perceptual distributions are clearly separated, whereas for two similar products, the two perceptual distributions are overlapping (the bigger the overlap, the closer the products).

In Figure 5.3, the perceptual distributions of the two products A and B are clearly separated: A and B are distinguishable.

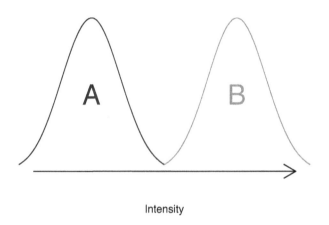

Intensity

FIGURE 5.3
Representation of the perceptual distribution of two distinguishable products, as considered in the Thurstonian approach.

In Figure 5.4, the perceptual distributions of the two products A and B overlap: A and B are confusable.

The standard deviation of these perceptual distributions is directly related to the variability within the products (*e.g.*, temperature, compounds distribution, *etc.*) and within the subjects (*e.g.*, memory, fatigue, adaptation, *etc.*). The more variable the products and/or the subjects, the larger the standard deviation associated with these perceptual distributions, and the higher the risks of overlap (*i.e.*, of confusion between products).

The aim of the Thurstonian approach is to calculate the true distance δ between products. This distance represents the distance between the means μ_A and μ_B of the perceptual distributions associated with product A and B.

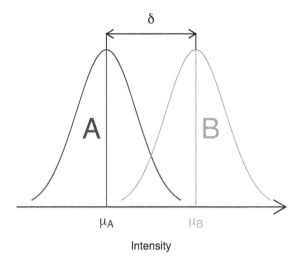

FIGURE 5.4
Representation of the perceptual distribution of two confusable products, as considered in the Thurstonian approach. In this case, the notion of distance δ between products is highlighted.

This distance δ is measured in standard deviations (σ_A and σ_B) associated with the perceptual distribution of A and B. As δ is unknown, its value is estimated. In practice, the estimation of δ is noted d'.

As opposed to the guessing model which links linearly p_d to p_g regardless of the protocol, the Thurstonian approach estimates the value of d' according to the protocol used. In the Thurstonian approach, the model that relates the proportion of correct answer p_c to d' depends on what is called a "psychometric function," and each protocol is defined by its own psychometric function. Hence, two protocols yielding the same proportion of correct answer yield different values of d'. It should be noted that these psychometric functions integrate the mental process guiding the decision rule that each subject follows while performing the task in their calculation. As a direct consequence, different protocols yield stable values of d' when the same products are involved: the Thurstonian model is not method-specific.

Let's illustrate an example of decision rule processed by a subject through the case of the triangle test. In a triangle test, two products A and one product B are presented to a subject who is then asked to detect the *odd* product amongst the three samples. During the testing process, the subjects measure internally distances between each pair of products and report the one that is the more distant from the other products. In other words, the product considered as the *odd* one is the product farther away from the two others. Illustrations of decisions in a triangle test are presented in Figure 5.5 and in

Figure 5.6. In Figure 5.5, the answer to the triangle test is correct since the *odd* product y is farther away from the two similar products x_1 and x_2.

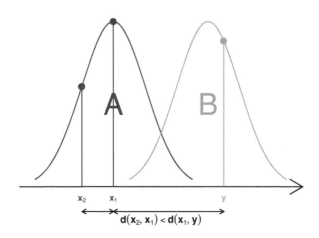

FIGURE 5.5
Schematization of an example of decision rule processed by a panelist in a triangle test leading to a correct answer to the task.

In Figure 5.6, the answer to the triangle test is wrong since the distance between the odd product y and x_1 is shorter than the distance between x_1 and x_2. In this example, x_2 would be selected by the subject as the odd product.

It is worth mentioning that since the instructions are different from one protocol to another, each discrimination test follows its own decision rule.

In Section 5.2.1, a triangle test was performed with 80 subjects in order to compare the current version of a cookie with the new prototype. Fourty-one correct answers were obtained in this test. To estimate the d' value between the current version of the cookie and the new prototype, the same function (*i.e.*, the **discrim** function from the sensR package) as previously stated is used:

```
> results <- discrim(correct=41,total=80,method="tetrad",statistic="exact",
+ test="difference")
> results
```

Estimates for the triangle discrimination protocol with 41 correct
answers in 80 trials. One-sided p-value and 95 % two-sided confidence
intervals are based on the 'exact' binomial test.

	Estimate	Std. Error	Lower	Upper
pc	0.5125	0.05588	0.39810	0.6259
pd	0.2687	0.08383	0.09716	0.4389
d-prime	1.5314	0.28926	0.86572	2.1081

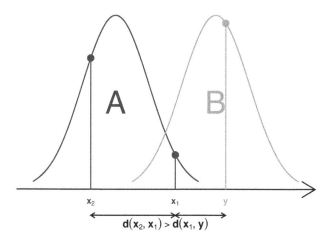

FIGURE 5.6
Schematization of an example of decision rule processed by a panelist in a triangle test leading to an incorrect answer to the task.

```
Result of difference test:
'exact' binomial test:  p-value = 0.0007063
Alternative hypothesis: d-prime is greater than 0
```

From this analysis, the distance between the two cookies is estimated at $d' = 1.53$. Since products are often considered as different when $d' \geq 1$ (arbitrary), it can be concluded here that the products are different.

Rather than using the triangle test, let's consider that a tetrad test was performed on these two variants of cookies. Although the tetrad test requires the subject to test four products instead of three (in the Tetrad test, two samples from one group and two from another are presented to the subjects who are asked to group them into two groups of two, based on similarity), the tetrad appears to be more powerful than the triangle test. This difference in power is directly related to the difference in decision rule involved in the two tasks. To justify this difference in power between protocols, let's first observe that the Thurstonian d' and the proportion of discriminators p_d both depend on one unique parameter in common: the proportion of correct answer p_c. Since p_c is linearly linked to p_d on the one hand, and is linked to d' on the other hand (through the psychometric function), it is easy to transit from one model to the other, *i.e.*, to find equivalences within each protocol between p_d and d'. In the sensR package, the **rescale** function provides directly these equivalences between models. By considering the previous threshold of $p_d = 20\%$ in a triangle test, let's define the equivalent value of d':

```
> rescale(pd=0.20,method="triangle")
```

```
Estimates for the triangle protocol:
        pc  pd  d.prime
1 0.4666667 0.2 1.287124
```

In the triangle test, a proportion of discriminators of 20% corresponds to a d' value of 1.29. This also corresponds to having a proportion of correct answers of about 47%. When applying the same parameter to the tetrad test, the following value of d' is obtained:

```
> rescale(pd=0.20,method="tetrad")
```

```
Estimates for the tetrad protocol:
        pc  pd   d.prime
1 0.4666667 0.2 0.9000829
```

In the tetrad test, a proportion of discriminators of 20% corresponds to a d' value of 0.90.

The differences in d' between the triangle test and the tetrad test can be interpreted as such: at p_d constant, the tetrad test allows detecting accurately smaller differences ($d'_{tetrad} = 0.90$ *versus* $d'_{triangle} = 1.29$) between products than the triangle test. In other words, the tetrad test is more powerful than the triangle test. In order to quantify the difference in power between the tetrad and the triangle test, let's position ourselves in a situation where the same difference (in terms of d') is evaluated in the two protocols. As we already know the power of the triangle test for a d' of 1.29, let's evaluate the power of the tetrad test in similar settings (*i.e.*, 80 subjects, $d' = 1.29$). Such power is obtained by using the **d.primePwr** function of the sensR package:

```
> pwr <- d.primePwr(d.primeA=1.29,sample.size=80,alpha=0.05,method="tetrad",
+ test="difference",statistic="exact")
> pwr
[1] 0.9959048
```

As noted here, the tetrad test is more powerful than the triangle test. Indeed, in the same settings, the triangle test is associated with a power of 74% whereas the tetrad test is associated with a power of 99%. A direct consequence of this relies in the fact that the tetrad test requires a smaller sample size than the triangle test to evaluate the same distance between products. This example also shows that the Thurstonian model is more accurate than the guessing model since the decision rule associated with each protocol is integrated within the data process. This is not the case with the guessing model which returns the same p_d for protocols associated with the same p_g, regardless of the different properties of the protocols (*i.e.*, difficulty, decision rules, *etc.*).

Tips: The rescale function of the sensR package

The guessing model is qualified as method-specific since p_d is linearly linked to p_g. Indeed, protocols associated with the same probability of guess p_g (*e.g.*, 3-AFC, tetrad, and triangle test) provide the same proportion of discriminators p_d, regardless of the degree of difficulty of these protocols.

A direct consequence of that is that at a fixed p_d threshold, different protocols might lead to different conclusions. The 3-AFC, which is an easier task (hence reaching higher p_c values), is easily associated with p_d values larger than 20%. In those cases, it would often be concluded that products are different. However, if the same products are compared with the triangle test, which is a more difficult task (hence reaching lower p_c values), this threshold of $p_d = 20\%$ would only be reached occasionally. Hence, no differences between the same products would often be concluded.

By using the **rescale** function of the sensR package, it can easily be seen that a threshold of $d'=1$ corresponds to a p_d value of 13% for the triangle test and of 45% for the 3-AFC test. To avoid any misinterpretation of the results when using the guessing model with different protocols, the threshold value of p_d should be readjusted in each case by using a fixed d' value as threshold.

```
> rescale(d.prime=1,method="triangle")

Estimates for the triangle protocol:
        pc       pd d.prime
1 0.4180467 0.12707       1

> rescale(d.prime=1,method="threeAFC")

Estimates for the threeAFC protocol:
        pc        pd d.prime
1 0.6337021 0.4505531       1
```

5.4   Exercises

Exercise 5.1 *Practicing the triangle test with more than two products*

The aim of this exercise is to analyze results from the triangle test: first, when two products are involved; then when four products are involved.

- Use the **triangle.design** function of the SensoMineR package to generate the design, with the following inputs: 2 products (*Pepsi Cola* and *Coca Cola*), 30 subjects, and 1 test per subject. Collect the data. Use the **triangle.test** function of the SensoMineR package to analyze the data. Compare the results previously obtained with those from the **triangle.pair.test** function of the SensoMineR package and those obtained from the sensR package. If you don't have any data, you can download the *triangle* data set on the book website.

- Use the **triangle.design** function of the SensoMineR package to generate the design, with the following inputs: 4 products (*Pepsi Cola, Coca Cola, Pepsi Cola light*, and *Coca Cola light*), 30 subjects, and 3 tests per subject. Collect the data. Use the **triangle.test** function of the SensoMineR package to analyze the data. If you don't have any data, you can download the *triangle_revisited* data set on the book website.

- Finally, let's consider 4 products (*Pepsi Cola, Coca Cola, Pepsi Cola light*, and *Coca Cola light*) that we want to test 2 by 2 from a hedonic point of view: which one of those 2 products do you prefer? Build an experimental design for such an experiment for 40 panelists. To do so, use the **optimaldesign** function of the crossdes package. Export this design into an Excel sheet.

Exercise 5.2 *Comparing the triangle test and the tetrad*

For health reasons, a cookie company decides to reduce the sugar content of one of his cookies. Once the variant has been created, Steve, the project manager, wants to ensure that the new variant is not detected as different from the original version by more than 15% of the population. To make sure that it is not the case, two studies are conducted, one in Europe, and one in the US.

- In Europe, a triangle test involving 80 consumers is performed. In this study, 32 panelists found a difference between the products. Using the **discrim** function of the sensR package, calculate the p_d value and conclude.

- In the US, only 50 consumers could be recruited. Since Steve was concerned about the power of such test, he decided to use another protocol which appears to be more powerful: the tetrad test. This test returned 23 correct answers. Knowing that the tetrad test has the same probability of guessing as the triangle test, compute the proportion of discriminators using the same procedure. Conclude.

- Run the same analysis again and let's compare the results in terms of d'. Using the **rescale** function of the sensR package, find the equivalent to a proportion of 15% of discriminators for the triangle test in the tetrad test. Is the new cookie suitable to replace the original product?

- Using the **discrimPwr** function of the sensR package, confirms that the tetrad test is more powerful than the triangle test.

Exercise 5.3 *Application of the Bradley-Terry model to two subgroups of subjects*

The experiment on the mascaras presented in Section 5.2.2 was originally designed to understand perception differences between female and male subjects. As women are the ones who use the product, we wanted to see whether what they like, and hopefully what they wear, was also corresponding to what male subjects like.

- Import the *mascara_gender.csv* file from the book website.

- Split the data imported into two groups, one for each gender. For each gender, create the contingency table summarizing how many times each product was preferred in each pair.

- Using the BradleyTerry2 package, compute the utilities for each gender. Compare the results.

- Similarly to Section 5.2.2, represent the *preference line* plot for both genders on the same graphic.

Exercise 5.4 *Assessment of the presentation order in the Bradley-Terry model*

This exercise aims at testing additional effect with the Bradley-Terry model. In this exercise, the focus is made on the presentation order and its impact on the results. By relating such procedure to sport (the paired comparison being the regular season in which all the teams are playing against each other), such effect assess the impact of playing home or away on the different results.

- Import the *mascara_total.csv* file from the book website.

- Run the Bradley-Terry model checking first for each product effect.

- Add two columns to the data set, one related to product A which only takes 1 (first order) and one related to product B which only takes 0 (last order). These two columns can be called *DesignOrder*. Such column can be constructed using the **data.frame** function.

- Update the model obtained previously by adding to the formula the *DesignOrder* effect. Compare the results of the new model with the original one: is the order of presentation having a significant effect on the evaluation of the products?

Exercise 5.5 *Application of the Bradley-Terry model to sensory attributes*

The data used to illustrate this exercise were collected by Elodie Dufeil, Audrey Emzivat, Alicia Gehin, and Nelly Le Cam during their final master's degree project[1]. We wanted to assess the feasibility of using consumers to get sensory profiles of complex products, in our case perfumes, using paired comparisons tests. For a given list of sensory attributes, consumers had to determine which of the two perfumes they were testing was the most ..., for each attribute: *e.g.*, which one of the two perfumes is the most heady?

- Import the *perfumes_paircomp.csv* file from the book website.

- With the summary function, describe the structure of the data set.

- Create a matrix with the products in rows, and the attributes in columns. To get the list of attributes, the **strsplit** function can be used (the character used "_" as `split`).

- For each attribute, select the two columns related to it, and create the data set as required by the **BTm** function of the BradleyTerry2 package.

- Run the Bradley-Terry model. Extract the utilities and store them in the matrix previously created.

- Once this matrix crossing the products in rows, the attributes in columns, and containing the abilities is obtained, run a PCA on this matrix, by setting the parameters according to your own point of view. Interpret the results.

5.5   Recommended readings

- Agresti, A. (2002). Categorical data analysis. 2nd edition. John Wiley & Sons.

- Bi, J., Ennis, D. M., & O'Mahony, M. (1997). How to estimate and use the variance of d' from difference tests. *Journal of Sensory Studies*, 12, (2), 87-104.

- Böckenholt, U. (2006). Visualizing individual differences in pairwise comparison data. *Food Quality and Preference*, 17, (3), 179-187.

---

[1]They were awarded for the 2013 edition of the Syntec trophy, whose aim is to reward the best master's degree thesis in the field of market research and public opinion polling. Their work was a methodological work with many different objectives, one of them aiming at evaluating the coherence between the sensory perception and the marketing perception of a product space.

- Bradley, R. A., & Terry, M. E. (1952). Rank analysis of incomplete block designs: the method of paired comparisons. *Biometrika*, 39(3-4), 324-345.

- Christensen, R. H. B. (2013). Statistical methodology for sensory discrimination tests and its implementation in sensR. Retrieved from `http://cran.r-project.org/web/packages/sensR/vignettes/methodology.pdf`.

- Christensen, R. H. B., & Brockhoff, P. B. (2010). sensR: an R-package for Thurstonian modeling of discrete sensory data. R package version 1.2-19. `http://www.cran.r-project.org/package=sensR`.

- Courcoux, P., & Semenou, M. (1997). Preference data analysis using a paired comparison model. *Food Quality and Preference*, 8, (5-6), 353-358.

- Ennis, D. M. (1993). The power of sensory discrimination methods. *Journal of Sensory Studies*, 8, (4), 353-370.

- Ennis, D. M., & Ennis, J. M. (2012). Accounting for no difference/preference responses or ties in choice experiments. *Food Quality and Preference*, 23, (1), 13-17.

- Ennis, J. M., Ennis, D., M., Yip, D., & O'Mahony, M. (1998). Thurstonian models for variants of the method of tetrads. *British Journal of Mathematical and Statistical Psychology*, 51, (2), 205-215.

- Ennis, J. M., & Jesionka, V. (2011). The power of sensory discrimination methods revisited. *Journal of Sensory Studies*, 26, (5), 371-382.

- Ennis, J. M., Rousseau, B., & Ennis, D. M. (2014). Sensory difference tests as measurement instruments: a review of recent advances. *Journal of Sensory Studies*, 29, (2), 89-102.

- Firth, D. (2005). Bradley-Terry models in R. *Journal of Statistical Software*, 12, 1-12. `http://www.jstatsoft.org/v12/i01`.

- Firth, D. (2010). Qvcalc: quasi-variances for factor effects in statistical models. R package version 0.8-7, `http://CRAN.R-project.org/package=qvcalc`.

- Gabrielsen, G. (2000). Paired comparisons and designed experiments. *Food Quality and Preference*, 11, (1), 51-61.

- Goos, P., & Großmann, H. (2011). Optimal design of factorial paired comparison experiments in the presence of within-pair order effects. *Food Quality and Preference*, 22, (2), 198-204.

- Næs, T., Brockhoff, P. B., & Tomic, O. (2010). Statistics for sensory and consumer science. John Wiley & Sons.

- Sailer, O. (2013). The "crossdes" package. Retrieved from `http://cran.r-project.org/web/packages/crossdes/crossdes.pdf`.

- Thurstone, L. L. (1927). A law of comparative judgment. *Psychological Review*, 34, (4), 273-286.

- Turner, H., & Firth, D. (2012). Bradley-Terry models in R: the BradleyTerry2 package. Retrieved from `http://cran.r-project.org/web/packages/BradleyTerry2/vignettes/BradleyTerry.pdf`.

- Worch, T., & Delcher, R. (2013). A practical guideline for discrimination testing combining both the proportion of discriminators and Thurstonian approaches. *Journal of Sensory Studies*, 28, (5), 396-404.

6

When products are grouped into homogeneous clusters

CONTENTS

"... data are usually aggregated into a matrix of distance between products. Unfortunately, in doing so, we lose the information related to the variability between subjects. This can easily be circumvent by directly analyzing the raw data with Multiple Correspondence Analysis (MCA). Moreover, similarly to what has been exposed in Chapter 2, MCA can be coupled with bootstrap techniques to get confidence ellipses around products. In practice, the analysis of sorting task data can be done by using the **fast** *function of the SensoMineR package."*

6.1   Data, sensory issues, and notations

In 1996, Daws reported: "A commonly used data collection technique is the method of free sorting. In the most general case, a group of n subjects are presented with a set of N objects and are asked to place "similar" objects together and to separate "dissimilar" ones. The collection of exhaustive and disjoint subsets produced by each subject will be called a sorting or a partition. The method of free sorting has been widely used as a data collection strategy for several reasons. The task is simple and relatively easy for subjects to carry out. Researchers (*e.g.*, Best and Ornstein, 1986) have been able to use it with children as young as four. The method is cognitively engaging (Miller, 1969), and subjects report enjoying the task."

As often, in most sensory experiments, the stimuli are products, and sensory scientists are interested in understanding how they are perceived. When providing a partition on the products (disjoint subsets of products), each subject is qualitatively assessing the products based on their similarities and differences, according to the group they belong to. In that sense, products can be considered as statistical units and subjects as variables. With respect to this point of view, the data that are collected can be put in a matrix $X = (x_{ij})$ of dimension $I \times J$, where I denotes the number of products and J the number of subjects. At the intersection of one row and one column, x_{ij} denotes the group to which product i belongs for subject j. Depending on the versions, whether subjects are asked to simply group products or to group products and verbalize the groups once obtained, x_{ij} can simply denote the number of the group the stimulus i belongs to or in the second situation, the "bag of words" used to qualify the group (*cf.* Chapter 4).

Initially, data are collected to understand similarities and differences within a set of products. Therefore, we expect from the analysis of such data that it provides (at least) a representation of the product space that would reflect the similarities and the differences provided by all subjects. From a cognitive perspective, we also want to understand how subjects have built their own partition with respect to the representation of the products provided by all subjects. Hence, we expect from the analysis that it provides an integrated representation of the subjects function of the way they have sorted the products.

Originally, such data were analyzed by first calculating a similarity matrix crossing the products both in rows and columns: at the intersection of row i and column i', the number of subjects who associated products i and i' in the same group. From this similarity matrix, a dissimilarity matrix was usually derived and Multidimensional Scaling (MDS) was then applied (*cf.* Exercises, Section 6.4). As data are aggregated over subjects for the construction of the dissimilarity matrix, a drawback of the method is that the individual information related to the subjects is "lost." To circumvent this problem, a methodology based on the raw data is proposed. Without transformation of any kind, the raw data matrix X is analyzed using Multiple Correspondence Analysis (MCA), a *principal component method* dedicated to qualitative variables.

This chapter is divided into three main parts: the first one dealing with the study of the product space itself, the second with the study of the product space and its interpretation when we have at our disposal verbalization data, and finally the third part dedicated to the subjects with respect to consensual representation of the products. In Section 6.3, *For experienced users*, hierarchical sorting task is presented.

6.2   In practice

6.2.1   How can I approach *sorting* data?

The most natural way to approach this way of collecting data starts from the observation of the subjects. Each subject is placing stimuli into categories; in other words, each stimulus is described by the group it belongs to, for a given subject. Consequently, a stimulus can be assimilated to a statistical unit of interest; a subject can be assimilated to a categorical variable that would sort stimuli according to their own categories (or groups). Depending on the task, subjects may describe the groups they have created.

In order to understand this point of view and the way data are getting analyzed, let's first import the data set *perfumes_sorting.csv* from the book website with the **read.table** function, and let's have a quick look at it with the **summary** and the **dim** functions. These data were collected thanks to Marine Cadoret during her PhD thesis: the stimuli used for her experiment are the same twelve perfumes as the ones used throughout the first part of the book.

```
> sorting <- read.table("perfumes_sorting.csv",header=TRUE,sep="\t",row.names=1)
> summary(sorting)
S1                                       S2           S3
  A:3   soft;discreet;refined           :3   D                    :1
  B:3   neutral;soap;classical          :2   discreet;chemical:2
  C:1   acid;violent;soft               :1   soap;hands           :1
  D:3   ammonia                         :1   soft;character   :4
  E:1   bathroom;low-quality            :1   very;discreet    :3
  F:1   flower;light;acid;soft;prickly:1   very;strong          :1
        (Other)                         :3
> dim(sorting)
[1] 12 60
```

Based on the **dim** function, the *sorting* data set contains 12 rows and 60 columns. In other words, there are 12 statistical units (the perfumes) described by 60 variables (the subjects). As expected, the variables of the data set are considered as categorical: for instance, the levels of the third variable *S3* are "*D*," "*discreet;chemical*," "*soap;hands*," "*soft;character*," "*very;discreet*," "*very;strong*."

Let's print the three first columns of the data set.

```
> sorting[,1:3]
                        S1                              S2               S3
Angel                   D              violent;pepper discreet;chemical
Aromatics Elixir        D                       ammonia      very;strong
Chanel N5               C              bathroom;low-quality      soap;hands
Cinéma                  B              soft;discreet;refined  soft;character
Coco Mademoiselle       B              soft;discreet;refined  soft;character
J'adore EP              A                  acid;violent;soft   very;discreet
J'adore ET              A              neutral;soap;classical  soft;character
L'instant               F       flower;light;acid;soft;prickly   very;discreet
Lolita Lempicka         E pepper;balanced;breed;soft;prickly  soft;character
```

Pleasures	A	neutral;soap;classical	very;discreet
Pure Poison	B	soft;discreet;refined	discreet;chemical
Shalimar	D	strong;raw	D

As previously explained and as shown in the output above, the statistical individual of interest is the stimulus. The structure of the data set is such that rows correspond to stimuli, and columns to subjects. Each subject is assimilated to a categorical variable that would sort stimuli according to their own groups. When subjects are asked to describe the groups they have formed, it can be noticed that they have more or less difficulty verbalizing their groups: *S1* did not verbalize at all, whereas *S2* verbalized all groups, and *S3* verbalized all groups except one.

More formally, the intersection of row i and column j corresponds to the index x_{ij} of the group in which product i belongs to according to subject j, as is the case with *S1*. When verbalization data are also provided, x_{ij} corresponds to the sequence of words subject j used to describe product i. As it will be presented later, this information (the words used to describe the groups) is of utmost importance when interpreting the data.

Naturally, to calculate a distance between two perfumes i and i' according to a given subject j, it is necessary to recode the variable j into dummy variables, which take the value 0 or 1 to indicate whether a product belongs to a category or not. For *S1* for instance, six dummy variables (*i.e.*, one per group) would be created. Naively, to calculate the distance between i and i', the Euclidean distance is applied on these dummy variables. For *S1*, *Angel* and *Shalimar* are at a distance of 0, as these perfumes belong to the same group; whereas, *Angel* and *Pleasures* are at a distance of 2, as these perfumes belong to two different groups.

If we had to calculate the distance over all the subjects, the most simple thing to do would be to add the individual distances that have just been calculated. For our three first subjects, it's easy to see that the distance between *Angel* and *Shalimar* is equal to 4, whereas the distance between *Angel* and *Pleasures* is equal to 6.

This naive approach can be "improved" by taking into account the number of stimuli per group. Let k denote the index of the dummy variables, K the total number of dummy variables, and I_k the number of products in the group associated with the dummy variable k. The idea is to stress on the differences for the very small groups for which products have been isolated. For *S1* for instance, the distance between *Angel* and *Pleasure* should be smaller than the distance between *Angel* and *Lolita Lempicka*, as *Lolita Lempicka* is the only perfume of its group, and *de facto* seems to have been isolated. This is not the case for *Pleasure*, which belongs to a group of three perfumes.

More formally, let's now consider the recoded data of dummy variables in which the intersection of one row i and one column k, denoted x_{ik}, is equal to 1 when i belongs to the group associated with the dummy variable k, and 0 otherwise. The distance between two products i and i' as considered from now on can be expressed the following way:

$$d^2(i, i') = \frac{I}{J} \sum_{k=1}^{K} \frac{(x_{ik} - x_{i'k})^2}{I_k}.$$

As we can see, the differences between two products are penalized by the fact that one of the two products belongs to a small group.

In the following section, the origin of the ratio I/J is explained. Additionally, the procedure on how a representation of the product space can be obtained using this distance is shown.

6.2.2 How can I get a representation of the product space?

To obtain a representation of the product space, the point of view presented here is the one introduced by Cadoret *et al.* in their paper (Cadoret, M., Lê, S., & Pagès, J., 2009). This point of view is based on MCA, which is the reference method for exploring a set of qualitative variables. In that sense, MCA can be seen as an extension of CA to the case where more than two qualitative variables are considered.

Regarding the structure of the data set to be analyzed (rows correspond to statistical units, columns correspond to variables, *cf.* Section 6.2.1), MCA can also be seen as a variation of PCA applied to qualitative data. In practice, in MCA, the original qualitative data set is first transformed into a table of dummy variables (or indicator variables), in which each dummy variable represents one level of one of the categorical variables of the original data set. In other words, the number of dummy variables generated corresponds to the total number of categories across variables present in the data set (in our context, this corresponds to the total number of groups generated over all subjects).

The distances between rows (*i.e.*, products) on the one hand, and columns (*i.e.*, dummy variables associated with groups provided by subjects) on the other hand, are then computed based on the previously defined formula (*cf.* Section 6.2.1). As in all *principal component methods* (or dimension reduction techniques), the idea of MCA is to find the dimensions that best represent distances amongst rows and amongst columns. In our particular case, two products are close if they have been grouped together by a high number of subjects.

Ultimately, MCA can be defined as a CA applied to our dummy variables data set. As such, the representations of the rows and the columns (in other words, of the individuals and the categories) can be superimposed. This feature is of utmost importance when interpreting the data.

Remark. The ratio I/J in the distance between two rows is a direct consequence of the Chi-square distance formula (*cf.* Chapter 4) applied to the table of dummy variables.

The analysis of sorting data can be performed "manually" using the **MCA** function from the FactoMineR package. However, we propose to use the **fast** function from the SensoMineR package, as this function provides additional results that are specific to sorting data. After loading the SensoMineR package, run the function **fast** on the *sorting* data set.

```
> library(SensoMineR)
> res.fast <- fast(sorting,sep.words=";",graph=FALSE)
> names(res.fast)
 [1] "eig"        "var"        "ind"        "group"      "acm"        "call"
 [7] "cooccur"    "reord"      "cramer"     "textual"    "descriptor"
```

The outputs provided by the **fast** function are divided into 11 objects (shown with the **names** function). Some of these outputs are directly related to the **MCA** function, whereas some others are specific to the **fast** function. The information related to the perfumes is placed in the object named **res.fast$ind**, which is divided into three parts: the coordinates on each dimension (**res.fastindcoord**), the contribution to the construction of each dimension (**res.fastindcontrib**), and the quality of representation on each dimension (**res.fastindcos2**). These results are the classical outputs of the **MCA** function.

```
> names(res.fast$ind)
[1] "coord"    "contrib" "cos2"
> res.fast$ind$coord
                        Dim 1        Dim 2        Dim 3        Dim 4        Dim 5
Angel                0.3954526   1.51711384   0.17272573 -0.865171134 -0.17817147
Aromatics Elixir     1.4836495  -0.22323266  -0.21394964 -0.550797798 -0.21623535
Chanel N5            1.3843400  -0.91069536   1.38741051  0.726426471 -0.01251468
Cinéma              -0.4476777   0.41025707   0.07157530  0.477447918  0.13971909
Coco Mademoiselle   -0.6544809   0.26099364   0.02100746  0.807177380  0.47342100
J'adore EP          -0.7067233  -0.62592308   0.17146765 -0.694520407 -0.22017087
J'adore ET          -0.6981726  -0.63594939   0.19467690 -0.508679141 -0.44473781
L'instant           -0.4703961   0.13728699  -0.58883593  1.155987278 -1.08229473
Lolita Lempicka     -0.1601233   1.39968241   0.69870786 -0.008770055  0.09073011
Pleasures           -0.7233064  -0.85839579  -0.11835628 -0.611310184 -0.34486916
Pure Poison         -0.4326015  -0.41923774  -0.15796241 -0.051669157  1.55758168
Shalimar             1.0300395  -0.05189995  -1.63846714  0.123878830  0.23754219
```

Additionally to the results related to the products, results related to the verbalization are also given. These results are either provided by the **MCA** function (**res.fast$var**), or specific to the **fast** function (**res.fast$textual** and **res.fast$descriptor**). As we will see later, these results are of utmost importance to get a full understanding of the product space.

Additionally to the numerical outputs, the **fast** function also provides graphical outputs. For instance, the representation of the product space can be obtained by using the **plot.fast** function of the SensoMineR package. To facilitate the visualization of the products (but not their interpretation, as we would require the verbalization), we propose to hide the words by using the `invisible` argument.

```
> plot.fast(res.fast,invisible="var")
```

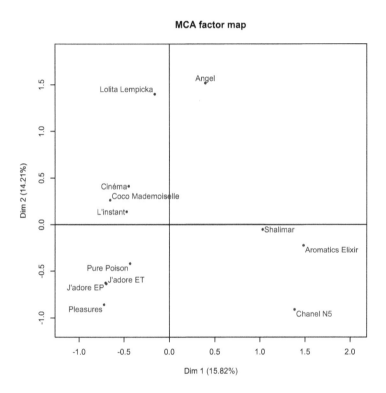

FIGURE 6.1
Representation of the perfumes on the first two dimensions resulting from an MCA, using the **fast** function (*sorting* data set).

According to Figure 6.1, the first dimension opposes *Shalimar, Aromatics Elixir* to *Cinéma, Coco Mademoiselle*, and *L'instant*. The second dimension opposes *Angel* and *Lolita Lempicka* to *Pleasures* and *Chanel N5*. Unless one has an accurate expertise on the products, such representation is not directly interpretable without any verbalization. By default, the graphical representations provided by the **fast** function are based on the first two principal dimensions of variability of the product space. The `axes` argument in the **plot.fast** function allows changing the dimensions on which graphs are plotted. To plot the product space on the third and fourth dimensions (results are not shown here), the following code is used:

```
> res.fast <- plot.fast(res.fast,axes=c(3,4),invisible="var")
```

Besides providing a representation of the product space, the **fast** function also enhances its representation of the products with confidence ellipses. The ellipses are obtained using re-sampling techniques. Practically, the function generates virtual panels of subjects by randomly drawing with replacement subjects from the original panel. For each virtual panel, a new product configuration is calculated and projected onto the original representation obtained from the original panel. By default, the number of panels generated is set to 200 (it can be modified by changing the B argument). Confidence ellipses are then drawn around each product by englobing 95% of the projections provided by the virtual panels (*cf.* Figure 6.2). This level of confidence of the ellipses can be modified by changing the `alpha` argument, set to 0.05 by default.

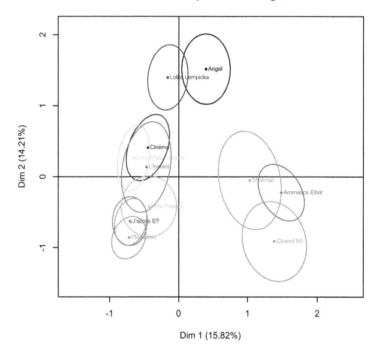

FIGURE 6.2
Representation of the perfumes and their confidence ellipses on the first two dimensions resulting from an MCA, using the **fast** function (*sorting* data set).

Tips: Handling graphical outputs

Besides being a powerful software for statistical computations, R is also an excellent tool for data visualization. However, a good control of the graphical tool is required to avoid overwriting one graphic with another one. If a graphical device is open, and you would like to generate another one in another window, it is possible by simply using the **dev.new** command. Similarly, the **dev.off** function closes the last graphical window open.

```
> dev.new()
> plot(sin,-pi,2*pi)
> dev.new()
> plot(cos,-pi,2*pi)
> dev.off()
```

Note that in some cases, the use of the **dev.new** function is not necessary since some R functions generate automatically graphical outputs in new windows. For example, the generic **plot** functions, that are associated with the principal component methods from the FactoMineR package (*e.g.*, the **plot.PCA** function used to plot the results from the **PCA** function, or similarly the **plot.CA** function used to plot results from the **CA** function, *etc.*) all present an argument called `new.plot` that controls the opening of new windows. If `new.plot=TRUE`, the resulting graphic is opened in a new window; otherwise not.

Instead of generating new graphical outputs directly on the screen, they can be directly printed within a "physical" file, on your computer. Such a file can be saved as *jpeg, pdf, tif, etc.* To do so, the **jpeg**, **pdf**, **tif**, *etc.* functions are used (*cf.* Appendix). Note that to finalize the creation of such physical file, the **dev.off** function should be used to close it.

```
> jpeg("MyGraphInR.jpg",width=800,height=600)
> plot(sin,-pi,2*pi)
> dev.off()
```

If you look in your working directory (run `getwd()` to get the address), a new file called "MyGraphInR.jpg" with your plot has been generated.

As you may have experienced already, and unless stated otherwise (for instance by using the argument `graph=FALSE`), some functions open many graphical outputs. It is the case of the **fast** function presented in this chapter. To close all the graphics at once, the **graphics.off** function can be used.

```
> res.fast <- fast(sorting,sep.words=";",graph=TRUE)
> graphics.off()
```

6.2.3 How can I fully interpret the product space?

As mentioned previously, the representation of the stimuli obtained by the sorting task is obtained thanks to MCA. The results of this analysis are stored in the object named `res.fast$acm`.

Tips: On using `res.fast$acm`

As mentioned previously, the core of our methodology is based on MCA. Internally, the **fast** function is a particular adaptation of the **MCA** function to sorting data. Besides providing specific results to sorting data, the **fast** function also stores all the results from the **MCA** function (of the FactoMineR package) directly in the object called `res.fast$acm`. Hence, some of the results are duplicated: for instance, we can easily see that the objects `res.fast$ind$coord` and `res$acm$ind$coord` are identical tables.

The reason for this duplication lies in the advantage in storing the raw results of the **MCA** function for further analysis. For instance, with respect to that output, it is easy to get clusters of products, or an automatic description of the dimensions, by using respectively the **HCPC** function and the **dimdesc** function on `res.fast$acm`, as has been shown in previous chapters (these functions do not apply on the object `res.fast`).

Besides providing a representation of the product space, MCA also represents the modalities on the same space. When verbalization is done, the "bags of words" are used to identify the groups panelists created, as this information is of utmost importance to interpret the results. However, since many subjects are providing many groups, this information can quickly be too cumbersome, making the interpretation tedious (*cf.* Figure 6.3), even after decreasing the size of the characters on screen (by using the `cex` argument).

```
> plot.MCA(res.fast$acm,choix="ind",invisible="ind",cex=0.7,col.var="grey30")
```

To facilitate the interpretation, the graphic can be improved by hiding or highlighting the results of some particular subjects only. For instance, the groups created by *S2* and *S24* can be highlighted through the `selectMod` argument (*cf.* Figure 6.4). This argument requires first selecting, within `res.fast$acm$var$coord`, the rows that correspond to the subjects of interest.

```
> S2.select <- grep("S2_",rownames(res.fast$acm$var$coord))
> S24.select <- grep("S24_",rownames(res.fast$acm$var$coord))
> plot.MCA(res.fast$acm,choix="ind",selectMod=c(S2.select,S24.select),
+ col.ind="grey60",col.var="grey30",cex=0.7)
```

Another important argument called `habillage` is also very useful as it colors the groups of one particular subject. Let's consider *S30*: to visualize the different groups that this panelist created, the following code is used (results not shown here):

MCA factor map

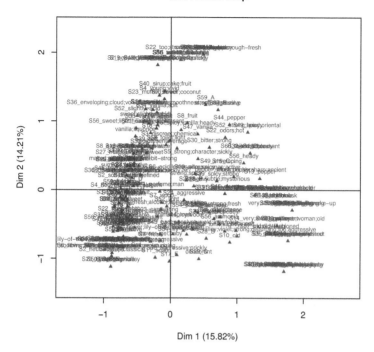

Dim 1 (15.82%)

FIGURE 6.3

Representation of the words used to describe the groups of perfumes on the first two dimensions resulting from an MCA, using the **plot.MCA** function (*sorting* data set).

```
> plot.MCA(res.fast$acm,choix="ind",invisible="var",habillage="S30")
```

Thanks to the coloring, it can be seen that *S30* created groups that match well (but not perfectly) the overall product representation.

Note that these graphics can also be obtained using the **plot.fast** function applied on the *res.fast* object. But since **plot.MCA** has more options, we prefer to apply **plot.MCA** on the object `res.fast$acm`.

These graphics can be completed by the automatic description of the dimensions using the **dimdesc** function of the FactoMineR package. This function provides the list of variables and the list of categories that are significantly related to each dimension, at a given threshold. The list of variables is easily exploitable, as it is sorted with respect to the *p-value* in ascending order. The list of categories is structured according to two poles, the categories that are significantly linked to the positive side of the dimension and the categories

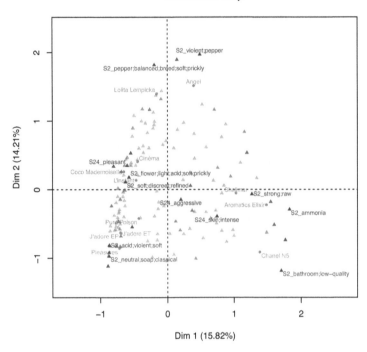

FIGURE 6.4
Representation of the words used by *S2* and *S24* to describe their groups on the first two dimensions resulting from an MCA, using the **plot.MCA** function (*sorting* data set).

that are significantly linked to the negative side of the dimension. In other words, the **dimdesc** function returns the categories that can be used to interpret the positive side of the dimension and the ones that can be used to interpret the negative side of the dimension.

```
> res.dimdesc <- dimdesc(res.fast$acm)
> names(res.dimdesc)
[1] "Dim 1" "Dim 2" "Dim 3"
> res.dimdesc$'Dim 1'
$quali
          R2        p.value
S13 0.9539951 1.081860e-05
S39 0.9712749 1.767325e-05
S14 0.9784508 6.408430e-05
S45 0.9526411 1.001886e-04
S35 0.9869866 1.493216e-04
...     ...          ...
```

```
$category
                                           Estimate      p.value
S39_spicy                                  1.2643794 4.242248e-06
S14_very;strong;perfume;woman;old          1.3561350 1.378966e-05
S35_pepper;winter;chestnut                 1.3321739 3.514317e-05
S45_unpleasant                             1.2767410 4.260275e-05
S51_solvent                                1.3063334 5.157908e-05
S13_ammonia                                0.7147522 6.388686e-04
S35_cream;sun                              0.9282186 8.018334e-04
     ...                                       ...         ...
S2_neutral;soap;classical                 -0.9033964 1.042970e-03
S39_verdure                               -0.6318149 9.065839e-04
S3_very;discreet                          -1.0927862 8.448786e-04
S35_flower;vegetable                      -0.6986721 7.707658e-04
S40_soft                                  -1.0407975 5.897788e-04
S40_flower                                -1.0539914 5.512265e-04
S35_fruit;flower;light;sweet              -0.8042688 3.992915e-04
S39_light                                 -0.6617273 1.630004e-04
S13_D                                     -1.1325721 1.115863e-05
```

Here, the positive side of the first dimension is characterized by stimuli that were qualified as *spicy, strong, unpleasant*, whereas the negative side is characterized by stimuli that were qualified as *light, discreet, soft*. With this output, the way subjects verbalize can easily be studied through the way words have been used. Is there some kind of a consensus amongst subjects in terms of words usage?

Thanks to the **dimdesc** function, the dimensions obtained by MCA can be interpreted. The next natural step would be to understand the product space at a product level. In other words, we expect that the analysis provides a description of the products with respect to the words provided by the subjects to describe their groups. Such extremely important information can be found in the `res.fast$textual` object. For a deeper understanding on how it works, please refer to Chapter 4; as in this part, categories are analyzed as free text comments.

```
> names(res.fast$textual)
 [1] "Angel"             "Aromatics Elixir"  "Chanel N5"        "Cinéma"
 [5] "Coco Mademoiselle" "J'adore EP"        "J'adore ET"       "L'instant"
 [9] "Lolita Lempicka"   "Pleasures"         "Pure Poison"      "Shalimar"
```

Let's take a closer look to the description of the products *Chanel N5* and *Lolita Lempicka*.

```
> res.fast$textual$'Chanel N5'
       Intern %  glob % Intern freq Glob freq      p.value     v.test
soap  10.588235 2.162719           9        21 0.0000489968  4.060361
sweet  1.176471 6.385170           1        62 0.0414884843 -2.038615
fruit  0.000000 5.973223           0        58 0.0083073661 -2.639315

> res.fast$textual$'Lolita Lempicka'
```

	Intern %	glob %	Intern freq	Glob freq	p.value	v.test
sweet	18.390805	6.385170	16	62	8.273369e-05	3.936340
vanilla	8.045977	3.398558	7	33	4.509772e-02	2.003742
flower	1.149425	7.826982	1	76	1.077786e-02	-2.549820

Based on this output, it appears that *Lolita Lempicka* has been described with the words *sweet, vanilla* and not by *flower*. We can also see that *soap* was significantly associated with *Chanel N5*.

If you remember, this output corresponds to the results of the **descfreq** function presented in Chapter 4.

6.2.4 How can I understand the data from a panel perspective?

As explained previously, the point of view adopted here consists in assimilating each subject to a qualitative variable. The coordinates of the stimuli on a dimension can also be assimilated to a variable, in this case a quantitative one. Based on this observation, it seems interesting to assess the link between the partition provided by a subject and the dimensions issued from all subjects. To do so, for each dimension s and each subject j, the "usual" *eta-squared* between a quantitative variable and a qualitative one is calculated:

$$\eta^2(F_s, j) = \frac{SS_{Between}}{SS_{Total}}.$$

By construction, this indicator lies between 0 and 1. Hence, subjects can be represented through their partition within a square of width/height 1: on a given dimension, the closer to 1 the indicator associated with a panelist, and the closer the partition of that panelist to the overall representation of that given dimension.

Figure 6.5 shows that the groups created by *S2* are similar to the main structure of variability issued from the whole panel of subjects (*i.e.*, to the results of the MCA). On the other hand, the groups provided by *S24* do not match the overall results.

```
> plot.fast(res.fast,choix="group")
> sorting[,c(2,24)]
```

	S2	S24
Angel	violent;pepper	aggressive
Aromatics Elixir	ammonia	skin;intense
Chanel N5	bathroom;low-quality	aggressive
Cinéma	soft;discreet;refined	pleasant
Coco Mademoiselle	soft;discreet;refined	pleasant
J'adore EP	acid;violent;soft	skin;intense
J'adore ET	neutral;soap;classical	aggressive
L'instant	flower;light;acid;soft;prickly	pleasant
Lolita Lempicka	pepper;balanced;breed;soft;prickly	pleasant
Pleasures	neutral;soap;classical	pleasant
Pure Poison	soft;discreet;refined	aggressive
Shalimar	strong;raw	skin;intense

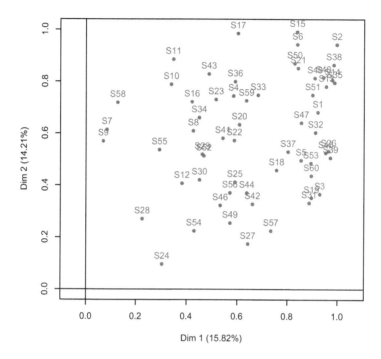

FIGURE 6.5
Representation of the subjects on the first two dimensions resulting from an
MCA, using the **plot.fast** function (*sorting* data set).

Remark. It can be shown that performing MCA on a set of qualitative variables is equivalent to performing MFA on the same set, by considering groups
of one variable only. This property is very important, and we will detail in
Chapter 7 how this representation of the subjects is obtained in MFA.

6.3 For experienced users: The hierarchical sorting task

Should I group this product with these ones, or should I leave it apart? This
hesitation reflects the most frequent question subjects ask themselves while
performing a sorting task. To avoid such hesitation (amongst others reasons),
a variation of the sorting task has been proposed. In this variant, although
the goal of the method stays unchanged (*i.e.*, obtaining groups of similar

products), the process to obtain these groups is different. Rather than defining groups of products in one shot, subjects are asked to create them in successive steps. In practice, each subject creates a first sequence of groups by dividing the whole product space into subsets of products. Each subset is then divided into smaller subsets, and so on until no more subgroups can be made. Subjects are free to stop their divisive process whenever they want. In fact, this variant is a succession of sorting tasks, which are following a certain hierarchy, hence its name hierarchical sorting task (HST). Since the subdivision of the groups is done from the higher level to the lower level of the hierarchy, this process is qualified as "top-down."

Another variant of the HST, in which subjects "crawl up" along the hierarchy, exists (the "bottom up" approach). In this second variant, subjects are first asked to separate the products in as many groups as possible. Once done, they are then asked to merge the closer groups together, and to repeat this procedure until the remaining groups are all too different to be combined. Although the sensory implications from these two variants slightly differ, both can be analyzed using the same statistical methodology. In this section, we focus our attention on the "top-down" approach of HST.

Whilst in a sorting task, each subject provides one unique partition; in an HST, each subject provides a sequence of nested partitions, each partition of the sequence corresponding to one of the steps. Similarly to the sorting task (*cf.* Section 6.1), the products are considered as statistical units, and the partitions provided by the subjects are considered as variables. Let I denote the number of products, J the number of subjects performing HST, K_j the number of partitions associated with subject j, and $\sum_j K_j = K$ the total number of variables. With respect to this point of view, the data collected can be put in a matrix $X = (x_{ij})$ of dimension $I \times K$. At the intersection of one row and one column, we have the group to which a product belongs, for one partition of one subject. Note that in HST, for a subject j, the distance between two products i and i' is not binary (like in the sorting task), but linked to the level K_j of the hierarchy where i and i' are regrouped.

Since this table only contains qualitative variables, it can be analyzed by MCA. However, a closer look at the data shows that the matrix X is structured into J submatrices X_j of dimension $I \times K_j$, each of them corresponding to the sequence of nested partitions provided by one subject j. Such structure is not respected by MCA, the hierarchal sort provided by subject j being considered as K_j independent partitions. To take into account the structure on the variables, *i.e.*, by considering the total matrix as a juxtaposition of submatrices (one submatrix or group corresponding to the entire hierarchical sort of one subject), Multiple Factor Analysis (MFA) is used.

As mentioned in Chapter 3, MFA is an exploratory multivariate analysis that applies to data tables in which the same set of individuals (*i.e.*, the products) is described by several sets of variables (*i.e.*, the hierarchical sort) structured in groups (*i.e.*, the subjects). Intrinsically, MFA balances the influence of each group in the construction of the final product space. In our case, this corresponds to balancing the influence of the J subjects, regardless of the number of levels (K_j) each of them provided. An important feature of MFA lies in the fact that it can deal with groups of different types including qualitative variables.

To illustrate this methodology, we use a study in which 89 children (aged from 7 to 10 years old) were asked to sort 16 cards. These cards represent simple geometric forms and were obtained using a 2^{7-3} fractional factorial design involving 7 factors of 2 levels (*cf.* Figure 6.6):

- Shape (round or square)

- Color of shape (blue or green)

- Color of background (yellow or orange)

- Size of shape (small or big)

- Position of shape (top or bottom, distinctive by a red line under the cards)

- Background pattern (plain or shaded)

- Contour pattern (continuous line or dotted line)

The data of this study were collected by Virginie Jesionka and Marie Durand during their master's degree studies.

A subset of the data (restricted to 30 children) can be found in the file *cards_HierarSort.csv*, which can be downloaded from the book website. For a better understanding of the task, the resulting data, and the way they have been coded, refer to Figure 6.7.

After importing the data, use the **summary** function to check the data. Note that since the first column of the data set corresponds to the names of the row, the `row.names` argument is set to 1.

```
> cards_hst <- read.table("cards_HierarSort.csv",header=TRUE,sep=";",row.names=1)
> summary(cards_hst[,1:5])
      S1L1           S1L2           S2L1           S2L2           S2L3
 Min.   :1.0    Min.   :1.00   Min.   :1.0    Min.   :1.00   Min.   :1.00
 1st Qu.:1.0    1st Qu.:2.75   1st Qu.:1.0    1st Qu.:1.75   1st Qu.:2.75
 Median :1.5    Median :4.50   Median :1.5    Median :2.50   Median :4.50
 Mean   :1.5    Mean   :4.50   Mean   :1.5    Mean   :2.50   Mean   :4.50
 3rd Qu.:2.0    3rd Qu.:6.25   3rd Qu.:2.0    3rd Qu.:3.25   3rd Qu.:6.25
 Max.   :2.0    Max.   :8.00   Max.   :2.0    Max.   :4.00   Max.   :8.00
```

The summary of *cards_hst* (restricted to its first five variables) shows that the variables are considered as quantitative, while we expect them to be qualitative. To change this, the **as.factor** function is applied to each column. To facilitate the task, let's automate it within a **for** loop, running from 1 to the number of columns in the data set.

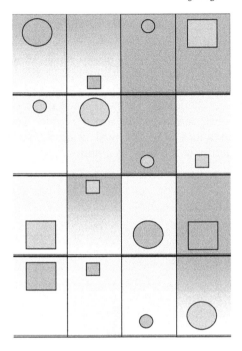

FIGURE 6.6
Representation of the 16 cards presented to the children.

```
> for (i in 1:ncol(cards_hst)) cards_hst[,i] <- as.factor(cards_hst[,i])
> summary(cards_hst[,1:5])
 S1L1        S1L2    S2L1  S2L2        S2L3
 1:8   1      :2   1:8   1:4   1        :2
 2:8   2      :2   2:8   2:4   2        :2
       3      :2         3:4   3        :2
       4      :2         4:4   4        :2
       5      :2               5        :2
       6      :2               6        :2
       (Other):4               (Other):4
```

To evaluate the structure of the data set, let's print the column names of the data set. The **colnames** function is used.

```
> colnames(cards_hst)
 [1] "S1L1"  "S1L2"  "S2L1"  "S2L2"  "S2L3"  "S3L1"  "S3L2"  "S3L3"  "S4L1"
[10] "S4L2"  "S5L1"  "S5L2"  "S6L1"  "S6L2"  "S6L3"  "S6L4"  "S7L1"  "S7L2"
[19] "S8L1"  "S8L2"  "S8L3"  "S9L1"  "S9L2"  "S10L1" "S11L1" "S11L2" "S11L3"
[28] "S12L1" "S12L2" "S13L1" "S13L2" "S13L3" "S14L1" "S14L2" "S14L3" "S15L1"
[37] "S15L2" "S15L3" "S16L1" "S16L2" "S17L1" "S17L2" "S17L3" "S18L1" "S18L2"
[46] "S18L3" "S19L1" "S19L2" "S20L1" "S20L2" "S20L3" "S21L1" "S21L2" "S21L3"
[55] "S22L1" "S22L2" "S22L3" "S23L1" "S23L2" "S23L3" "S24L1" "S24L2" "S24L3"
[64] "S25L1" "S25L2" "S25L3" "S26L1" "S26L2" "S26L3" "S27L1" "S27L2" "S27L3"
[73] "S28L1" "S28L2" "S28L3" "S29L1" "S29L2" "S29L3" "S30L1" "S30L2" "S30L3"
```

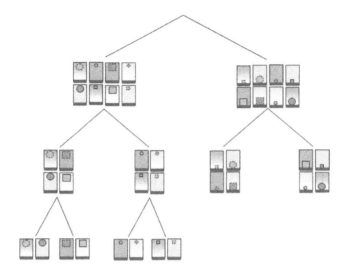

FIGURE 6.7
Illustration of a possible hierarchy of groups obtained by hierarchical sorting task.

As can be seen, the first child (*i.e.*, *S1*) did a partition in two levels (*i.e.*, *S1L1* and *S1L2*), whereas the second child (*i.e.*, *S2*) did it in three (*i.e.*, *S2L1*, *S2L2*, and *S2L3*). By going through all the outputs, a vector containing the number of partitions (or equivalently of steps) for each child is generated (to do so, we used the **c** function). This information, which reflects the structure of the data, is necessary for further analysis. The **length** function is then used to count the number of subjects in the data set.

```
> group.cards <- c(2,3,3,2,2,4,2,3,2,1,3,2,3,3,3,2,3,3,2,3,3,3,3,3,3,3,3,3,3,3)
> length(group.cards)
[1] 30
```

To verify that the groups are made correctly, we can check that the sum of the values in `group.cards` corresponds to the number of columns present in *cards_hst*.

```
> sum(group.cards)
[1] 81
> ncol(cards_hst)
[1] 81
```

As mentioned previously, this structured data set is analyzed by MFA on qualitative groups. Hence, the **MFA** function of the FactoMineR package is used. In this function, the important arguments to define are related to the groups of variables: `group` provides information about the structure of the data set (this information is contained in `group.cards`), while `type` defines the nature of each group (in this case `"n"` for all the groups since the groups are all qualitative; `"n"` stands for *nominal*). Finally, a name can be given to each group through the argument `name.group`.

```
> res.mfa <- MFA(cards_hst,group=group.cards,type=rep("n",30),
+ name.group=paste("S",1:30,sep=""),graph=FALSE)
> plot.MFA(res.mfa,choix="ind",invisible="quali",habillage="none")
> plot.MFA(res.mfa,choix="group",habillage="none")
```

The representation of the product space obtained (*cf.* Figure 6.8) shows that the most important criterion used by the children to separate the cards is related to the shape of the form (from left to right: circles *versus* squares). The second most important criterion is related to the size of the form (up to down: small *versus* large).

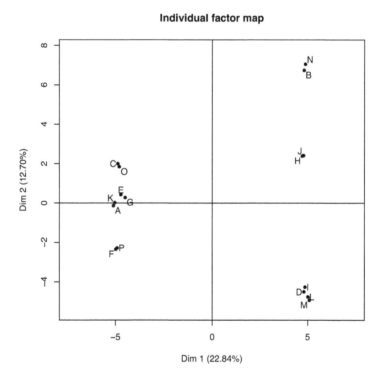

FIGURE 6.8
Representation of the cards on the first two dimensions resulting from an MFA, using the **plot.MFA** function (*cards_hst* data set).

The group representation (*cf.* Figure 6.9) highlights how each child is re-lated to the MFA space: most children (*e.g.*, *S5*, *S9*, and *S24*) being on the left side (close to the value 1), the shape is a major criterion in the generation of the groups for most of them. For some others (*e.g.*, *S12*), the shape of the forms was not determinant in the creation of the groups, whereas the size of the forms was. Finally, it can be noted that *S11* was neither influenced by the shape nor by the size of the forms to generate the groups.

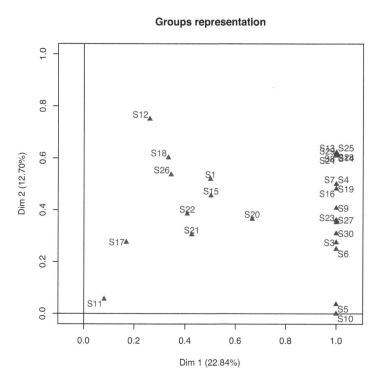

FIGURE 6.9
Representation of the children on the first two dimensions resulting from an MFA, using the **plot.MFA** function (*cards_hst* data set).

To visualize the way a child grouped the products in a certain level, the `habillage` argument of the **plot.MFA** function is used. Let's highlight the cards of the groups of the second child at the first level (this information is positioned at the third column of the data set, hence the `habillage` argument is set to 3).

```
> plot.MFA(res.mfa,invisible="quali",habillage=3,col.hab=c("black","red"),
+ title="Cards colored according to level 1 of subject 2")
```

By playing with the value of `habillage`, it is possible to highlight the groupings done at different levels for the different children. In this case, it can be seen that the first separation made by child 2 is related to the shape of the forms (*cf.* Figure 6.10).

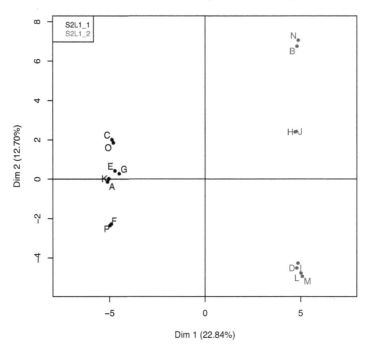

FIGURE 6.10
Representation of the cards highlighting the first level of separation proposed by subject 2 on the first two dimensions resulting from an MFA, using the **plot.MFA** function (*cards_hst* data set).

Alternatively to the **MFA** function, which is a general function, the **fahst** function (SensoMineR package) can be used to analyze HST data. Similarly to **fast** and **MCA** for sorting data, this function is based on the **MFA** function, and provides additional outputs that are specific to HST data.

```
> res.fahst <- fahst(cards_hst,group=group.cards,
+ name.group=paste("S",1:30,sep=""),graph=FALSE)
> names(res.fahst)
[1] "eig"    "var"    "ind"    "group" "call"
```

Besides providing some visual representation of some descriptive statistics (*e.g.*, histogram of the number of products within each group, or of the number of levels per child), the **fahst** function provides a "level and trajectory" representation. This representation shows the steps children followed to sort the cards (*cf.* Figure 6.11). To generate this representation, the **plot.fahst** function with the arguments `choix="level"` and `traj=TRUE` is used.

```
> plot.fahst(res.fahst,choix="level",traj=TRUE)
```

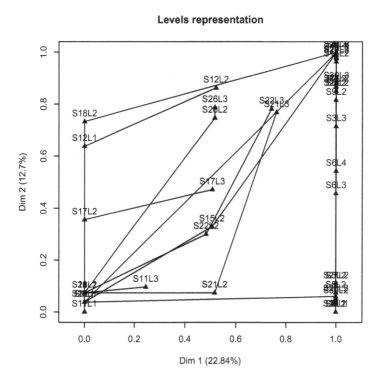

FIGURE 6.11
Representation of the levels of hierarchy and the trajectories of the children on the first two dimensions resulting from an MFA, using the **plot.fahst** function (*cards_hst* data set).

On this trajectory representation, it can be seen that *S12L1* has a quite high coordinate on the second dimension and a null coordinate on the first dimension. The first criterion used by *S12* to separate the cards is related to the size of the forms. In the second level, this child used the shape to separate the cards, since *S12L2* has a higher coordinate on the first dimension. When looking at *S18*, it can be seen that *S18L1* has a null coordinate on both dimension 1 and dimension 2. The first criterion used by this child to separate the cards is different from the size and shape of the forms. However, in the second level, *S18L2* has a high coordinate on dimension 2 and a null coordinate on dimension 1. This child used the size of the form to separate the products. Finally, in a third step, they used the shape to separate the products, as *S18L3* has high coordinates on both dimension 1 and dimension 2. Finally, it can be seen that other children followed other tracks. For example, *S21* first separated the cards according to some different criteria than shape or form (*S21L1*), before using, respectively, the shape (*S21L2*) and the form (*S21L3*) to sort the cards.

6.4 Exercises

Exercise 6.1 *Getting a co-occurrence matrix from the raw data*

This exercise is particularly important as, in order to get a representation of the space product, other approaches are based on distance matrices between stimuli. It's crucial to know how to obtain such matrices from what we've called the raw data (of course the reverse operation is not possible, hence the importance of keeping the raw data).

- Import the *perfumes_sorting.csv* file from the book website.

- For confident R-users, try to construct the matrix of co-occurrence yourself. Once done, run the **fast** function from the SensoMineR package on the imported data set, and compare your results with the one provided by this function.

- For the others, run the **fast** function of the SensoMineR package to the imported data set. Search within the outputs of the **fast** function the matrix of co-occurrence.

- Once extracted, export this matrix to Excel.

Exercise 6.2 *Representation of the products using Correspondence Analysis (CA) and multidimensional scaling (MDS)*

In this exercise, we will use the matrix of co-occurrence generated in the Exercise 6.1 to obtain a representation of the product space with CA on the one hand, and with MDS on the other hand.

- Import (or use) the matrix of co-occurrence created in the Exercise 6.1 to your R session.

- Analyse your dissimilarity matrix using the **CA** function of the FactoMineR package.

- Transform this matrix of co-occurrence to a dissimilarity matrix. Run the MDS on the matrix of dissimilarity: to do so, use the **smacofSym** function of the smacof package (we consider here 3 dimensions and metric=TRUE). To obtain orthogonal dimensions, run a PCA on the object $conf thus obtained.

Exercise 6.3 *Advertisements: how do you perform compared to our panel?*

This experiment was designed to assess the importance of different factors in an advertisement. To do so, we designed 16 different advertisements based on 6 different factors. The aim of this exercise is to understand the importance of supplementary information (in our case, individual and qualitative variables).

- Go to the book website, print the advertisement, and do the sorting task.

- Add your results to the *advertisement.csv* file that you can also download from the book website.

- Import *advertisement.csv* and *advertisement_design.csv* from the book website.

- Merge these two files together, and run the analysis, by assigning the right important to the different information. You can either use the **fast** function, or the **MCA** function. Interpret the results.

Exercise 6.4 *Sorting on incomplete data*

To understand the impact of incomplete data on sorting, we made an experiment on subjects who tested only a subset of the 12 products.

- Import the data set *perfumes_incomplete.csv* from the book website. By looking at the data set, how many products were evaluated by each subject?

- Use and adapt the code proposed in Exercise 6.1 to generate the co-occurrence matrix. Conclude regarding the design of experiment.

- The strategy we adopt to treat missing data is to regroup the missing products within each subject into a group called "Missing". Using the **replace** function, replace the empty cells by the word "Missing".

- Run the **fast** function on this "completed" data set. Interpret the results.

6.5   Recommended readings

- Abdi, H., & Valentin, D. (2007). Multiple correspondence analysis. In N. J. Salkind (Ed.), Encyclopedia of measurement and statistics, pp. 651-657, Thousand Oaks, CA: Sage.

- Abdi, H., Valentin, D., Chollet, S., & Chrea, C. (2007). Analyzing assessors and products in sorting tasks: DISTATIS, theory and applications. *Food Quality and Preference*, 18, (4), 627-640.

- Bécue-Bertaut, M., & Lê, S. (2011). Analysis of multilingual labeled sorting tasks: application to a cross-cultural study in wine industry. *Journal of Sensory Studies*, 26, (5), 299-310.

- Best, D. L., & Ornstein, P. A. (1986). Children's generation and communication of mnemonic organizational strategies. *Developmental Psychology*, 22, 845-853.

- Cadoret, M., & Husson, F. (2013). Construction and evaluation of confidence ellipses applied at sensory data. *Food Quality and Preference*, 28, (1), 106-115.

- Cadoret, M., Lê, S., & Pagès, J. (2009). A Factorial Approach for Sorting Task data (FAST). *Food Quality and Preference*, 20, (6), 410-417.

- Cadoret., M., Lê, S., & Pagès, J. (2011). Statistical analysis of hierarchical sorting data. *Journal of Sensory Studies*, 26, (22), 96-105.

- Carroll, J. D. (1972). Individual differences and multidimensional scaling in R. Shepard, A. Romney, & S. Nerlove (Eds.), Multidimensional scaling: Theory and applications in the behavioral sciences. New York, Academic Press, 105-155.

- Chollet, S., Lelièvre, M., Abdi, H., & Valentin, D. (2011). Sort and beer: everything you wanted to know about the sorting task but did not dare to ask. *Food Quality and Preference*, 22, (6), 507-520.

- Daws, J. T. (1996). The analysis of free-sorting data: beyond pairwise co-occurrences. *Journal of Classification*, 13, (1), 57-80.

- Escofier, B., & Pagès, J. (1994). Multiple factor analysis (AFMULT package). *Computational Statistics & Data Analysis*, 18, (1), 121-140.

- Husson, F., Lê, S., & Pagès, J. (2010). *Exploratory multivariate analysis by example using R*. Chapman & Hall/CRC Computer Science & Data Analysis.

- Lawless, H. T., Sheng, N., & Knoops, S. S. C. P. (1995). Multidimensional-scaling of sorting data applied to cheese perception. *Food Quality and Preference*, 6, (2), 91-98.

- Lelièvre, M., Chollet, S., Abdi, H., & Valentin, D. (2008). What is the validity of the sorting task for describing beers? A study using trained and untrained assessors. *Food Quality and Preference*, 19, (8), 697-703.

- Miller, G. A. (1969). A psychological method to investigate verbal concepts. *Journal of Mathematical Psychology*, 6, 169-191.

- Santosa, M., Abdi, H., & Guinard, J.-X. (2010). A modified sorting task to investigate consumer perceptions of extra virgin olive oils. *Food Quality and Preference*, 21, (7), 881-892.

7

When products are positioned onto a projective map

CONTENTS

"...the shape of the map is of utmost importance. With the Napping®, by positioning products onto a rectangle of dimension 60 cm by 40 cm, subjects are unconsciously prioritizing their two most spontaneous axes of variability between products. Thanks to Multiple Factor Analysis (MFA), it is possible to extract from the data the common dimensions of variability, as well as the specific ones, between the products depicted by the subjects. When applying MFA, it is crucial not to standardize the data, in order to keep the relative importance of the axes. Such analysis can be performed using the **MFA** *function of the FactoMineR package."*

7.1   Data, sensory issues, and notations

In 1994, Risvik *et al.* published an innovative paper entitled *"Projective Mapping: a tool for sensory analysis and consumer research."* As written in their abstract, "this paper looks at a third and alternative method of producing a two-dimensional, perceptual map utilizing a projective-type method whereby individual assessors themselves are required to place products on the space according to the similarities and differences they perceive." The idea of positioning stimuli on a sheet of paper based on their perceived similarities arose in sensory analysis, as well as the idea of statistically analyzing this type of information. In their paper, the authors proposed to use Generalized Procrustes Analysis (GPA) to analyze such data.

In 2005, Pagès introduced the Napping® in "*Collection and analysis of perceived product inter-distances using multiple factor analysis: application to the study of 10 white wines from the Loire Valley.*" As specified in this paper, the protocol consists in positioning products (more generally, stimuli) on a rectangle, according to their resemblances and differences: two products are close if they are perceived as similar; on the contrary, two products are distant as they are perceived as different.

With respect to the way data are collected, the two methods - projective mapping and Napping - look similar. But both the context in which each method was conceived and their respective motivations are very much different. In a sense, Risvik and Pagès converged from two different starting points. From our point of view, Napping embraces both the way data are collected and analyzed. In Napping, the specified dimensions of the rectangle, on which stimuli are positioned, render the method unique (60 *cm* by 40 *cm*). The main feature of this method is to reveal and to order the first two dimensions of variability for each subject: as it is rectangular (and not square nor round), the subject implicitly opposes products with respect to the first dimension, at first, and with respect to the second dimension, in a second time. Additionally, Napping provides the possibility to describe products, or groups of products, using free text comments. This feature is of utmost importance, as without this information the interpretation of the sensory dimensions of the product space is impossible.

Naturally, in both projective mapping and Napping, the statistical units of interest are the products. They are measured in function of their positions (*i.e.*, their coordinates) on the rectangle. Arbitrarily, the origin of the basis is situated in the left-bottom corner of the rectangle: by doing so, the coordinates on the x-axis lie between 0 and 60, and between 0 and 40 on the y-axis. If I denotes the number of products and J denotes the number of subjects, the data set $X = (x_{ik})$ of dimension $(I, 2J)$ is such that x_{ik} presents the coordinate of product i on the x-axis (for $k = 2j - 1$), or on the y-axis (for $k = 2j$) for subject j. This data set is a multivariate quantitative data set, in the sense that variables (*i.e.*, coordinates) are quantitative. In addition, this data set is naturally structured in groups, more precisely in J groups of two variables, each group corresponding to one subject; *de facto* it can also be considered as a multiple data table.

Similarly to the previous tasks, we expect from the statistical analysis of such data that it provides a representation of the products (stimuli) based on their similarities and differences. Such product space should be obtained based on the J subjects' point of view, each of them being equally balanced (one subject should not be dominating). Typically, the statistical analysis of such data set should take into account the "natural" partition on the variables, as well as the structure of the data (*i.e.*, the ratio between the length and the height of the rectangle). Additionally, we expect to have a representation of the subjects that would be interpreted jointly with the former representation of the products, and that would show how the individual stimuli representations provided by the subjects coincide. Finally, if additional data are available (*e.g.*, free text comments), we would like to be able to relate the product space to this additional information to deepen the interpretation of the product space.

Similarly to Chapter 3, Multiple Factor Analysis is considered for the analysis of such data. Indeed, MFA was conceived precisely for the purpose that each group of variables is equally balanced within the global analysis. To keep the specified dimensions of the rectangle unchanged within the analysis, a particularity of MFA consisting in performing unstandardized Principal Component Analysis (PCA) on each separate group, is presented in this chapter. Additional (direct and indirect) outputs of the MFA (*e.g.*, the group representation, the *RV* coefficient) are also presented here.

Tips: Introducing the notion of *projective test*

For a better understanding of the intrinsic nature of the data collected during a Napping task, and their potential in terms of interpretability, let's review the notion of projective test.

"In psychology, a projective test is a personality test designed to let a person respond to ambiguous stimuli, presumably revealing, hidden emotions and internal conflicts. This is sometimes contrasted with a so-called "objective test" in which responses are analyzed according to a universal standard (for example, a multiple choice exam). The responses to projective tests are analyzed for their meanings rather than being based on presuppositions about meaning, as is the case with objective tests. Projective tests have their origins in psychoanalytic psychology, which argues that humans have conscious and unconscious attitudes and motivations that are beyond or hidden from conscious awareness." (`http://en.wikipedia.org/wiki/Projective_test`)

In Marketing, projective techniques "are unstructured prompts or stimulus that encourage the respondent to project their underlying motivations, beliefs, attitudes, or feelings onto an ambiguous situation. (...) Examples of projective techniques include:

- word association - say the first word that comes to mind after hearing a word - only some of the words in the list are test words that the researcher is interested in, the rest of them being fillers, which can be useful when testing brand names;

- sentence completion - respondents are given incomplete sentences and asked to complete them;

- thematic apperception tests - respondents are shown a picture (or series of pictures) and asked to make up a story about the picture(s);

- third-person technique - a verbal or visual representation of an individual and their situation is presented to the respondent - the respondent is asked to relate the attitudes or feelings of that person - researchers assume that talking in the third person will minimize the social pressure to give standard or politically correct responses." (http://en.wikipedia.org/wiki/Qualitative_marketing_research)

Remark. The appellation Napping takes its origin from the French word *nappe*, which means *tablecloth* in English. The suffix *"ing"* was added to this French word to express the action of constructing a *nappe* (typical of French sense of humor). The concept of tablecloth will be referred hereafter as *"rectangle."*

7.2   In practice

7.2.1   How can I approach Napping® data?

The purpose of this section is to explain how Napping data can be statistically approached. In this section, the importance of the weighting in MFA, a crucial feature that has already been mentioned in Chapter 3, is discussed.

The data set used to illustrate this methodology is stored in the file entitled *perfumes_napping.csv*, which can be downloaded from the book website. These data were also collected by Mélanie Cousin, Maëlle Penven, Mathilde Philippe, and Marie Toularhoat as part of their master's degree project. After importing the data in your R session, use the **summary** function to have an overview of the data set (here, the 10 first columns are shown).

```
> napping <- read.table("perfumes_napping.csv",header=TRUE,sep=",",
+ dec=".",row.names=1)
> summary(napping[,1:10])
      X1              Y1              X2              Y2              X3
Min.   : 3.30   Min.   : 8.20   Min.   : 2.00   Min.   : 3.70   Min.   : 6.00
1st Qu.:15.47   1st Qu.:18.95   1st Qu.:26.55   1st Qu.:10.78   1st Qu.:11.07
Median :20.35   Median :31.85   Median :29.50   Median :20.55   Median :26.55
Mean   :27.13   Mean   :26.64   Mean   :29.58   Mean   :17.85   Mean   :25.65
3rd Qu.:44.33   3rd Qu.:33.48   3rd Qu.:34.98   3rd Qu.:22.50   3rd Qu.:36.25
Max.   :57.50   Max.   :37.80   Max.   :56.00   Max.   :37.20   Max.   :54.50
      Y3              X4              Y4              X5              Y5
Min.   : 4.70   Min.   : 7.10   Min.   : 5.60   Min.   : 2.20   Min.   : 3.70
1st Qu.:20.00   1st Qu.:15.32   1st Qu.:13.72   1st Qu.:17.25   1st Qu.:12.25
Median :29.20   Median :25.80   Median :20.30   Median :25.40   Median :18.10
Mean   :23.15   Mean   :28.32   Mean   :19.45   Mean   :22.33   Mean   :19.32
3rd Qu.:29.32   3rd Qu.:41.62   3rd Qu.:25.02   3rd Qu.:29.12   3rd Qu.:23.35
Max.   :29.60   Max.   :55.90   Max.   :35.20   Max.   :42.00   Max.   :38.00
```

The first natural idea that comes to mind when seeing such data is to represent them within the rectangle provided by each subject. In other words, the idea is to plot the perfumes on a rectangle of dimension 60×40. To do so, the **rectangle** function is used. As this function does not belong to any package, this is a good occasion to introduce you to the writing of a function in R (*cf.* Appendix and the corresponding Tips).

Tips: Plotting Napping® data

As shown in the following code, the **rectangle** function is based on three main graphical functions: **plot**, **points**, and **text**. The idea is to first represent the frame of the graphical output, then the perfumes using dots; finally the labels of the perfumes are displayed. To use the function, we introduce two parameters, the data set on which the analysis is performed and the index of the subject we would like to represent. Of course the initial structure of the data set is important as it should be similar to the one of a Napping-type data set, as explained in the introduction.

```
> rectangle <- function(data,i){
+      plot(data[,((i-1)*2+1):(i*2)],col="blue",xlim=c(0,60),
+      ylim=c(0,40),xlab="",ylab="",
+      main=paste("Napping: Subject ",i,sep=""),type="n",asp=1)
       points(data[,((i-1)*2+1):(i*2)],col="blue",pch=20)
       text(data[,((i-1)*2+1):(i*2)],
+      label=rownames(data),col="blue",pos=3,offset=0.2)
+ }
```

Let's apply the **rectangle** function to represent *Subject 38*. The following code is then used:

```
> rectangle(napping,38)
```

Figure 7.1 shows that *Subject 38* has separated *Chanel N5, Shalimar, Aromatics Elixir*, from *J'adore EP* and *J'adore ET* along the length of the rectangle. Although this separation is quite clear, the length of the rectangle may not always be the main dimension of variability amongst the products, although it is expected from the task. Indeed, it depends on the way the subjects performed the task. To get such information, a *natural* step consists in performing PCA on each rectangle.

Let's analyze these data using standardized PCA first (`scale.unit=TRUE`, by default), before applying an unstandardized PCA (`scale.unit=FALSE`).

```
> library(FactoMineR)
> res.pca38.std <- PCA(napping[,75:76],scale.unit=TRUE,graph=FALSE)
> res.pca38.unstd <- PCA(napping[,75:76],scale.unit=FALSE,graph=FALSE)
> par(mfrow=c(1,2))
> plot.PCA(res.pca38.std,choix="ind",title="Napping: Subject 38
```

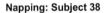

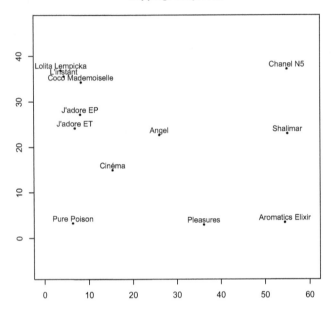

FIGURE 7.1
Representation of the Napping configuration provided by *Subject 38* using the
rectangle function (*napping* data set).

```
+ (Standardized)")
> plot.PCA(res.pca38.unstd,choix="ind",title="Napping: Subject 38
+ (Unstandardized)")
```

As shown by Figure 7.2, and as expected, the two representations of the
product space are different. By standardizing the data, the length and the
height of the rectangle are given the same importance, which is not in ac-
cordance with the spirit of Napping. Intrinsically, by using Napping, subjects
favor the length of the rectangle when positioning the products. Hence, the
representation of the perfumes using PCA when data are unstandardized is
much closer (actually, it is very similar) to the *genuine* representation (*cf.*
Figure 7.1) than when data are standardized.

Hence, to respect the intrinsic nature of the data, unstandardized PCA
should be performed on Napping data.

Let's compare the rectangles of subjects 38 and 57 by applying the **summary** function on the *napping* data set restricted to the sole coordinates of
these two subjects.

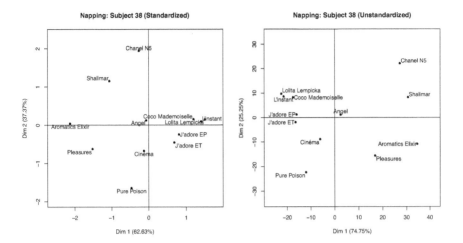

FIGURE 7.2
Representation of the perfumes on the first two dimensions resulting from a PCA on the standardized data (left) and on the unstandardized data (right), using the **plot.PCA** function (*Subject 38, napping* data set).

```
> summary(napping[,c(75:76,113:114)])
      X38              Y38              X57              Y57
 Min.   : 3.800   Min.   : 3.00   Min.   : 6.20   Min.   : 8.20
 1st Qu.: 6.775   1st Qu.:12.12   1st Qu.: 8.20   1st Qu.:16.27
 Median :11.850   Median :23.60   Median :18.00   Median :19.75
 Mean   :23.317   Mean   :22.16   Mean   :24.74   Mean   :19.47
 3rd Qu.:40.675   3rd Qu.:34.62   3rd Qu.:36.83   3rd Qu.:23.75
 Max.   :55.000   Max.   :37.20   Max.   :56.79   Max.   :32.00
```

Apparently, these two subjects are very similar in the way they used the *x*-axis and the *y*-axis, at least in terms of range.

If we run a PCA on the unstandardized data of *Subject 57*, it appears that the resulting representation of the product space (*cf.* Figure 7.3) shows some similarity to the one of *Subject 38*. This is, for instance, the case of the opposition between *Aromatics Elixir, Chanel N5,* and *J'adore EP, J'adore ET*. However, some differences can also be observed as, for instance, the opposition between *J'adore EP* and *J'adore ET* on the second dimension, for the sole *Subject 57*.

```
> res.pca57.unstd <- PCA(napping[,113:114],scale.unit=FALSE,graph=FALSE)
> plot.PCA(res.pca57.unstd,choix="ind",title="Napping: Subject 57
+ (Unstandardized)")
```

However, a closer look at the numerical indicators provided by the **PCA** function shows that the two subjects are not comparable anymore, at least with respect to the variance of their first dimension.

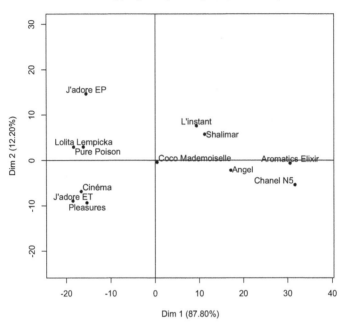

FIGURE 7.3
Representation of the perfumes on the first two dimensions resulting from a
PCA on the unstandardized data, using the **plot.PCA** function (*Subject 57,
napping* data set).

```
> res.pca38.unstd$eig
        eigenvalue percentage of variance cumulative percentage of variance
comp 1    428.1210                74.75043                         74.75043
comp 2    144.6128                25.24957                        100.00000
> res.pca57.unstd$eig
        eigenvalue percentage of variance cumulative percentage of variance
comp 1   343.33539                 87.8003                          87.8003
comp 2    47.70584                 12.1997                         100.0000
```

To compare these two analyses, one idea could be to balance the first
dimension of each separate analysis, without changing the structure amongst
products induced by each analysis. To do so, a homothetic transformation is
applied on each individual configuration: the data of each subject are divided
by the square root of the first eigenvalue of each separate analysis, *i.e.*, by
the standard deviation of the first dimension of each separate analysis. After
performing such transformation, let's re-run an unstandardized PCA on each
rectangle.

```
> res.pca38.unstd <- PCA(napping[,75:76]/sqrt(428.1210),scale.unit=FALSE,
+ graph=FALSE)
> res.pca38.unstd$eig
        eigenvalue percentage of variance cumulative percentage of variance
comp 1  1.0000000                 87.8003                           87.8003
comp 2  0.1389482                 12.1997                          100.0000

> res.pca57.unstd <- PCA(napping[,113:114]/sqrt(343.33539),scale.unit=FALSE,
+ graph=FALSE)
> res.pca57.unstd$eig
        eigenvalue percentage of variance cumulative percentage of variance
comp 1  1.000000                  74.75043                          74.75043
comp 2  0.337785                  25.24957                         100.00000
```

Once transformed, let's compare the two unstandardized PCA representations:

```
> par(mfrow=c(1,2))
> plot.PCA(res.pca38.unstd,choix="ind",title="Napping: Subject 38
+ (Unstandardized, Transformed)")
> plot.PCA(res.pca57.unstd,choix="ind",title="Napping: Subject 57
+ (Unstandardized, Transformed)")
```

Both graphically (*cf.* Figure 7.4) and numerically, the two rectangles are comparable. This is a direct consequence of the transformation performed, which corresponds exactly to the particular weighting procedure considered in MFA.

7.2.2   How can I represent the product space on a map?

As mentioned previously, as a result of the Napping, each subject provides two vectors of coordinates of dimension $I \times 1$ each (one for the x-axis, one for the y-axis), where I denotes the number of stimuli to be positioned on the rectangle. Hence, the final data set to be analyzed, denoted X, is obtained by merging the J couples of vectors of coordinates, where J denotes the number of subjects. In other words, X can be seen as a data set structured into J groups of two variables each. Typically, the statistical analysis of such data set X should take into account the *natural* partition on the variables.

Multiple Factor Analysis (MFA, Escofier & Pagès, 1994) was conceived precisely for this purpose, with the recurring idea of balancing the part of each group of variables within a global analysis. When variables are quantitative, MFA can be defined as an extension of PCA, that takes into account a group structure on the variables, and that balances the part of each group. In that sense, MFA can be seen as a weighted PCA.

Technically, when performing MFA on Napping data, the separate analyses of each group of coordinates are not standardized (*i.e.*, variables are not scaled to unit variance). Otherwise, the analyst would lose the information about the relative importance of the dimensions of variability amongst the stimuli, which is basically why Napping was originally proposed on a rectangle. As we will

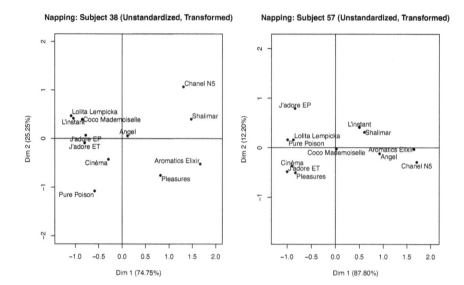

FIGURE 7.4
Representation of the perfumes for *Subject 38* (left) and *Subject 57* (right) on
the first two dimensions resulting from a PCA on the unstandardized data,
using the **plot.PCA** function (*napping* data set).

see, when analyzing the data, one has to specify this fundamental option, as
it is closely related to the intrinsic nature of the data.

To obtain a representation of the products based on the rectangles provided by all the subjects, the following command involving the **MFA** function
is used. As explained earlier, this product space is obtained after balancing
the subjects. In this particular case, it is of utmost importance to set the `type`
parameter of the **MFA** function to `"c"`, as data shouldn't be scaled to unit variance (`"c"` stands for *centered*). Based on the intrinsic structure of the data, the
data set is a succession of 60 groups of 2 variables each (`group=rep(2,60)`).

```
> res.napping <- MFA(napping,group=rep(2,60),type=rep("c",60),
+ name.group=paste("S",1:60,sep=""),graph=FALSE)
> plot.MFA(res.napping,choix="ind",habillage="none")
```

The product space in the context of Napping data is presented in Figure 7.5. This space opposes, on the first dimension, *Aromatics Elixir*, *Chanel
N5*, and *Shalimar* to the rest of the products. The second dimension opposes
Angel and *Lolita Lempicka* to *Pleasures* and *Chanel N5*. However, the sole
representation of the products does not speak for itself unless the analyst has
a great expertise on the products (hence the need of integrating supplementary
information, if available).

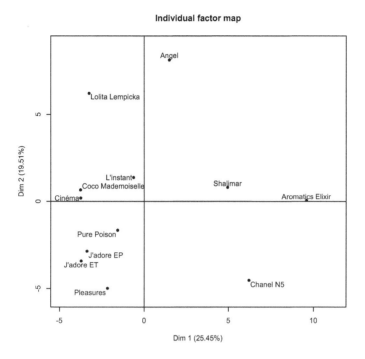

FIGURE 7.5
Representation of the perfumes on the first two dimensions resulting from an MFA on the unstandardized groups of variables, using the **plot.MFA** function (*napping* data set).

7.2.3 How can I interpret the product space with the verbalization data?

Supplementary information is essential for interpreting a Napping factorial plane, as the relative positioning of the stimuli (usually the products) is generally not self-contained. Hence, the idea to supplement the original information from Napping (*i.e.*, the only coordinates of the stimuli) by adding comments on the stimuli arises. In practice, once the subjects placed the stimuli on their rectangle, they are then asked to verbalize[1] the criteria used to group or separate the stimuli.

In practice, this information is aggregated over the subjects and coded into a matrix, where the rows correspond to the stimuli and the columns to the words used. In this matrix, the intersection of row i and column j

---

[1]Literally to express their thoughts, feelings, and emotions in words.

defines the number of times that the word j was used to qualify the stimulus i (*cf.* Chapter 4). This matrix is intrinsically a contingency table, but as the information contained in this matrix is quantitative its columns can also be considered as continuous variables. In this section, we will limit ourselves to this latter point of view, as it is how data were originally analyzed.

Such matrix is usually considered as a supplementary group in the MFA. In other words, this matrix is added to the matrix of coordinates and is treated as a supplementary group, whereas the groups of coordinates are treated as active groups.

Let's recall that when comments are provided, the raw data are often organized in a matrix in which each row corresponds to a free comment, and in which each row is described by three variables, one specifying the stimulus, one specifying the subject, and one with the free comment. As described in Chapter 4, such a table can automatically be transformed into a contingency table with the **textual** function of the FactoMineR package.

To illustrate this procedure, let's consider the *perfumes_comments.csv* file presented in Chapter 4. After importing the data, let's generate the contingency table using the **textual** function.

```
> comments <- read.table("perfumes_comment.csv",header=TRUE,sep="\t",
+ quote="\"")
> res.textual <- textual(comments,num.text=3,contingence.by=1,sep.word=";")
> res.textual$cont.table[,1:7]
```

	a bit strong	acidic	adult	aggressive	airduster	airy	alcohol
Angel	0	0	0	5	0	0	2
Aromatics Elixir	0	0	0	4	0	0	5
Chanel N5	0	1	0	2	0	0	2
Cinéma	1	1	0	1	0	0	0
Coco Mademoiselle	0	1	1	0	0	0	0
J'adore EP	0	1	0	0	0	1	0
J'adore ET	0	0	1	0	0	1	0
L'instant	1	0	1	0	0	0	0
Lolita Lempicka	0	1	0	0	0	0	0
Pleasures	0	1	1	0	0	1	0
Pure Poison	0	2	0	0	0	1	1
Shalimar	0	0	0	5	1	0	4

In order to remove the words rarely used, we restrict the contingency to the words whose occurrences are higher or equal to 5.

```
> words_select <- res.textual$cont.table[,apply(res.textual$cont.table,
+ 2,sum)>=5]
> colnames(words_select)
 [1] "acidic"       "aggressive"     "alcohol"       "amber"          "black pepper"
 [6] "candy"        "caramel"        "character"     "citrus"         "classic"
[11] "cologne"      "concentrated"   "discrete"      "disinfectant"   "elderly"
[16] "female"       "flower"         "flowery"       "forest"         "fresh"
[21] "freshness"    "fruity"         "grand mother"  "green"          "heady"
[26] "heavy"        "intense"        "lavender"      "lemony"         "light"
[31] "male"         "man"            "medicine"      "musk"           "nature"
[36] "neutral"      "old"            "oriental"      "perfume"        "perfumed"
[41] "pleasant"     "powder"         "powerful"      "rose"           "sensitivity"
```

[46] "sensual"	"shampoo"	"shower gel"	"sickening"	"soap"
[51] "soft"	"softness"	"sour"	"spicy"	"spring"
[56] "strength"	"strong"	"subtle"	"summer"	"sweet"
[61] "toiletry"	"toilets"	"vanilla"	"vegetal"	"warm"
[66] "woman"	"woody"	"young"		

Finally, from the original 250 words, 68 words are kept in the restricted contingency table.

In order to integrate the words into the representation of the product space provided by MFA, the two data sets *napping* and *words_select* need to be combined. This can be done using the **cbind** function, or alternatively the **merge** function. Once combined, MFA as described previously is performed. In this analysis, the groups corresponding to the coordinates of the rectangles provided by the subjects are considered as active, whereas the group of variables that corresponds to the words is considered as an illustrative group. In other words, the group related to the words do not contribute to the construction of the axes, but is used for the interpretation of the product space. But before performing such analysis, the contingency table should be transformed into row-profiles (*cf.* Chapter 4). Such transformation is done by dividing the frequency by the column margin.

To facilitate the reading of the graphics, let's limit the variable representation to the sole representation of the words. To do so, the **plot.MFA** function is used, with particular attention on its `select` parameter.

```
> row_profile <- words_select/apply(words_select,MARGIN=1,FUN=sum)
> napping_words <- cbind(napping,row_profile[rownames(napping),])
> res.mfa <- MFA(napping_words,group=c(rep(2,60),68),type=rep("c",61),
+ num.group.sup=61,name.group=c(paste("S",1:60,sep=""),"Words"),graph=FALSE)
> plot.MFA(res.mfa,choix="var",select=colnames(row_profile),cex=0.8)
```

The representation of the product space (*cf.* Figure 7.5) can now be jointly interpreted with the representation of the words (*cf.* Figure 7.6), like in any PCA. On the negative side of the first dimension, the products (*J'adore EP* and *J'adore ET*, for instance) are described by the words *fruity, soft, pleasant, etc.*, while the products located on the positive side of the first dimension (*Aromatics Elixir* and *Shalimar*, for instance) are described with the words *strong, cologne, elderly, etc.* On the second dimension, *Angel* and *Lolita Lempicka* are described with words like *sickening* and *vanilla*, whereas *Pleasures* and *Chanel N5* were described with words like *powder, lavender, soap*, and *disinfectant*.

Tips: Combining matrices or data frames

To merge the matrix of coordinates of the rectangles (*i.e.*, *napping* data set) with the contingency table crossing the products in rows and the words used to describe them in columns (*i.e.*, *row_profile* data set), the **cbind** function is used. This function aims at combining tables by columns in the simplest way,

Correlation circle

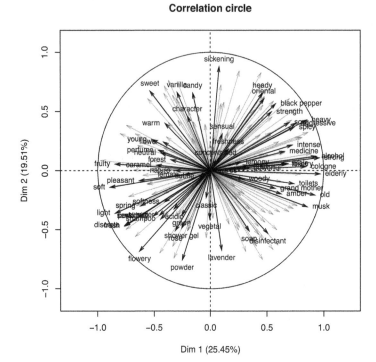

FIGURE 7.6
Representation of the words, used as supplementary elements, on the first two dimensions resulting from an MFA on the unstandardized groups of variables, using the **plot.MFA** function (*napping_words* data set).

i.e., the first row of the first matrix with the first row of the second matrix. However, this function has a drawback: it does not check automatically the names of the rows in the two matrices before combining them.

In order to prevent any inversion of rows between the two matrices, we force the table *row_profile* to have the same rows' order as *napping*. To do so, the following piece of code is used: **[rownames(napping),]**. Another solution to prevent any possible inversion during the combination of matrices consists in using the **merge** function. This function aims at merging two data frames or matrices using common columns or row names. Since in this case the two matrices should be merged in row, the by parameter should be set as 0 or equivalently row.names.

```
> napping_words <- merge(x=napping,y=row_profile,by=0)
```

```
> head(napping_words[,1:5])
          Row.names   X1   Y1   X2   Y2
1             Angel 43.2 34.7 56.0 22.2
2   Aromatics Elixir 52.0 33.1 15.5  3.7
3         Chanel N5 33.0 34.6  2.0 20.0
4            Cinéma 18.8 31.9 30.5  9.2
5 Coco Mademoiselle 19.8 31.8 29.2 11.3
6         J'adore EP  5.5 18.8 18.6 21.1
```

By looking at a subset of the *napping_words* (function **head**), one can remark that the information used to merge the two matrices together (here the row names) is added in the resulting matrix as a new column (the first one, called *Row.names*). To correct for that, we use the first column of *napping_words* to define the names of the rows (using the **rownames** function) before deleting that column from the matrix.

```
> rownames(napping_words) <- napping_words[,1]
> napping_words <- napping_words[,-1]
> head(napping_words[,1:5])
                    X1   Y1   X2   Y2   X3
Angel             43.2 34.7 56.0 22.2 23.8
Aromatics Elixir  52.0 33.1 15.5  3.7 54.5
Chanel N5         33.0 34.6  2.0 20.0  6.3
Cinéma            18.8 31.9 30.5  9.2 38.8
Coco Mademoiselle 19.8 31.8 29.2 11.3 45.8
J'adore EP         5.5 18.8 18.6 21.1  9.2
```

As can be shown here, the resulting matrix *napping_words* is identical to the one obtained previously using the **cbind** function. Note, however, that the **merge** function requires identical names in the two matrices to achieve its combination. Such requirement is also necessary with the **cbind** function when the order of the rows is controlled.

For more information concerning the **merge** function, please look at its help file using the following code.

```
> ?merge
```

As the representation of the words is overcrowded, an automatic description of the dimensions provided by MFA is performed. To do so, the **dimdesc** function of the FactoMineR package is applied on `res.mfa`. This function provides a list containing the variables whose correlation coefficients with the components are significant (according to a given threshold, to be specified with the `proba` parameter). To reduce the amount of the outputs, only the words are shown here.

```
> res.dimdesc <- dimdesc(res.mfa,proba=0.05)
> select.words <- which(rownames(res.dimdesc$Dim.1$quanti) %in%
+ colnames(row_profile))
```

```
> res.dimdesc$Dim.1$quanti[select.words,]
            correlation        p.value
elderly       0.9663368  3.217379e-07
strong        0.9579942  9.597197e-07
alcohol       0.9338704  8.907438e-06
old           0.9218128  2.015934e-05
musk          0.8539574  4.072882e-04
heavy         0.8467769  5.112298e-04
cologne       0.8410392  6.081786e-04
aggressive    0.7532915  4.674430e-03
toilets       0.7360478  6.347904e-03
spicy         0.7359291  6.360791e-03
intense       0.7186988  8.450342e-03
sour          0.6950108  1.211098e-02
male          0.6815834  1.464650e-02
toiletry      0.6702873  1.706599e-02
medicine      0.6512529  2.178874e-02
amber         0.6249114  2.980647e-02
spring       -0.6197992  3.157919e-02
pleasant     -0.6294760  2.828520e-02
flowery      -0.6295407  2.826406e-02
fresh        -0.7568259  4.377358e-03
discrete     -0.7600999  4.115132e-03
fruity       -0.8424754  5.826784e-04
light        -0.8428464  5.762277e-04
soft         -0.8883968  1.127435e-04
```

As illustrated above, the **dimdesc** function provides for each dimension a sorted list of variables. In our example, the five words the most positively correlated with the first dimension are *elderly, strong, alcohol, old*, and *musk*. Similarly, the five words the most negatively correlated with the first dimension are *soft, light, fruity, discrete*, and *fresh*. Let's recall that this dimension opposes perfumes such as *Aromatics Elixir, Shalimar*, and *Chanel N5* to perfumes such as *Cinéma* and *J'adore* (*cf.* Figure 7.5).

Remark. The analysis of the words by considering the matrix as a contingency table is quite straightforward, as the only argument to modify when calling the **MFA** function concerns the `type` parameter: `"f"` stands for *frequencies*, as the matrix is considered as a contingency table. In this case, words are represented on the representation of the perfumes.

```
> res.mfa <- MFA(napping_words,group=c(rep(2,60),68),type=c(rep("c",60),"f"),
+ num.group.sup=61,name.group=c(paste("S",1:60,sep=""),"Words"),graph=FALSE)
```

7.2.4 How can I represent the consumers, and how can I explain the product representation through their individual rectangles?

Let's consider a multiple data table, $X = [X_1| \ldots |X_J]$, on which MFA is performed. In MFA, the representation of the groups of variables is obtained by projecting the matrices of scalar products $W_j = X_j X_j'$ (of dimension $I \times I$) onto $F_s F_s'$, where (F_s) is the sequence of vectors of coordinates of individuals issued from the MFA on X.

In our case, as each group of variables corresponds to the rectangle of one subject, representing the groups consists in indirectly representing the subjects. This representation is obtained using the following code:

```
> plot.MFA(res.mfa,choix="group",habillage="none",
+ col.hab=c(rep("black",60),"grey40"))
```

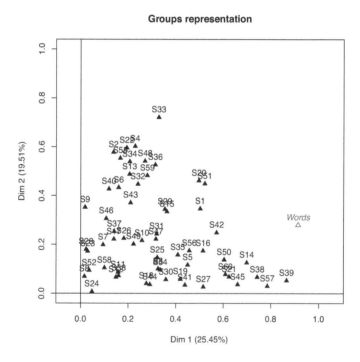

FIGURE 7.7
Representation of the subjects on the first two dimensions resulting from an MFA on the unstandardized groups of variables, using the **plot.MFA** function (*napping_words* data set).

This representation has some important features. The coordinate of a group j on an axis of rank s is equal to $L_g(F_s, X_j)$ (*cf.* Chapter 3, Section 3.2.3). Consequently, and because of the particular weighting of each group in MFA, the coordinates of a group lie between 0 and 1. By definition of the L_g measure, the coordinate of a group on an axis of rank s is equal to 1 if F_s

corresponds to the first dimension of the PCA performed on the sole group X_j.

In this representation, two subjects are close if they induce similar structures on the statistical individuals, in our case the perfumes. In other words, two subjects are all the more close that they perceive the product space similarly.

According to Figure 7.7, *Subject 57* has a rather high coordinate on the first dimension. It means that the first dimension, F_1, of the common configuration obtained with the MFA of all subjects represents an important dimension of variability for *Subject 57*. Indeed, Figure 7.5 shows an opposition between *Aromatics Elixir* and *Chanel N5* to the rest of the perfumes on the first dimension of the MFA, which is also the case for the first dimension of *Subject 57* as shown by Figure 7.3.

This representation of the groups also shows how the group of *Words* fit within the MFA. Here, it appears that the first dimension of the MFA represents an important dimension of variability of *Words* ($L_g(F_1, Words) \simeq 1$). This is, however, not the case for the second dimension since the $L_g(F_2, Words)$ is only of around 0.3.

Let's have a closer look at *Subject 33*, who has a rather high coordinate on the second dimension (*cf.* Figure 7.7). To do so, let's perform an unstandardized PCA on his rectangle. Like previously, this can be done using the following code:

```
> res.pca33.unstd <- PCA(napping[,65:66],scale.unit=FALSE,graph=FALSE)
> plot.PCA(res.pca33.unstd,choix="ind",title="Napping: Subject 33
+ (Unstandardized)")
```

As shown in Figure 7.8, *Subject 33* has mainly opposed *Lolita Lempicka* and *Angel* to *Chanel N5* and *Pure Poison*. This opposition corresponds to the opposition of the products observed on the second dimension of the product space provided by the MFA (*cf.* Figure 7.5). In that sense, the second dimension of MFA corresponds to an important dimension of variability for *Subject 33*, hence the relatively high coordinate of this subject on the second dimension in the group representation (*cf.* Figure 7.7).

Such group representation can be completed with an important index introduced by Escoufier (1970), called the *RV* coefficient. This useful coefficient measures the link between two groups of variables, X and Y. By definition, the *RV* coefficient between X and Y is expressed by the following formula:

$$RV(X, Y) = \frac{tr(XX'YY')}{\sqrt{tr(XX'^2)tr(YY'^2)}}.$$

This coefficient takes values between 0 (each variable of X is uncorrelated to each variable of Y) and 1 (the configurations of the individuals induced by X and Y are homothetic).

Based on this coefficient, the relationship between configurations of subjects can easily be evaluated. The table of the *RV* coefficient is automatically

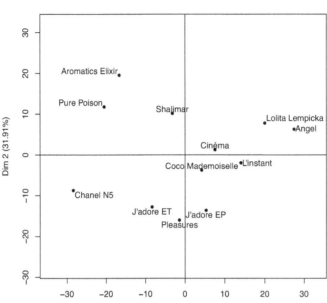

FIGURE 7.8
Representation of the perfumes on the first two dimensions resulting from a PCA on the unstandardized data, using the **plot.PCA** function (*Subject 33, napping* data set).

provided by the **MFA** function, within the object `res.mfa$group$RV`. Let's have a close look at a subset of this table by looking at the particular subjects *S24*, *S4*, *S33*, *S57*, and *S39*.

```
> round(res.mfa$group$RV[c(4,24,33,39,57),c(4,24,33,39,57)],3)
        S4    S24    S33    S39    S57
S4   1.000 0.027 0.750 0.166 0.066
S24  0.027 1.000 0.057 0.095 0.059
S33  0.750 0.057 1.000 0.134 0.109
S39  0.166 0.095 0.134 1.000 0.823
S57  0.066 0.059 0.109 0.823 1.000
```

As expected (*cf.* Figure 7.7), *S39* and *S57* are close ($RV = 0.82$) since the first dimension of the MFA corresponds to an important dimension of variability for these two subjects. Similarly, *S4* and *S33* are close ($RV = 0.75$) since the second dimension of the MFA corresponds to an important dimension of variability for these two subjects. On the contrary, *S24* is linked to none of the other panelists as their separation of the products is unique (and is related to none of the first two dimensions).

7.3 For experienced users: The sorted Napping®

It is through the observation of subjects during the verbalization phase that came the idea of the sorted Napping. It appeared that most of the subjects were not verbalizing product by product, but instead of an individual description of the products, subjects were providing a description of the products at a more global level, *i.e.*, with respect to groups of products.

Actually, when they were asked to verbalize, subjects were naturally making groups of products by circling them with their pen, and then were describing the groups: hence, the denomination sorted Napping (derived from sorting task).

This behavioral pattern may have several explanations. One of them could be that the intrinsic nature of Napping is holistic, and that's precisely what we want. Napping naturally reveals the main sensory dimensions of variability amongst the products; but when it comes to describing the products, subjects have to switch to a more analytical way of thinking: putting words on something that was latent and that has just been revealed may be difficult for the subjects. Hence, the recurrent behavioral pattern that has been observed and that has shown that subjects were more likely to verbalize at a cluster of products level. Ultimately, sorted Napping® can be assimilated to Napping followed by a verbalization phase.

The rationale behind the sorted Napping is to take into account simultaneously both types of information, that from the coordinates, and that from the comments on the groups of products. Naturally, the product is considered as the statistical unit of interest on which we collect, for a given subject, continuous data related to its position, and qualitative data related to the way it has been perceived. Hence, to each subject we can associate a data set composed of three variables: two for the coordinates and one for the comments (*cf.* Chapter 4).

This latter information is absolutely meaningful as it reinforces the notion of distance between products, based on Napping only. The data combining both sorting and Napping are stored in the *perfumes_sorted_napping.csv* file on the book website. After importing the data, let's have a look at the 6 first columns.

```
> sorted_napping <- read.table("perfumes_sorted_napping.csv",header=TRUE,
+ sep="\t",dec=".",row.names=1)
```

```
> sorted_napping[,1:6]
                    X1   Y1 S1   X2   Y2                                        S2
Angel             43.2 34.7 D  56.0 22.2                            violent;pepper
Aromatics Elixir  52.0 33.1 D  15.5  3.7                                   ammonia
Chanel N5         33.0 34.6 C   2.0 20.0                       bathroom;low-quality
Cinéma            18.8 31.9 B  30.5  9.2                       soft;discreet;refined
Coco Mademoiselle 19.8 31.8 B  29.2 11.3                       soft;discreet;refined
J'adore EP         5.5 18.8 A  18.6 21.1                          acid;violent;soft
J'adore ET         3.3 24.7 A  29.5 22.5                    neutral;soap;classical
L'instant         20.9  8.2 F  41.5 23.7          flower;light;acid;soft;prickly
Lolita Lempicka   47.7 19.0 E  40.3  7.4 pepper;balanced;breed;soft;prickly
Pleasures          4.7 12.6 A  29.5 22.5                    neutral;soap;classical
Pure Poison       19.2 32.5 B  29.2 13.4                       soft;discreet;refined
Shalimar          57.5 37.8 D  33.2 37.2                                 strong;raw
```

The core of the methodology proposed by Pagès *et al.* to obtain a representation of the product space lies on a Hierarchical Multiple Factor Analysis (HMFA), in which at the first level of the hierarchy of variables, the part of each subject is balanced, and at the second level of the hierarchy, the part of each type of data is balanced. In other words, for each subject an MFA is first performed on two groups of variables: a first one constituted of two continuous variables (the coordinates) and a second one constituted of one categorical variable only (the comments). As for the Napping the PCA performed on the first group of continuous variables is unstandardized in order to keep the relative importance of the *x*-axis over the *y*-axis. As for the analysis of sorting Task data, MCA is performed on the second group of data (constituted of one qualitative variable) within each subject. These analyses provide J matrices of coordinates of the products. The second step of the HMFA consists in running an MFA on these J matrices in order to balance the part of each subject. Once again, the PCA performed on each of these matrices is unstandardized in order to keep the relative importance of the dimensions.

In our case the following hierarchy on the variables needs to be considered:

```
> hierar <- list(rep(c(2,1),60),rep(2,60))
> hierar
[[1]]
  [1] 2 1 2 1 2 1 2 1 2 1 2 1 2 1 2 1 2 1 2 1 2 1 2 1 2 1 2 1 2 1 2 1 2 1
 [35] 2 1 2 1 2 1 2 1 2 1 2 1 2 1 2 1 2 1 2 1 2 1 2 1 2 1 2 1 2 1 2 1 2 1
 [69] 2 1 2 1 2 1 2 1 2 1 2 1 2 1 2 1 2 1 2 1 2 1 2 1 2 1 2 1 2 1 2 1 2 1
[103] 2 1 2 1 2 1 2 1 2 1 2 1 2 1 2 1

[[2]]
 [1] 2 2 2 2 2 2 2 2 2 2 2 2 2 2 2 2 2 2 2 2 2 2 2 2 2 2 2 2 2 2 2 2 2 2 2
[36] 2 2 2 2 2 2 2 2 2 2 2 2 2 2 2 2 2 2 2 2 2 2 2 2
```

The first level of the hierarchy is composed of groups of two variables and groups of one variable, consecutively. The second level of the hierarchy is composed of groups of two groups, consecutively.

Then, an HMFA is performed on the data set using the **HMFA** function of the FactoMineR package (*cf.* Figure 7.9). To do so, once again it is very important to specify properly the parameter **type**.

```
> res.hmfa <- HMFA(sorted_napping,H=hierar,type=rep(c("c","n"),60))
```

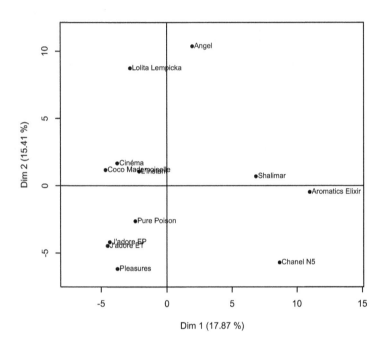

FIGURE 7.9
Representation of the perfumes on the first two dimensions resulting from an
HMFA, using the **HMFA** function (*sorted_napping* data set).

To simplify the analysis of sorted Napping data and to improve one's
understanding with specific "sensory" outputs, the use of the **fasnt** function
of the SensoMineR package is recommended. Its main parameters are **don**, the
name of the data set to analyze, and **first**, which indicates which one of the
Napping or the sorting data comes first in the data set.

```
> res.fasnt <- fasnt(sorted_napping,first="nappe",sep.word=";")
```

```
> names(res.fasnt)
[1] "eig"        "ind"        "quali.var"  "quanti.var" "group"
[6] "indicator"  "textual"    "call"
```

Beyond the representation of the products, the **fasnt** function provides also
an individual description of each product, the possibility to generate virtual
panels in order to represent confidence ellipses around the products (*cf.* Figure 7.10), and a representation of the subjects (*cf.* Figure 7.11). Numerically,

besides providing the results of HMFA, **fasnt** also provides results specific to sorted Napping, such as the analysis of the words (`$textual`) and some particular indicators (`$indicator`).

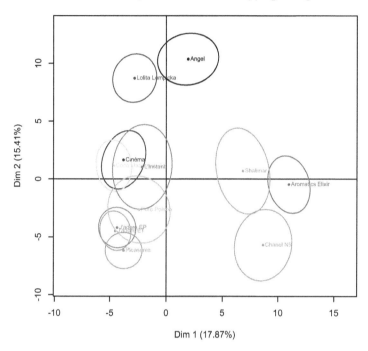

Confidence ellipses for the sorted napping configuration

FIGURE 7.10

Representation of the perfumes and the confidence ellipses on the first two dimensions resulting from HMFA, using the **fasnt** function (*sorted_napping* data set).

7.4 Exercises

Exercise 7.1 *Comparing sorting and Napping® with (Hierarchical) Multiple Factor Analysis*

The aim of this exercise is to compare the representations of the product space obtained by sorting task on the one hand, and by Napping on the other hand.

Subjects representation

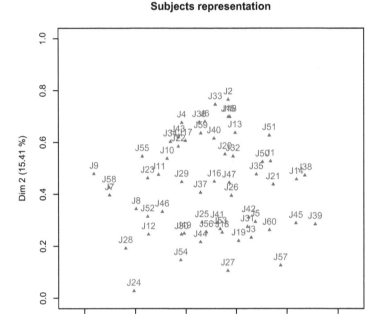

FIGURE 7.11
Representation of the subjects on the first two dimensions resulting from
HMFA, using the **fasnt** function (*sorted_napping* data set).

To do so, we will first apply MFA on the coordinates of the products issued
from the analysis of the sorting data on the one hand, and of the Napping
data on the other hand. In a second step, we will combine the analysis of the
two data sets within the global analysis through HMFA.

- First import the *perfumes_sorting.csv* file from the book website. Analyse
  this data set using the MCA procedure presented in Chapter 6. Extract all
  the possible dimensions.

- Then, import the *perfumes_napping.csv* file from the book website. Analyse
  this data set using the MFA procedure presented in Chapter 7. Extract all
  the possible dimensions.

- After renaming them appropriately, combine these two sets of coordinates
  in one unique table. Run a MFA on these two groups by setting up the
  correct parameters.

- Finally, combine the raw sorting data to the raw Napping data. Run the HMFA using the correct setting, and compare the results obtained previously.

Exercise 7.2 *Napping® and the INDSCAL model with the* smacof *package and the* SensoMineR *package*

The aim of this exercise is to illustrate the use of the INDSCAL model on the data obtained by Napping.

- Import the *perfumes_napping.csv* data from the book website.

- From this data set, build a list of dissimilarity matrices, in other words, a list containing objects of class *dist*.

- Use the **smacofIndDiff** function of the smacof package to perform the INDSCAL model.

- Use the **indscal** function of the SensoMineR package to perform the INDSCAL model. Compare both results.

Exercise 7.3 *Projective mapping and Generalized Procrustes Analysis*

The aim of this exercise is to illustrate the use of Generalized Procrustes Analysis (GPA) on Napping data.

- To do so, open an R session and load the FactoMineR package.

- Import the *perfumes_napping.csv* data from the book website.

- Perform a Generalized Procrustes Analysis with the **GPA** function of the FactoMineR package. Comment on the results.

Exercise 7.4 *Getting clusters of emotions with Napping® and Hierarchical Ascendant Classification*

In practice, Napping is mainly used to evaluate distances between products. However, its usage is not only limited to products (mainly words), and can also involve music, pictures, packaging, or concepts or words. In this study, we wanted to understand how French people would group feelings and emotional terms together.

- Import the *Napping_France.csv* file from the book website.

- Using the **MFA** function of the FactoMineR package, generate the word space. Interpret the results.

- By applying the **HCPC** function on your word space, make clusters of similar terms. Conclude.

7.5 Recommended readings

- Cousin, M., Penven, M., Toularhoat, M., Philippe, M., Cadoret, M., & Lê S. (2008). Confrontation of products spaces based on consumers' spontaneous data and experts' conventional profile. Application to the sensory evaluation of 12 luxury fragrances. 10th European Symposium on Statistical Methods for the Food Industry, Louvain-la-Neuve, Belgium, January 2008.

- Cousin, M., Penven, M., Toularhoat, M., Philippe, M., Cadoret, M., & Lê S. (2008). Construction of a products space from consumers' data (Napping and ultra flash profile). Application to the sensory evaluation of 12 luxury fragrances. 10th European Symposium on Statistical Methods for the Food Industry, Louvain-la-Neuve, Belgium, January 2008.

- Escofier, B., & Pagès, J. (1994). Multiple factor analysis (AFMULT package). *Computational Statistics & Data Analysis*, 18, (1), 121-140.

- Escoufier, Y. (1970). Echantillonage dans une population de variables aléatoires réelles. *Publ. Inst. Stat. Univ. Paris*, 19, 1-47.

- Escoufier, Y. (1973). Le traitement des variables vectorielles. *Biometrics*, 29, 751-760.

- Gazano, G., Ballay, S., Eladan, N., & Sieffermann, J.-M. (2005). Flash profile and fragrance research: the world of perfume in the consumer's words. ESOMAR Fragrance Research Conference, New-York, NY, 15-17 May 2005.

- Gower, J.C. (1975). Generalized procrustes analysis. *Psychometrika*, 40, 33-50.

- http://en.wikipedia.org/wiki/Projective_test

- Pagès, J. (2005). Collection and analysis of perceived product inter-distances using multiple factor analysis: application to the study of 10 white wines from the Loire Valley Origina. *Food Quality and Preference*, 16, (7), 642-649.

- Pagès, J., Cadoret, M., & Lê, S. (2010). The sorted napping: a new holistic approach in sensory evaluation. *Journal of Sensory Studies*, 25, (5), 637-658.

- Perrin, L., Symoneaux, R., Maître, I., Asselin, C., Jourjon, F., & Pagès, J. (2008). Comparison of three sensory methods for use with the Napping procedure: case of ten wines from Loire valley. *Food Quality and Preference*, 19, (1), 1-11.

- Risvik, E., McEwan, J. A., Colwill, J. S., Rogers, R., & Lyon, D. H. (1994). Projective mapping: a tool for sensory analysis and consumer research. *Food Quality and Preference*, 5, (4), 263-269.

- Risvik, E., McEwan, J. A., & Rødbotten, M. (1997). Evaluation of sensory profiling and projective mapping data. *Food Quality and Preference*, 8, (1), 63-71.

Part III

Affective descriptive approaches

Affective descriptive approaches

This is it. This last part concludes our sensory journey. After having devoted two thirds of the book to the comprehension of our product space, we are now moving on towards the notion of consumers and their so-called drivers of liking, in other words, the Holy Grail. Obviously, this third part is related to the two previous ones, as the issue of drivers of liking, as it is usually tackled, is linked to the characteristics of the products. Needless to say, a good understanding of the first two parts would ensure a good understanding of this last crucial part.

By definition, *affective* means "relating to moods, feelings, and attitudes." In this chapter, we are particularly dealing with measures that are directly related to pleasure, the so-called "hedonic[1] data." Here, two types of such measures are considered: a first one, emblematic of what has been usually collected and analyzed so far, that could be named the *pure* hedonic score; and a second one, emblematic of what is actually done, that could be named the *mixed*[2] hedonic score. These two types of hedonic measures correspond to a progressive evolution of the way hedonic data were collected and studied.

For the first type of measures, not much is expected from consumers. The only thing sensory scientists believed they could ask them is whether they would like a product or not, and how much they would like it. The analysis of such data is usually performed in two steps: a first one that consists in studying the hedonic scores solely, and a second one that consists in connecting hedonic data to sensory data. Unfortunately, this connection is somehow artificial, as the data to be linked are stemming from two independent sources. For the second type of measures, consumers are given much more responsibilities, as the hedonic information they provide is directly linked to the sensory characteristics of the products. This information is not only unique, it is crucial in a product's development phase.

Similarly to the first part, the first two chapters of this third part are dedicated to a standard every sensory scientist must understand and master to move on. Chapter 8 highlights the kind of information that can be expected from the analysis of *pure* hedonic data. Notably, we propose ways to find the products that have been preferred and to define homogenous clusters of consumers in terms of preferences. Chapter 9 presents the different strategies

---

[1]Hedonic is defined in the dictionary as "relating to or considered in terms of pleasant (or unpleasant) sensations."

[2]Here, the word *mixed* refers to both notions of sensory and hedonic through the notion of ideal.

that can be used to connect the hedonic information to the sensory information statistically. In particular, two standards, the so-called internal preference mapping and external preference mapping, are presented. The last chapter of this third part, Chapter 10, is dedicated to particular tasks that directly or indirectly involve the notion of ideal, within the sensory methodology. This final chapter focuses its attention on the Just About Right task (JAR) and the Ideal Profile Method (IPM).

As mentioned in the *Preface*, in terms of statistics, this last part is by far the most complicated one to embrace. One reason is that the statistical methods used are more advanced, compared to the ones that figure in the first two parts. The other reason is that these statistical methods are combined to provide a methodology. For example, in Chapter 8, it is shown how to deal with specific *post hoc* tests. In Chapter 9, various multiple regression models are used to explain hedonic scores by sensory attributes. Additionally, it is shown how the results issued from these models can be projected on a graphical representation of the products issued from a Principal Component Analysis (PCA). Finally, in Chapter 10, amongst others, results from regression models are combined with exploratory multivariate methods to assess the quality of Ideal Profile data.

8

When products are solely assessed by liking

CONTENTS

"...the main goal of the analysis consists in defining the products that are preferred, in overall. Since hedonic ratings are personal (it takes all to make a world), no rules (or agreement) can be expected from the consumers hedonic assessments. Still, it is important to look at general trends across consumers, and it is common to search for homogenous clusters of consumers, with respect to their preferences. To do so, cluster analysis (e.g., HAC) is required. In practice homogeneous clusters of consumers are defined using the **HCPC** *function of the* FactoMineR *package."*

8.1 Data, sensory issues, and notations

Products are developed to be sold to so-called consumers, supposedly the main reason being that consumers like them and are ready to pay for them. Hence, the data we are dealing with in this part are crucial, as they determine up to a certain point whether a product can be successful or not (without taking into account any other aspects of the product, such as marketing, price, *etc.*). As these data are so important, we can usually consider, without loss of generality, that consumers have tasted all the products of the product space, and have scored them on what is called a hedonic scale (the most famous one being the

9-point Likert scale), hence associating liking scores to products. Sometimes, as we will see in Section 8.3, *For experienced users*, consumers are asked to score the products on more or less specific liking questions.

Let I denote the number of products and J the number of consumers performing the test. Without loss of generality (and also since it is the most frequent case), let's consider that each consumer has scored all I products on overall liking. As in Chapter 1 and in Chapter 2, it is quite natural to consider as statistical unit the hedonic assessment of each product by each consumer. Hence, the data set to be analyzed is composed of $I \times J$ rows and 3 columns, two categorical ones with the product and the consumer information, and one numerical one with the hedonic scores.

Due to their intrinsic nature, the data collected present some important specificities: such data are both product and consumer related. Based on the point of view adopted, the same hedonic data are going to be structured in different ways, depending on the objectives of the study. Therefore, we will consider different data sets with different formats (different numbers of rows and columns) of the same data originally collected. Each data set, depending on its format (in other words, depending on the choice of the statistical unit of interest), answers to specific questions through specific statistical analyses.

If the attention is on understanding the data from a product perspective, the scientist wants to understand how products are liked by consumers. In this case, the statistical unit is the product, and the goal of the analysis is hence to understand if some products are more liked, or, as we will see later, preferred to other products. Ideally, the analysis should also provide a map of the products, two products being close if they have induced the same "liking" behavior to consumers.

However since the data collected are related to the "affective", and since it takes all sorts to make a world (*i.e.*, consumers have different liking behaviors towards the products), such data can also be studied from a consumer perspective. In this situation, the statistical unit of interest is the consumer and the analysis should provide information regarding the heterogeneity of the panel of consumers in terms of liking behavior. Besides providing information regarding the groups of consumers (*i.e.*, to which group belongs each consumer?), the analysis should also provide a map of consumers based on their liking behavior (two consumers being close if they liked more the same products).

For these two specific situations (product-specific and consumer-specific), the data need to be restructured into two equivalent matrices, one being the transposed version of the other. When the analysis is product-oriented, this matrix crosses the products in rows and the consumers in columns. At the intersection of row i and column j is stored the hedonic score x_{ij} provided by the consumer j to product i. When the analysis is consumer-oriented, this matrix presents the consumers in rows and the products in columns.

Before focusing on either the product or the consumer specifically, it is important to first discuss about the intrinsic nature of the data collected: are we really directly interested in the liking scores measured? This question is important in order to correctly interpret results issued from the statistical analyses of such data. Hence, it is the first section of this chapter.

In the following section, the hedonic measure is assessed from a product perspective. In particular, we will identify the product that was the best rated by the consumers through ANOVA extended with the notion of *post hoc* tests[1]. From a multivariate point of view, the principles of *internal preference mapping* (MDPref) will be presented.

As hedonic data are collected based on consumers' opinions, Section 8.2.3 will focus on how to get homogeneous clusters of consumers. In other words, we will see how to segment a consumer market. This feature is, in practice, of utmost importance, as it is the basis for any optimization procedure (*cf.* Chapter 9 and Chapter 10 for more information). Section 8.3, *For experienced users*, is a natural extension of the cluster analysis (presented in Section 8.2.3) as it shows how external information can be used to characterize the different clusters of consumers obtained.

8.2 In practice

8.2.1 How can I approach *hedonic* data?

To understand the intrinsic nature of the data, let's import the file *per-fumes_liking_small.csv* from the book website. This data set consists in 6 perfumes assessed by 2 consumers, each row corresponding to the assessment of one perfume by one consumer.

```
> small <- read.table("perfumes_liking_small.csv",header=TRUE,sep=",",dec=".",
+ quote="\"")
> small
   Panelist        Product Liking
1         A          Angel    6.0
2         A Aromatics Elixir    6.5
3         A       Chanel N5    7.0
4         A       J'adore EP    9.0
5         A       J'adore ET    8.5
6         A        Shalimar    5.5
7         B          Angel    2.0
8         B Aromatics Elixir    4.0
9         B       Chanel N5    3.0
10        B       J'adore EP    6.0
11        B       J'adore ET    7.0
12        B        Shalimar    1.0
```

---

[1] Although they have not been presented in Chapter 2, *post hoc* tests can also be applied to sensory variables using the exact same procedure as presented in Section 8.2.2.

Regarding these data, the main question one may want to answer is whether differences in liking between products are observed: are the intrinsic differences between products impacting their liking assessments? To answer this question, let's represent *abusively* the data using a box plot per product. As in Chapter 1, the **boxplot** function is used.

```
> boxplot(Liking~Product,data=small,col="lightgray")
```

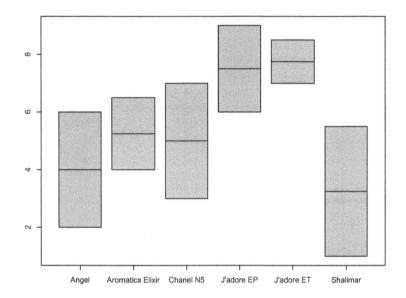

FIGURE 8.1
Representation of the hedonic scores per product, using the **boxplot** function (*small* data set).

Figure 8.1 shows that, although *J'adore EP* and *J'adore ET* seem to be associated with higher liking scores, there is no obvious evidence of a strong *Product* effect here, because of the outer products (*e.g.*, *Chanel N5* and *Angel* for instance).

Nevertheless, if we compare in the same way the two consumers (*cf.* Figure 8.2), it seems that consumer *A* gave rather high scores, whereas consumer *B* gave rather low scores. It seems obvious that *A* and *B* behaved completely differently. Does it mean that *A* liked a lot of things, whereas *B* did not? Or, does it mean that *A* and *B* used the scale of notation differently?

```
> boxplot(Liking~Panelist,data=small,col="lightgray")
```

To confirm this difference between consumers, a *t*-test is performed. This test statistically compares the means of consumer *A* and consumer *B*. Such analysis can be done using the **t.test** function:

```
> t.test(Liking~Panelist,var.equal=TRUE,data=small)
```

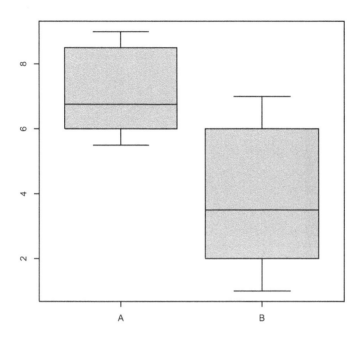

FIGURE 8.2
Representation of the hedonic scores per consumer, using the **boxplot** function
(*small* data set).

```
Two Sample t-test

data:  Liking by Panelist
t = 2.9448, df = 10, p-value = 0.01467
alternative hypothesis: true difference in means is not equal to 0
95 percent confidence interval:
 0.7909022 5.7090978
sample estimates:
mean in group A mean in group B
       7.083333        3.833333
```

As shown in the previous output, we reject the null hypothesis under which
there are no differences between A and B. In other words, the differences
between A and B have to be taken into account somehow in the interpretation
of Figure 8.1, as they are significant ($p\text{-}value=0.014$).

Let's now have a look at the data from a different perspective, by representing each consumer by a broken line, using the **graphinter** function of the SensoMineR package.

```
> library(SensoMineR)
> graphinter(small,col.p=2,col.j=1,firstvar=3,numr=1,numc=1)
```

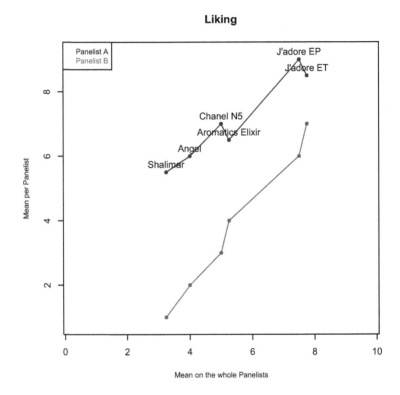

FIGURE 8.3
Visualization of the hedonic scores given by the panelists, using the **graphinter** function (*small* data set).

As shown in Figure 8.3, although the two consumers provided different scores to the products (consumer *A* scoring higher than consumer *B*), both seem to evaluate the products in a similar way, at least in terms of ranking. Indeed, both consumers appreciated *J'adore EP* and *J'adore ET* the most, and *Shalimar* the least. This finally reveals a strong *Product* effect.

With these elements in mind, let's now try to find the proper model to understand the liking scores in function of the factors involved in the experiment.

First, let's consider the most natural and simple ANOVA model:

$$Y_{ij} \sim \mu + \alpha_i + \epsilon_{ij},$$

where

- Y_{ij} is the hedonic score given by consumer j to product i;

- α_i is the ith coefficient associated with the *Product* effect;

- and ϵ_{ij} denotes the error term.

As in Chapter 1, the errors are assumed to be normally distributed with mean zero and constant variance, σ^2, and also to be independent:

- $\forall(i,j), \epsilon_{ij} \sim N(0, \sigma^2)$;

- $\forall(i,j) \neq (i',j'), \text{Cov}(\epsilon_{ij}, \epsilon_{i'j'}) = 0$.

As explained in Chapter 2, two points of view are usually considered: the first one where there is no reference product (this point of view is similar to the one adopted in PCA when data are centered) and the second one, where products are compared to a reference product. In the case where there is no reference, the user has to specify the contrasts used to estimate the coefficients of the ANOVA model. In that case, the contrasts can be expressed the following way:

$$\sum_{i=1}^{I} \alpha_i = 0.$$

The **AovSum** function of the FactoMineR estimates the coefficients of the ANOVA model without any prior information on the products, *i.e.*, by considering that $\sum \alpha_i = 0$.

```
> res <- AovSum(Liking~Product,data=small)
> res
$Ftest
              SS df    MS F value Pr(>F)
Product   33.354  5 6.6708  1.1477 0.4284
Residuals 34.875  6 5.8125


$Ttest
                              Estimate Std. Error    t value      Pr(>|t|)
(Intercept)                  5.4583333  0.6959705  7.8427649 0.0002271191
Product - Angel             -1.4583333  1.5562374 -0.9370892 0.3848816812
Product - Aromatics Elixir  -0.2083333  1.5562374 -0.1338699 0.8978828625
Product - Chanel N5         -0.4583333  1.5562374 -0.2945138 0.7782892698
Product - J'adore EP         2.0416667  1.5562374  1.3119249 0.2375023888
Product - J'adore ET         2.2916667  1.5562374  1.4725688 0.1912903369
Product - Shalimar          -2.2083333  1.5562374 -1.4190208 0.2056908169
```

As shown by the *F*-test, the *Product* effect is not significant. It seems that the products were not appreciated significantly differently. Considering the differences between the consumers, one can hypothesize that the *Product* effect is masked by the *Panelist* effect.

Let's now consider the following ANOVA model in which the *Panelist* effect was added:

$$Y_{ij} \sim \mu + \alpha_i + \beta_j + \epsilon_{ij},$$

where

- Y_{ij} is the hedonic score given by consumer j to product i;

- α_i is the ith coefficient associated with the *Product* effect;

- β_j is the jth coefficient associated with the *Panelist* effect;

- and ϵ_{ij} denotes the error term.

Here again, the errors are assumed to be normally distributed with mean zero and constant variance, σ^2, and also to be independent:

- $\forall(i,j), \epsilon_{ij} \sim N(0, \sigma^2)$;

- $\forall(i,j) \neq (i',j'), \mathrm{Cov}(\epsilon_{ij}, \epsilon_{i'j'}) = 0.$

Such results are obtained using the same procedure, after introducing the *Panelist* effect in the ANOVA model:

```
> res <- AovSum(Liking~Product+Panelist,data=small)
> res
$Ftest
              SS df     MS F value    Pr(>F)
Product   33.354  5  6.671  10.464 0.0110869 *
Panelist  31.687  1 31.687  49.706 0.0008871 ***
Residuals  3.188  5  0.638
---
Signif. codes:  0 '***' 0.001 '**' 0.01 '*' 0.05 '.' 0.1 ' ' 1

$Ttest
                              Estimate Std. Error    t value     Pr(>|t|)
(Intercept)                  5.4583333  0.2304886 23.6815750 2.500285e-06
Product - Angel             -1.4583333  0.5153882 -2.8295823 3.669274e-02
Product - Aromatics Elixir  -0.2083333  0.5153882 -0.4042260 7.027568e-01
Product - Chanel N5         -0.4583333  0.5153882 -0.8892973 4.145759e-01
Product - J'adore EP         2.0416667  0.5153882  3.9614152 1.072767e-02
Product - J'adore ET         2.2916667  0.5153882  4.4464865 6.724226e-03
Product - Shalimar          -2.2083333  0.5153882 -4.2847960 7.827707e-03
Panelist - A                 1.6250000  0.2304886  7.0502399 8.870514e-04
Panelist - B                -1.6250000  0.2304886 -7.0502399 8.870514e-04
```

Regarding this second model, the *Product* effect is now highly significant (*p-value*=0.011), highlighting the fact that the products were appreciated significantly differently. If we now look at the results of the *t*-test, we can see that all the α_i coefficients associated with the products are significantly different from 0 (except for *Aromatics Elixir* and *Chanel N5*). With $\alpha_i > 0$, *J'adore EP* and *J'adore ET* are the two most preferred products. Oppositely, *Shalimar* is the least preferred product (with $\alpha_i < 0$).

The addition of the *Panelist* effect in our ANOVA model has for consequence a semantic change when interpreting the results: although liking scores were originally collected, the products are now interpreted in terms of preferences.

Let's see what happens if we are now considering an exploratory multivariate point of view. As presented in Section 8.1, let $X = (x_{ij})$ denote a matrix of hedonic scores crossing the products in rows, and the consumers in columns, x_{ij} being the score given by consumer j to product i. This matrix is *natural* in the sense that products are evaluated by consumers, and *de facto* can be considered as our statistical units of interest when the consumers are the variables. Each consumer j is associated here to a vector of I hedonic scores.

To each consumer j, one can associate a vector of I hedonic scores. This vector may be analyzed without any prior transformation or mean centered. In the latter case, x_{ij} is transformed the following way:

$$\forall i, j \ x_{ij} \longleftarrow x_{ij} - \bar{x}_j,$$

where $\bar{x}_j$ denotes the average hedonic score over the products for consumer j, $\bar{x}_j = \frac{1}{I} \sum x_{ij}$.

These two situations correspond to two distinct point of views: in the first case, the interest is in the raw liking scores, whereas in the second case, the interest is not in how much products are liked, but in how they are preferred. From a practical point of view, adopting one or the other point of view changes considerably the analyses and their interpretation. When liking scores are used, it is not always clear how results should be interpreted from a consumer point of view, as the data mix confusingly the notion of accepting/rejecting products and the notion of the use of the scale: when consumers score all the products with relatively high scores, does that mean that they like all the products very much, or does that simply mean that they are enthusiastic and tend to use a higher part of the scale? For that matter, preference data seem more stable (*preference is preference*) as such *noise* is removed by construction. However, the interpretation loses this important notion of acceptability: a product that is preferred by a consumer can still be rejected in overall.

To have a better understanding of these two situations, let's now consider that our statistical unit of interest is the consumer. The data set considered is the transposed of X, and we want to represent the consumers with respect to their hedonic scores. Such representation is based on a distance that expresses a multivariate point of view, as consumers have tested several products (each product corresponds to one dimension). Actually, we can consider a consumer as a vector of dimension I.

When data are not transformed, this representation of the consumers is such that two consumers are all the more close that they like the same products (and dislike the same products, as well), and all the more distant that they don't like the same products. When data are centered, this representation of the consumers is such that two consumers are all the more close that they have the same preferences, and all the more distant that they don't have the same preferences.

As mentioned earlier, the data set to be analyzed corresponds to a matrix of dimension $J \times I$, in which rows correspond to consumers, columns correspond to products. Since we want to give the same weight in the analysis to all the products, a standardized PCA is performed on this matrix.

Let's import the full data set stored in *perfumes_liking.csv* from the book website.

```
> liking <- read.table("perfumes_liking.csv",header=TRUE,sep=",",
+ quote="\"")
> liking$consumer <- as.factor(liking$consumer)
> summary(liking)
     consumer                  product          liking
 171    :  12   Angel             :103   Min.   :1.000
 553    :  12   Aromatics Elixir  :103   1st Qu.:5.000
 991    :  12   Chanel N5         :103   Median :6.000
 1344   :  12   Cinéma            :103   Mean   :5.688
 1661   :  12   Coco Mademoiselle :103   3rd Qu.:7.000
 1679   :  12   J'adore EP        :103   Max.   :9.000
 (Other):1164   (Other)           :618
```

As shown by the outputs of the **summary** function, this data set needs to be restructured. This can be done using the following code:

```
> consumer <- levels(liking$consumer)
> nbconso <- length(consumer)
> product <- levels(liking$product)
> nbprod <- length(product)
> newmat <- matrix(0,nbconso,0)
> rownames(newmat) <- consumer
> for (p in 1:nbprod){
+     data.p <- liking[liking$product==product[p],]
+     data.add <- as.matrix(data.p[,3])
+     rownames(data.add) <- data.p[,1]
+     newmat <- cbind(newmat,data.add[rownames(newmat),])
+ }
> colnames(newmat) <- product
> head(newmat)
     Angel Aromatics Elixir Chanel N5 Cinéma Coco Mademoiselle J'adore EP
171      3                3         7      6                 6          5
553      3                5         1      4                 7          8
991      7                6         8      6                 7          7
1344     5                6         8      8                 6          9
1661     5                4         1      7                 5          5
1679     5                4         5      6                 5          6
```

	J'adore ET	L'instant	Lolita Lempicka	Pleasures	Pure Poison	Shalimar
171	6	3	3	5	6	3
553	6	4	3	6	5	3
991	8	9	6	8	7	6
1344	7	6	8	7	7	8
1661	7	7	7	7	6	3
1679	5	3	5	7	5	2

Let's represent the consumers according to the first point of view. In this case, we do not consider any transformation on the data, and we apply a PCA on this table. Let us remind that for this PCA, two consumers are all the more close that they gave the same scores to the products, in other words, that they have similar liking profiles (they like the same products and also dislike the same products). For instance, if we consider the following consumers 4640, 1755, and 8840, we expect to find the first two really close on the first factorial plane (it seems that they liked everything), and very far from the third one.

```
> newmat[c("4640","1755","8840"),]
```

	Angel	Aromatics	Elixir	Chanel N5	Cinéma	Coco Mademoiselle	J'adore EP
4640	8	8	8	8		7	8
1755	9	9	8	6		8	6
8840	1	1	1	6		2	7

	J'adore ET	L'instant	Lolita Lempicka	Pleasures	Pure Poison	Shalimar
4640	7	9	7	7	8	8
1755	6	9	8	8	9	9
8840	7	1	1	5	8	1

If we now run a PCA on the data set using the following code, we can actually see this opposition between *4640* and *1755*, on the one hand, and *8840*, on the other hand (*cf.* Figure 8.4).

```
> raw.pca <- PCA(newmat,graph=FALSE)
> plot.PCA(raw.pca,choix="ind",title="When liking is not centered")
```

From this analysis, the most surprising results are provided by the variable representation which highlights what is usually called a *size effect* (*cf.* Figure 8.5).

```
> plot.PCA(raw.pca,choix="var",title="When liking is not centered")
```

This *size effect* highlights the fact that the variables of the data set are all positively correlated. In terms of consumers and liking, it means that some consumers like everything, and some others dislike everything. In other words, the first dimension is interpreted as a gradient of liking that opposes consumers who like everything, from consumers who dislike everything[1].

Such *size effect* can be directly highlighted by projecting the average liking score provided by the consumers as a supplementary variable (*cf.* Figure 8.5). Such results can be obtained using the following code:

---

[1] Again, this notion of gradient of liking is only completely true under the assumption that all consumers are using the scale similarly.

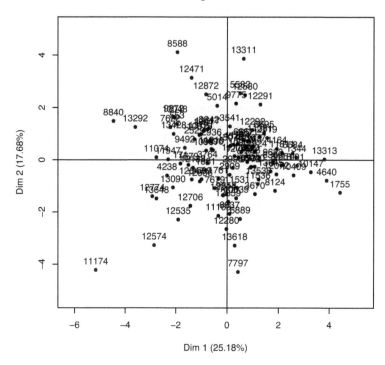

FIGURE 8.4
Representation of the consumers on the first two dimensions resulting from a PCA, using the **plot.PCA** function (*newmat* data set).

```
> newmat2 <- cbind(newmat,as.matrix(apply(newmat,1,mean)))
> colnames(newmat2)[13] <- "Average"
> raw.pca2 <- PCA(newmat2,quanti.sup=13,graph=FALSE)
> plot.PCA(raw.pca2,choix="var")
```

The second dimension is what is called a *shape effect*. Consumers are differentiated along the second axis, in the sense that amongst the consumers who like everything, some consumers really like products such as *J'adore EP*, *J'adore ET*, and some others really like products such as *Shalimar* and *Chanel N5*. It seems that this segmentation of the consumers is not really relevant, and the information it provides is not really useful from a practical point of view.

Let's now consider the second point of view for which the data set is mean centered by rows, *i.e.*, the data we are dealing with are not pure liking scores anymore, but preferences. To center the data by rows, the **apply** function and the **sweep** function are used:

Variables factor map (PCA)

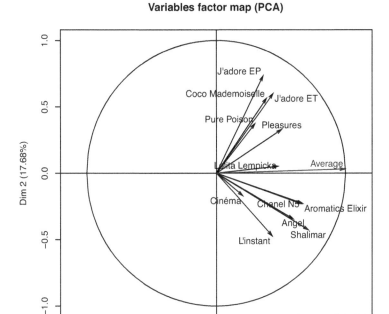

FIGURE 8.5
Representation of the products as variables on the first two dimensions resulting from a PCA, using the **plot.PCA** function (*newmat* data set).

```
> means <- apply(newmat,1,mean)
> round(means,2)[1:11]
   171     553     991    1344    1661    1679    1755    1761    1801    1956    2119
  4.67    4.58    7.08    7.08    5.33    4.83    7.92    5.42    5.17    6.08    6.50
> newmat_centered <- sweep(newmat,1,means,FUN="-")
> round(apply(newmat_centered,1,mean))[1:11]
   171     553     991    1344    1661    1679    1755    1761    1801    1956    2119
     0       0       0       0       0       0       0       0       0       0       0
```

Tips: Using the scale function

Alternatively to the combination of the **apply** function with the **sweep** function, the **scale** function can be used. This function automatically centers
and/or scales the columns of a matrix. Since we want to center (not scale) the
rows, one solution consists in:

- transposing the original table;

- centering the table using the scale function (we set the option scale=FALSE);

- transposing the resulting matrix again.

In the previous example, this would consist in using the following code:

```
> newmat_centered2 <- t(newmat)
> newmat_centered2 <- scale(newmat_centered2,scale=FALSE)
> newmat_centered2 <- t(newmat_centered2)
> head(newmat_centered2)
```

As expected, the two procedures provide the exact same results.

As we can see, data have been centered by rows; we can now apply a PCA on this transformed data set.

```
> center.pca <- PCA(newmat_centered,graph=FALSE)
> plot.PCA(center.pca,choix="var",title="When liking is centered")
```

In this new analysis, the first dimension opposes consumers who prefer perfumes such as *J'adore EP* and *J'adore ET*, to consumers who prefer perfumes such as *Shalimar* or *Chanel N5* (*cf.* Figure 8.6).

Remark. To understand PCA and the impact of pre-processing the data (*i.e.*, using preference data rather than raw liking data), we propose to look at further dimensions of the PCA liking space. More particularly, let's compare the variable representation obtained with the raw data on dimensions 2 and 3 with the variable representation obtained on the preference data on the first two dimensions.

```
plot.PCA(raw.pca,choix="var",axes=c(2,3))
```

As can be seen in Figure 8.6 and Figure 8.7, the two representations are very similar (we do not interpret the symmetry along the X or Y axis here). Such result is very common as after separating the consumers according to their differences in the use of the scale[2] (first dimension of the PCA on the raw data), their separation is made based on differences in liking behavior, *i.e.*, their preferences.

8.2.2   How can I identify the *best* product?

To identify the *best* product, two different approaches are considered: a univariate and a multivariate one. In the univariate approach, the usual approach

---

[2]From now on, the confusion between accepting/rejecting all the products and the use of the high/low part of the scale will be referred as "use of the scale."

When liking is centered

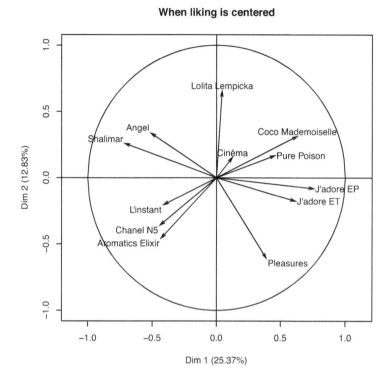

FIGURE 8.6
Representation of the variables using the **PCA** function, on the first two dimensions (*newmat_centered* data set).

involving ANOVA is considered. However, rather than using the *t*-test to evaluate the significance of the difference between each product and the overall mean in terms of liking, another procedure is considered. This procedure aims at comparing each pairs of product together, through what is called *post hoc* tests (also known as *a posteriori* tests). Here, the Fisher's LSD test is used. For the multivariate approach, the so-called *internal preference mapping* (also known as MDPref) is considered.

Let's consider the second ANOVA model introduced in the previous section, where *Liking* is explained by the *Product* effect and the *Panelist* effect. In the case where there is no reference, the user has to specify the contrasts used to estimate the coefficients of the ANOVA model. In that case, the contrasts

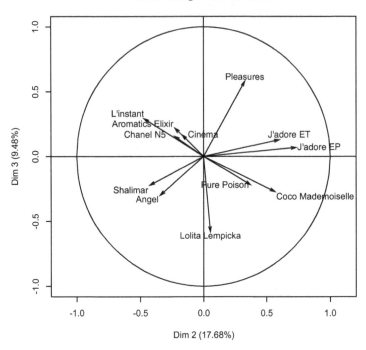

FIGURE 8.7
Representation of the products as variables on the second and the third dimensions resulting from a PCA, using the **plot.PCA** function (*newmat* data set).

can be expressed the following way:

$$\sum_{i=1}^{I} \alpha_i = 0, \sum_{j=1}^{J} \beta_j = 0.$$

After changing the contrasts in R using the **options** function, let's run a two-way ANOVA on the *liking* data set imported in the previous section. Such analysis can be obtained using the **lm** function.

```
> options(contrasts=c("contr.sum", "contr.sum"))
> model <- lm(liking~product+consumer,data=liking)
> anova(model)
```

```
Analysis of Variance Table

Response: liking
           Df   Sum Sq Mean Sq F value    Pr(>F)
product     11   806.83  73.348 27.2717 < 2.2e-16 ***
consumer   102   988.95   9.696  3.6049 < 2.2e-16 ***
Residuals 1122 3017.67    2.690
---
Signif. codes:  0 '***' 0.001 '**' 0.01 *' 0.05 '.' 0.1 ' ' 1

> summary(model)
Call:
lm(formula = liking ~ product + consumer, data = liking)

Residuals:
    Min      1Q  Median      3Q     Max
-4.4757 -1.0283  0.1145  1.1139  4.9167

Coefficients:
            Estimate Std. Error t value Pr(>|t|)
(Intercept)  5.68770    0.04665 121.929  < 2e-16 ***
product1    -1.15372    0.15471  -7.457 1.76e-13 ***
product2    -1.37702    0.15471  -8.901  < 2e-16 ***
product3    -0.53236    0.15471  -3.441 0.000601 ***
product4     0.72006    0.15471   4.654 3.64e-06 ***
product5     0.77832    0.15471   5.031 5.69e-07 ***
product6     0.89482    0.15471   5.784 9.46e-09 ***
product7     0.64239    0.15471   4.152 3.54e-05 ***
product8     0.13754    0.15471   0.889 0.374191
product9     0.16667    0.15471   1.077 0.281594
product10    0.28317    0.15471   1.830 0.067470 .
product11    0.64239    0.15471   4.152 3.54e-05 ***
consumer1   -1.02104    0.47112  -2.167 0.030425 *
consumer2   -1.10437    0.47112  -2.344 0.019245 *
consumer3    1.39563    0.47112   2.962 0.003117 **
consumer4    1.39563    0.47112   2.962 0.003117 **
...
---
Signif. codes:  0 '***' 0.001 '**' 0.01 '*' 0.05 '.' 0.1 ' ' 1

Residual standard error: 1.64 on 1122 degrees of freedom
Multiple R-squared:  0.3731,Adjusted R-squared:  0.3099
F-statistic: 5.909 on 113 and 1122 DF,  p-value: < 2.2e-16
```

One of the drawback of this output is that it is impossible to interpret the results unless one knows how the levels of the *Product* effect are organized. To do so, let's use the **levels** function.

```
> levels(liking$product)
 [1] "Angel"              "Aromatics Elixir"  "Chanel N5"
 [4] "Cinéma"             "Coco Mademoiselle" "J'adore EP"
 [7] "J'adore ET"         "L'instant"         "Lolita Lempicka"
[10] "Pleasures"          "Pure Poison"       "Shalimar"
```

The other drawback of this output is that the *t*-test for the last level of each factor included in the model has to be calculated manually, by deduction. According to the previous outputs, we can say that *Coco Mademoiselle*, *J'adore EP*, and *J'adore ET* were the most appreciated products.

Let's now have a look at the case where products are compared to one particular reference product. This point of view is the default option in R and the comparisons are systematically made with respect to the first category of the factor.

```
> options(contrasts=c("contr.treatment", "contr.poly"))
> model <- lm(liking~product+consumer,data=liking)
> anova(model)
Analysis of Variance Table

Response: liking
            Df  Sum Sq Mean Sq F value    Pr(>F)
product     11  806.83  73.348 27.2717 < 2.2e-16 ***
consumer   102  988.95   9.696  3.6049 < 2.2e-16 ***
Residuals 1122 3017.67   2.690
---
Signif. codes:  0 '***' 0.001 '**' 0.01 *' 0.05 '.' 0.1 ' ' 1

> summary(model)
Call:
lm(formula = liking ~ product + consumer, data = liking)

Residuals:
    Min      1Q  Median      3Q     Max
-4.4757 -1.0283  0.1145  1.1139  4.9167

Coefficients:
                           Estimate Std. Error t value Pr(>|t|)
(Intercept)               3.513e+00  4.981e-01   7.053 3.05e-12 ***
productAromatics Elixir  -2.233e-01  2.285e-01  -0.977 0.328713
productChanel N5          6.214e-01  2.285e-01   2.719 0.006649 **
productCinéma             1.874e+00  2.285e-01   8.199 6.55e-16 ***
productCoco Mademoiselle  1.932e+00  2.285e-01   8.454  < 2e-16 ***
productJ'adore EP         2.049e+00  2.285e-01   8.964  < 2e-16 ***
productJ'adore ET         1.796e+00  2.285e-01   7.860 8.99e-15 ***
productL'instant          1.291e+00  2.285e-01   5.650 2.03e-08 ***
productLolita Lempicka    1.320e+00  2.285e-01   5.778 9.79e-09 ***
productPleasures          1.437e+00  2.285e-01   6.288 4.61e-10 ***
productPure Poison        1.796e+00  2.285e-01   7.860 8.99e-15 ***
productShalimar          -4.854e-02  2.285e-01  -0.212 0.831817
consumer553              -8.333e-02  6.695e-01  -0.124 0.900968
consumer991               2.417e+00  6.695e-01   3.610 0.000320 ***
consumer1344              2.417e+00  6.695e-01   3.610 0.000320 ***
consumer1661              6.667e-01  6.695e-01   0.996 0.319592
...
---
Signif. codes:  0 '***' 0.001 '**' 0.01 '*' 0.05 '.' 0.1 ' ' 1

Residual standard error: 1.64 on 1122 degrees of freedom
Multiple R-squared:  0.3731,Adjusted R-squared:  0.3099
F-statistic: 5.909 on 113 and 1122 DF,  p-value: < 2.2e-16
```

As *Angel* is the first level of the *Product* effect, it does not appear in the output: by construction, its estimate is equal to 0, and all the other estimates are calculated with respect to *Angel* (as the reference product). The products that were more ($\alpha_i > 0$) or less ($\alpha_i < 0$) appreciated than Angel are easily detected.

Finally, now the ANOVA results are obtained, *a posteriori* tests can be performed. For multiple comparisons strategies (so-called *post hoc* tests), the agricolae package can be used. Here, we will focus particular attention on the Fisher's Least Significant Difference (Fisher's LSD) provided by the **LSD.test** function.

```
> options(contrasts=c("contr.sum","contr.sum"))
> model <- aov(liking~product+consumer,data=liking)
> library(agricolae)
> res.LSD <- LSD.test(model,"product",p.adj="none",group=TRUE,
+ main="Results of the Fisher LSD")
> names(res.LSD)
[1] "statistics" "parameters" "means"      "comparison" "groups"
```

The **LSD.test** function returns 5 main outputs. In the `statistics` outputs, when the data are perfectly balanced (each product has been assessed the exact same number of times), the last value provided is called LSD. This corresponds to the least significant difference required between two products to be significantly different. Since here, the LSD value is of 0.45, every pair of products with a difference in the average liking score larger than 0.45 is significant at 5% (this significance threshold can be adjusted using the `alpha` parameter).

The other output of interest is either `comparison` (if group=FALSE) or `groups` (if group=TRUE). In the first case, the results of each pair comparison is provided, whereas in the second case (presented here), the products are grouped together based on their significant differences: two products associated with the same letter are not significantly different at the level considered.

```
> res.LSD$statistics
      Mean       CV MSerror       LSD
  5.687702 28.83384 2.689544 0.4483865
> res.LSD$groups
                    trt   means  M
1  J'adore EP            6.582524  a
2  Coco Mademoiselle 6.466019  a
3  Cinéma               6.407767 ab
4  J'adore ET           6.330097 ab
5  Pure Poison          6.330097 ab
6  Pleasures            5.970874 bc
7  Lolita Lempicka   5.854369  c
8  L'instant            5.825243  c
9  Chanel N5            5.155340  d
10 Angel                4.533981  e
11 Shalimar             4.485437  e
12 Aromatics Elixir  4.310680  e
```

In this example, the results of the Fisher's LSD test show that *J'adore EP*, *Coco Mademoiselle*, *Cinéma*, *J'adore ET*, and *Pure Poison* are the significantly most liked products, whereas *Aromatics Elixir*, *Shalimar*, and *Angel* are significantly the least liked products.

Such results can also be graphically visualized with the **bar.group** function of the agricolae package. The following code generates Figure 8.8.

```
> bar.group(res.LSD$groups,ylim=c(0,10),density=4,border="black",cex.names=0.7)
```

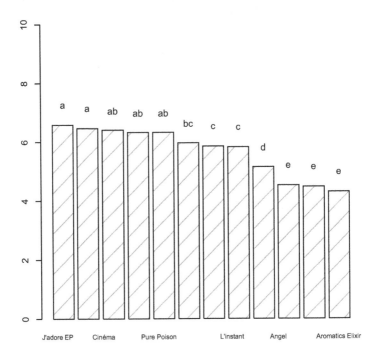

FIGURE 8.8
Representation of the perfumes resulting from an ANOVA and an LSD test, using the **bar.group** function (*liking* data set).

Tips: Different types of *post hoc* tests

Although the Fisher's LSD test is quite easy to use thanks to the LSD value it computes, it is quite criticized by statistician. The main reasons for these criticisms rely in the fact that, when 100 tests are performed, at least 5 of them tend to highlight significant differences amongst pairs of products, even if these differences do not exist. These 5 significant tests correspond to the 5% type I error one would accept.

To make the tests more accurate, other *post hoc* tests and *p-value* adjustment have been set up. Amongst the alternative *post hoc* tests, we can mention the Tukey's HSD. In terms of *p-value* adjustment, we can mention the Bonferroni adjustment.

All these tests are available in the agricolae package. For *p-value* adjustment, the p.adj parameter should be used (*cf.* ?p.adjust). Concerning alternative *post hoc* tests, the **HSD.test**, **SNK.test**, **duncan.test**, *etc.* functions could be used. Note that, all these functions work the same way as the **LSD.test** function. For an example of the use of the **HSD.test** function, refer to Exercise 8.2.

Finally, it is worth mentioning that all these tests are paired comparison tests, *i.e.*, all pairs of products are compared together. However, in some situations, the product space contains a product considered as the reference, and only the comparison of each product to that reference should be performed. The classical *post hoc* test used in this situation is the Dunnett's test. Such analysis is provided in the multcomp package, by using the **glht** function and by setting the linfct parameter as **mcp(product="Dunnett")**.

Let's now consider an exploratory multivariate point of view to identify the most liked products. From this point of view, as mentioned in Section 8.1, the statistical unit of interest is the product, the variables are the consumers. The idea of MDPref, is to get a representation of the products based on the hedonic scores: two products are all the more close that they are liked similarly. To do so, a standardized PCA is performed on X.

Remark. Since by construction in the PCA, the columns (*i.e.*, the consumers) are at least mean centered, no pre-processing of the data is required here. Consequently, two products are all the more close that they are preferred the same way by the consumers.

The data set we are interested here is the transposed version of the *newmat* data set generated previously. To transpose this matrix, the **t** function is used. Let's perform a PCA on this transposed matrix, followed by HAC on the dimensions resulting from the PCA:

```
> data.mdpref <- t(newmat)
> mdpref <- PCA(data.mdpref,ncp=Inf)
```

The first dimension of the PCA (*cf.* Figure 8.9) opposes *J'adore ET* and *J'adore EP* to *Shalimar*. In other words, consumers who preferred *J'adore EP* tend to also like more *J'adore ET*, and tend to appreciate less *Shalimar*, and vice versa. *J'adore ET* and *J'adore EP* seem to have been preferred similarly, but very differently from *Shalimar*.

The variable representation of the PCA (*cf.* Figure 8.10) adds the consumer information in the interpretation. Here, a strong agreement amongst consumers is observed, as most of the variables are positively correlated with

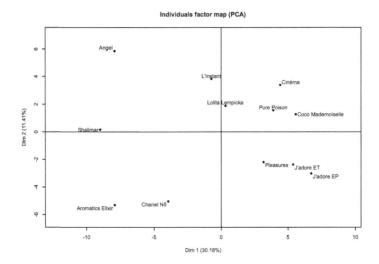

FIGURE 8.9
Representation of the perfumes on the first two dimensions resulting from
PCA, using the **PCA** function (*data.mdpref* data set).

the first dimension. This agreement confirms the strong *Product* effect already
shown by ANOVA. The conjoint interpretation of Figure 8.9 and Figure 8.10
allows to say that the perfumes with positive coordinates on the first dimension (*e.g.*, *J'adore ET* and *J'adore EP*) are the ones that are preferred by the
majority of the consumers.

   If we now run the **HCPC** function of the FactoMineR package on the outputs resulting from the PCA, we can obtain groups of products similar in
terms of the way they have been preferred (*cf.* Figure 8.11).

```
> prod.hcpc <- HCPC(mdpref)
```

   By cutting the dendrogram at the level proposed by the **HCPC** function (6
groups of products are considered), let's notice that the cluster corresponding
to the most appreciated perfumes (*cf.* Figure 8.11) is very similar to the one
proposed by the Fisher's LSD test.

8.2.3   How can I get homogeneous clusters of consumers?

Considering clusters of consumers is of utmost importance, in product optimization. Indeed, as mentioned in Section 8.1, liking scores are related to the
affective evaluation of the products, and since it takes all kinds to make a
world, it is important to take into account the variability of the consumers in
terms of preferences. Hence, to increase the accuracy of the analysis and of

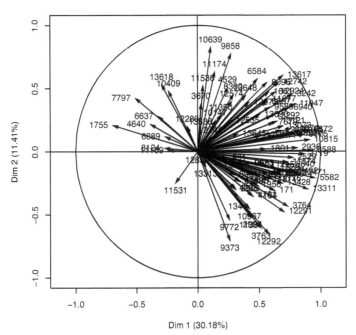

FIGURE 8.10
Representation of the consumers on the first two dimensions resulting from PCA, using the **PCA** function (*data.mdpref* data set).

the improvement guidelines, it is of utmost importance to determine the best product within homogeneous segments of consumers.

The variable representation in the MDPref already provides an idea of how homogeneous in terms of preference the panel of consumers is: if all the consumers (as variables) are highly positively correlated together, the consumers panel is quite homogeneous. However, if the consumers are defining clear separate structures, or are scattered all around the correlation circle, then it is necessary to segment the consumers since different preference patterns are observed.

Since the focus of this analysis is on the consumers, they are considered here as the statistical units of interest. *De facto*, the matrix considered in this part is the transposed of X as previously defined, with J rows and I columns. As evoked previously, data are mean centered by rows, and consumers are compared regarding the products they prefer. Such data set corresponds to the matrix *newmat_centered* defined previously. As in Section 8.2.2, let's evaluate the variability between consumers by performing a PCA on this data set. Here

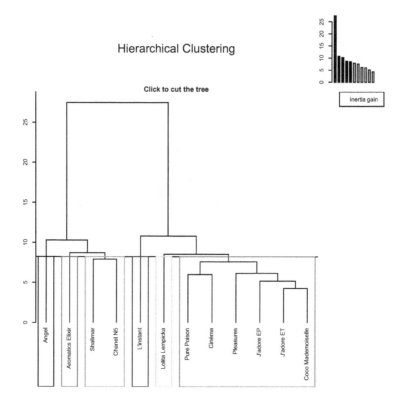

FIGURE 8.11
Representation of the perfumes according to the cluster they belong to (dendrogram) resulting from HAC, using the **HCPC** function (*data.mdpref* data set).

again, the results thus obtained can be completed by applying the **HCPC** function of the FactoMineR package on the consumer space thus obtained.

```
> center.pca <- PCA(newmat_centered,graph=FALSE)
> cons.hcpc <- HCPC(center.pca)
```

The **HCPC** function provides an interactive graphical output of the dendrogram that allows users to choose the number of clusters they want to consider (*cf.* Figure 8.12). Users have to click on the output depending on the number of clusters they want to consider. A "natural" cut is proposed by the function: once the number of groups is defined (here, we consider 2 groups of consumers although the **HCPC** function suggests 4), the function displays a representation of the consumers (*cf.* Figure 8.13). This representation corresponds to the first plane of the PCA already performed previously, in which the clusters of consumers are highlighted using a color code.

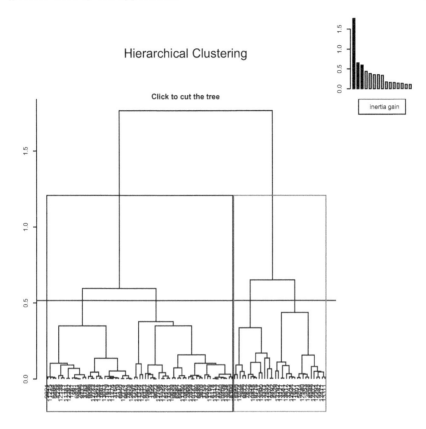

FIGURE 8.12
Representation of the consumers according to the cluster they belong to resulting from HAC, using the **HCPC** function (*newmat_centered* data set).

The conjoint interpretation of Figure 8.6 and Figure 8.13 allows to say that consumers from cluster 1 are the ones who preferred perfumes such as *Shalimar* and *Chanel N5*, by opposition to consumers from cluster 2 who preferred perfumes such as *Pleasures* and *J'adore ET*.

This visual interpretation of the clusters can be done *automatically* with the numerical outputs provided by the **HCPC** function. In particular the description of the groups by both the individuals and the variables of the data set.

```
> names(cons.hcpc)
[1] "data.clust" "desc.var"   "desc.axes"  "call"       "desc.ind"
```

The first object of the list, **res.hcpc$data.clust**, contains the original data set plus a new categorical variable that corresponds to the index of the

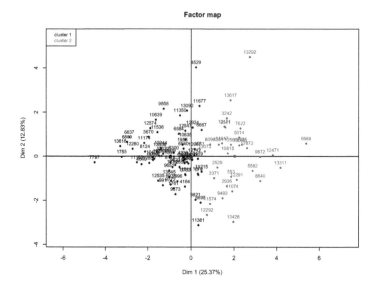

FIGURE 8.13

Representation of the consumers on the first two dimensions resulting from PCA, using the **HCPC** function (*newmat_centered* data set).

group a consumer belongs to. If we apply the **summary** function on this data set (restrained to the last variables) the number of consumers belonging to each group is shown.

```
> summary(cons.hcpc$data.clust[,10:13])
   Pleasures          Pure Poison          Shalimar          clust
 Min.   :-3.9167   Min.   :-2.91667   Min.   :-4.08333    1:70
 1st Qu.:-0.5833   1st Qu.:-0.08333   1st Qu.:-2.33333    2:33
 Median : 0.3333   Median : 0.83333   Median :-1.16667
 Mean   : 0.2832   Mean   : 0.64240   Mean   :-1.20226
 3rd Qu.: 1.2917   3rd Qu.: 1.45833   3rd Qu.: 0.04167
 Max.   : 3.2500   Max.   : 4.58333   Max.   : 2.41667
```

According to this output, there are two groups of consumers, a first one with 70 consumers, a second one with 33 consumers.

The second object of the list, res.hcpc$desc.var, contains a description of each cluster with respect to all the variables of the data set, whether they are active or illustrative, quantitative or qualitative.

The first object of the list, res.hcpc$desc.var$quanti$'1', corresponds to the description of the first cluster of consumers with respect to the quantitative variables of the data set.

```
> round(cons.hcpc$desc.var$quanti$'1',3)
          v.test Mean in category Overall mean sd in category
Shalimar   5.671           -0.567       -1.202          1.415
```

```
Angel                   4.552        -0.581       -1.154        1.614
Aromatics.Elixir        4.541        -0.867       -1.377        1.378
Chanel.N5               4.447        -0.024       -0.532        1.378
Pure.Poison            -3.844         0.276        0.642        1.278
Pleasures              -4.167        -0.124        0.283        1.266
J'adore.ET             -4.545         0.190        0.642        1.267
Coco.Mademoiselle      -5.019         0.305        0.778        1.085
J'adore.EP             -5.295         0.376        0.895        1.300
                     Overall sd p.value
Shalimar                1.649        0
Angel                   1.851        0
Aromatics.Elixir        1.653        0
Chanel.N5               1.682        0
Pure.Poison             1.401        0
Pleasures               1.436        0
J'adore.ET              1.462        0
Coco.Mademoiselle       1.388        0
J'adore.EP              1.441        0
```

In this output, a closer look at the *V-tests* in the first column confirms that consumers belonging to that cluster *significantly* preferred products such as *Shalimar, Angel,* and *Chanel N5* (compared to consumers from cluster 2).

```
> round(cons.hcpc$desc.var$quanti$'2',3)
                  v.test Mean in category Overall mean sd in category
J'adore.EP         5.295            1.995        0.895          1.055
Coco.Mademoiselle  5.019            1.783        0.778          1.425
J'adore.ET         4.545            1.601        0.642          1.385
Pleasures          4.167            1.146        0.283          1.395
Pure.Poison        3.844            1.419        0.642          1.333
Chanel.N5         -4.447           -1.611       -0.532          1.759
Aromatics.Elixir  -4.541           -2.460       -1.377          1.666
Angel             -4.552           -2.369       -1.154          1.731
Shalimar          -5.671           -2.551       -1.202          1.248
                  Overall sd p.value
J'adore.EP         1.441            0
Coco.Mademoiselle  1.388            0
J'adore.ET         1.462            0
Pleasures          1.436            0
Pure.Poison        1.401            0
Chanel.N5          1.682            0
Aromatics.Elixir   1.653            0
Angel              1.851            0
Shalimar           1.649            0
```

For cluster 2, the *V-tests* confirm that consumers belonging to that cluster *significantly* preferred products such as *Pleasures, J'adore ET,* and *J'adore EP* (compared to consumers from cluster 1).

Finally, the last object of the list, res.hcpc$desc.ind, contains a description of the clusters with respect to the individuals of the data set, from the most typical (at the center of the class) to the least typical (most distant to the center of the class).

```
> cons.hcpc$desc.ind
$para
cluster: 1
    12072        8118       13313        9651       12656
0.5324024 0.6499575 0.6651064 0.7249414 1.0367414
------------------------------------------------------------
cluster: 2
     5582       10815        5014        1801        9775
1.335172 1.564519 1.574419 1.652632 1.684653

$dist
cluster: 1
     7797       13618        6637        1755        6889
6.848122 5.855288 5.658590 5.599939 5.315219
------------------------------------------------------------
cluster: 2
     8588       13292       13311       12471       11074
6.402092 5.858666 5.435772 5.188151 5.145778
```

Since the results of the **HCPC** function helps comparing the clusters together, it seems interesting to complete these results with a visual output, which shows the liking/preference trends within and between clusters. To do so, we propose to compute for each cluster the average liking score per product. Hence, a first step consists in combining the cluster information to the raw hedonic scores stored in *newmat*. Such information can be obtained using the following code:

```
> clusters <- as.matrix(cons.hcpc$data.clust$clust)
> rownames(clusters) <- rownames(cons.hcpc$data.clust)
> colnames(clusters) <- "Cluster"
> cons.clust <- merge(x=newmat,y=clusters,by=0)
> rownames(cons.clust) <- cons.clust[,1]
> cons.clust <- cons.clust[,-1]
```

The average score by product for each cluster of consumers can be obtained with the **aggregate** function.

```
> cluster.mean <- aggregate(cons.clust[,1:12],by=list(cons.clust[,13]),
+ FUN="mean")[,-1]
> round(cluster.mean,2)
  Angel Aromatics Elixir Chanel N5 Cinéma Coco Mademoiselle J'adore EP
1  5.31            5.03   5.87      6.46              6.20         6.27
2  2.88            2.79   3.64      6.30              7.03         7.24
  J'adore ET L'instant Lolita Lempicka Pleasures Pure Poison Shalimar
1       6.09     6.23            6.01      5.77        6.17     5.33
2       6.85     4.97            5.52      6.39        6.67     2.70
```

Finally, we can represent graphically the differences in terms of liking between clusters. To do so, the following code is used:

```
> plot(1:12,cluster.mean[1,],ylim=c(1,9),xlab="",ylab="Average Liking Scores",
+ main="Mean by clusters",col=1,lwd=2,type="l",axes=FALSE,cex.lab=0.7)
> lines(1:12,cluster.mean[2,],lwd=2,col=2)
> axis(1,at=1:12,label=colnames(cluster.mean),cex.axis=0.55,las=2)
> axis(2,cex.axis=0.7)
> legend("topleft",legend=paste("Cluster",1:2),bty="n",col=1:2,lwd=2,cex=0.7)
```

Remark. Graphically, the comparison between clusters can only be done if both clusters use the scale of notation (in overall) similarly. Otherwise, the comparison is biased towards the clusters which uses a higher part of the scale. To make sure that the two clusters use the scale in the same way, the following code can be used:

```
> mean(as.matrix(cons.clust[cons.clust[,13]=="1",-13]))
[1] 5.895238
> mean(as.matrix(cons.clust[cons.clust[,13]=="2",-13]))
[1] 5.247475
```

As expected, the two clusters are using sensibly the same part of the scale: the liking scores can be compared here between clusters.

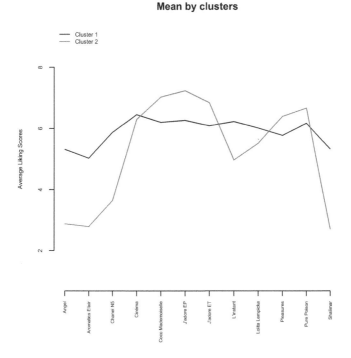

FIGURE 8.14
Representation of the average liking scores per product and by cluster, using the **plot** function (*cons.clust* data set).

From Figure 8.14, it can be seen that cluster 1 accepts more *Angel* and *Shalimar* than Cluster 2, which appreciates more *J'adore EP* (separation between broken lines). However, for cluster 1, *Angel* and *Shalimar* for instance are not the preferred products, since *Cinéma* is associated with the highest liking score.

8.3 For experienced users: Dealing with multiple hedonic variables and supplementary consumer data

In the present study, 80 consumers evaluated eight lipsticks, selected according to two important factors that were supposed to influence consumers' buyer behavior. The first one was the color of the lipstick: for this very important factor, we decided to choose four amongst the best-selling colors in the French market, *i.e.*, pink, red, brown, and prune. The second factor was whether the lipstick was glossy or matt.

From a hedonic point of view, each product was evaluated by each consumer according to five hedonic variables:

- How do you like the texture of the lipstick?

- How do you like the odor of the lipstick?

- How do you like the taste of the lipstick?

- How do you like the color of the lipstick?

- How do you like the lipstick?

In terms of usage and attitude questions, 39 questions related to make-up in general, and 18 questions related to lipstick in particular were asked.

- How do you apply your make-up?

- What were the reasons to use make-up for the first time?

- ...

- Why would you wear make-up?

- For whom would you wear lipstick?

Finally, 60 more questions related to the personality of the consumer were asked: from 1 (never) to 7 (always) would you say that you are jealous? sincere? *etc.*

The data were collected by Charlène Benniza, Lisa Defeyter, and Maud Leportier during their master's degree studies[1].

---

[1]They were awarded for their poster entitled *"Intimate Projective Mapping: Combining context and emotional states to understand consumer behavior,"* at the International Conference on Correspondence Analysis and Related Methods (CARME) in 2011. They also received the Prix spécial du jury for the 2011 edition of the Syntec trophy, which aim is to reward the best master's degree thesis in the field of market research and public opinion polling.

8.3.1 Dealing with multiple hedonic variables

One of the main feature of this data set is the fact that different hedonic aspects have been collected. At some point, these different hedonic variants have to be considered in the analysis. In this section, we will show how it can be done to get clusters of consumers such as two consumers belong to the same cluster if they have the same preferences. Regarding this problematic, the statistical units of interest are the consumers, and the variables are the hedonic variants. In other words, the core of this segmentation lies in analyzing a matrix of dimension 80×40, where 80 is the number of consumers and 40 the total number of hedonic scores given by a consumer.

For the analysis of such data set, at least two points of view can be adopted, each one being associated with a specific multiple data table structure. A first one, consisting in considering 8 submatrices of 5 columns, each submatrix corresponding to a lipstick, each column corresponding to one hedonic aspect. The second point of view consists in considering 5 submatrices of 8 columns, each submatrix corresponding to a hedonic aspect, each column corresponding to one lipstick.

As usual, MFA is used to balance the role of each group within a global analysis. The analysis of the first multiple data table by MFA is such as the role of each lipstick in the multivariate description of the consumers in terms of preferences is balanced. An MFA on the second multiple data table would balance the role of each hedonic aspect in the multivariate description of the consumers in terms of preferences. No option is better than another, but each one corresponds to a specific question. The second analysis is less common and functional than the first one, but it should provide rich and subtle insights on how consumers behave.

Let's focus on the first point of view, the most practical one in terms of product development, and let's see how to perform such analysis with R. To do so, let's import the *lipsticks.csv* file from the book website.

```
> lipsticks  <- read.table("lipsticks.csv",header=TRUE,sep=",",dec=".",
+ quote="\"")
```

After loading the FactoMineR package (if required), let's run a MFA on the liking scores. In this analysis, we consider 11 groups of variables, some of them being projected as supplementary. The 8 first groups are considered as active, whereas the 3 last groups are considered as illustrative (num.group.sup=9:11). The 8 first groups are made up of 5 variables, the 3 last groups are made up of 39, 18, and 60 variables, respectively (group=c(rep(5,8),39,18,60)). The variables of the 8 first groups are scaled to unit variance, as the ones from the last group; the variables from the 2 remaining groups are considered as categorical (type=c(rep("s",8),"n","n","s")).

```
> mfa.lipsticks <- MFA(lipsticks,type=c(rep("s",8),"n","n","s"),
+ group=c(rep(5,8),39,18,60),name.group=c("RedMatt","PinkMatt","PruneGlossy",
+ "RedGlossy","BrownGlossy","BrownMatt","PinkGlossy","PruneMatt",
+ "Usages make-up","Usages lipstick","Self description"),num.group.sup=9:11)
```

The interpretation of the results remains to be done, and we highly recommend the readers the do so by using all the MFA's outputs presented throughout the book.

8.3.2   Dealing with supplementary consumer data

In this part, for simplification purposes and also because it is the most common situation, we will limit ourselves to one hedonic aspect to get clusters of consumers: the color of the lipstick (since it is the most determinant one from our experience). Once clusters are obtained, we focus our attention on the interpretation of the clusters by adding supplementary consumer data, notably the usage and attitude questions, and the self-description questions.

To do so, we run a PCA on the data set *lipsticks* by considering as active variables the assessment of the color for each product, and by considering as illustrative categorical variables the usage and attitude questions (`quali.sup=9:65`), and as illustrative continuous variables the personality questions (`quanti.sup=66:125`). Note that since many variables are discarded from the analysis (only the hedonic variables specific to the color are selected), the first qualitative variables becomes the 9th column in our data set.

```
> pca.lipsticks <- PCA(lipsticks[,c(4,9,14,19,24,29,34,39,41:157)],
+ quali.sup=9:65,quanti.sup=66:125)
```

Then, we apply the **HCPC** function on the dimensions resulting from the PCA, to get homogeneous clusters of consumers. In this particular case, we set the parameter `proba` to 0.1, in order to be less restrictive regarding the tests used to automatically describe the clusters, and *de facto* to enlarge the description of our clusters. In this example, to determine the number of clusters, we used the *natural cut* proposed by the **HCPC** function (`nb.clust=-1`).

```
> res.hcpc <- HCPC(pca.lipsticks,nb.clust=-1,proba=0.1)
```

As the clusters are based on preferences and as preferences were naturally considered as quantitative variables, let's first have a look at the description of the clusters with respect to the quantitative variables. For illustration purposes, we will limit ourselves to the first two clusters.

```
> round(res.hcpc$desc.var$quanti$'1',3)
```

	v.test	Mean in category	Overall mean
ColorRedMatt	5.178	0.605	-1.573
ColorRedGlossy	5.150	1.105	-1.090
ColorPruneGlossy	3.991	0.912	-0.631
ColorPruneMatt	2.567	-1.011	-1.832
Sense.of.competition	2.168	4.346	3.763

Willingness.to.take.a.position	1.840	5.000	4.612
Unpredictable	1.805	3.983	3.557
Athletic	-1.701	3.099	3.582
ColorPinkMatt	-1.728	-1.318	-0.681
Calm.voice	-1.743	4.077	4.513
Avoid.being.vulgar	-1.927	4.385	4.912
Conscientious	-2.731	5.498	5.937
ColorBrownGlossy	-4.269	-1.549	0.324
ColorPinkGlossy	-4.462	-1.165	0.603
ColorBrownMatt	-4.857	-1.934	0.046

	sd in category	Overall sd	p.value
ColorRedMatt	2.433	2.593	0.000
ColorRedGlossy	1.927	2.629	0.000
ColorPruneGlossy	1.555	2.386	0.000
ColorPruneMatt	1.704	1.972	0.010
Sense.of.competition	1.568	1.660	0.030
Willingness.to.take.a.position	0.961	1.299	0.066
Unpredictable	1.521	1.456	0.071
Athletic	1.388	1.751	0.089
ColorPinkMatt	1.701	2.274	0.084
Calm.voice	1.466	1.541	0.081
Avoid.being.vulgar	1.841	1.690	0.054
Conscientious	1.151	0.992	0.006
ColorBrownGlossy	2.575	2.707	0.000
ColorPinkGlossy	1.957	2.443	0.000
ColorBrownMatt	1.846	2.514	0.000

In terms of preferences, consumers from cluster 1 seem to have mostly appreciated red and prune lipsticks: the glossiness of the lipstick does not seem to be that important compared to the color. Most importantly, we can see that questions related to the self-description of the consumer are characterizing that first cluster. This result is remarkable, as these questions were considered as illustrative variables, and did not take part actively in the calculation of the distances amongst consumers (based on preference data only). Consumers of this cluster, who rather like red and prune lipstick, depicted themselves as having a high sense of competition, as willing to take a position, and also as quite unpredictable, compared to the other consumers[2] (the mean of these variables for this cluster is significantly higher than the overall mean).

```
> round(res.hcpc$desc.var$quanti$'2',3)
```

	v.test	Mean in category	Overall mean	sd in category
ColorPinkMatt	6.299	1.709	-0.681	1.376
ColorPinkGlossy	3.080	1.859	0.603	1.679
Defend.their.opinions	-1.735	5.320	5.625	1.256
Gentle	-1.864	5.040	5.375	1.113
Quick.heal.wounds	-1.883	4.240	4.650	1.274
Theatrical	-1.933	2.520	3.112	1.769
Happy	-2.221	5.360	5.713	1.015
ColorRedGlossy	-2.438	-2.160	-1.090	2.342
Feminine	-2.514	5.280	5.671	1.114
ColorPruneGlossy	-2.562	-1.651	-0.631	2.486

---

[2]These results are all relative, in the sense that they have been obtained relatively to the other consumers from other clusters.

```
Strong.personality        -2.680             4.400        5.000        1.497
ColorRedMatt              -4.342            -3.451       -1.573        1.416
                      Overall sd p.value
ColorPinkMatt              2.274   0.000
ColorPinkGlossy            2.443   0.002
Defend.their.opinions      1.053   0.083
Gentle                     1.077   0.062
Quick.heal.wounds          1.305   0.060
Theatrical                 1.837   0.053
Happy                      0.951   0.026
ColorRedGlossy             2.629   0.015
Feminine                   0.932   0.012
ColorPruneGlossy           2.386   0.010
Strong.personality         1.342   0.007
ColorRedMatt               2.593   0.000
```

Consumers from cluster 2 seem to prefer pink lipsticks. These consumers depicted themselves as not having a strong personality, as not feminine, as not happy; they don't heal quickly when they are hurt, and they don't tend to defend their opinions. Once again, these results are remarkable as the personality questions were considered as illustrative and as the description of the consumers seems to be really consistent.

In terms of usage and attitude questions, the description of the clusters suggests that the consumers of cluster 2, who appeared to be *fragile*, *sensitive*, and certainly *discreet* also reported to use make-up to be *another woman*, to be *looked at*, and to *feel more confident*.

```
> round(res.hcpc$desc.var$category$'2',3)
                                                 Cla/Mod Mod/Cla Global
Why.make.up.to.be.another.woman=yes              100.000      12   3.75
Why.make.up.to.please=yes                         43.750      56  40.00
What.is.the.main.occasion=to hang out             50.000      36  22.50
Begin.make.up.to.be.looked.at=yes                 50.000      36  22.50
Why.lipstick.to.feel.more.confident=yes          100.000       8   2.50
How.do.you.apply.makeup.on.your.skin=very little  47.368      36  23.75
How.often.do.you.use.lip.pencil=+1 per week        0.000       0   7.50
Why.lipstick.to.feel.more.confident=no            29.487      92  97.50
Begin.make.up.to.be.looked.at=no                  25.806      64  77.50
Why.make.up.to.please=no                          22.917      44  60.00
Why.make.up.to.be.another.woman=no                28.571      88  96.25
How.often.do.you.use.foundation=each day          13.793      16  36.25
                                                 p.value v.test
Why.make.up.to.be.another.woman=yes                0.028  2.197
Why.make.up.to.please=yes                          0.057  1.906
What.is.the.main.occasion=to hang out              0.065  1.845
Begin.make.up.to.be.looked.at=yes                  0.065  1.845
Why.lipstick.to.feel.more.confident=yes            0.095  1.670
How.do.you.apply.makeup.on.your.skin=very little   0.099  1.650
How.often.do.you.use.lip.pencil=+1 per week        0.096 -1.662
Why.lipstick.to.feel.more.confident=no             0.095 -1.670
Begin.make.up.to.be.looked.at=no                   0.065 -1.845
Why.make.up.to.please=no                           0.057 -1.906
Why.make.up.to.be.another.woman=no                 0.028 -2.197
How.often.do.you.use.foundation=each day           0.011 -2.540
```

These interpretations are certainly fragile (as consumers from cluster 2), but need to be meditated due to their high consistency.

8.4 Exercises

Exercise 8.1 *On the impact of external information on hedonic scores*

This exercise addresses an important issue in consumer behavior: the impact of emotions on liking, and most particularly one of the most important emotion for a human being, *Love*. To answer this question we brought in 77 couples in love (one man, one woman). For each couple, the core of the experiment was based on the assessment of 10 luxury perfumes for men. This assessment was conducted in four steps:

1. A first one where the man was asked to give a hedonic judgement on a set of ten perfumes.

2. A second one where the woman was also asked to give a hedonic judgement on the same set of ten perfumes, but labelled with different digit numbers than the one used for the man.

3. A third one where the woman explains to the man her preferences with respect to her labels.

4. Finally, a fourth step where the man has to assess for a second time the set of ten perfumes but with the labels used for the woman and most important with the liking scores and the comments given by the woman.

At no time the man is supposed to know that he's assessing twice the same perfumes, nor the woman that she's assessing the same products as the man. The idea of this exercise is to assess the impact of one's beloved on their hedonic scores.

- Import the *perfumes_before_after.csv* file from the book website.

- Select the subset of data related to the men only. Make sure that the selection of subset is done correctly by updating the levels of the *consumer* variable.

- Identify the ANOVA model that will allow you to answer to the question of the impact of external information on the hedonic scores.

- Apply this model and run the ANOVA. For simplicity, we recommend you to use the **AovSum** function of the FactoMineR to run your ANOVA model. Comment on your results.

Exercise 8.2 *What gender has to do with liking?*

As a follow up to Exercise 8.1, we are now interested in understanding the differences in liking between gender.

- Import the *perfumes_before_after.csv* file from the book website.

- Since we are interested in the first evaluation of the product, first select a subset of the data that only corresponds to the evaluation before.

- Split this subset into two groups, each group corresponding to one gender. Update in each subset the consumer levels.

- For each gender, run a two-way ANOVA. Using the agricolae package, run *post hoc* tests using the **LSD.test** function, and the **HSD.test** function.

- Within a gender, compare the results between *post hoc* tests. Conclude.

- Within a *post hoc* test, compare the results between genders. Conclude.

- Finally, using the first subset of data generated (including before only), run a nested ANOVA model including the product effect, the gender effect, the consumer effect within the gender (noted gender/consumer in R), and the interaction between product and gender. Conclude.

Exercise 8.3 *When data are missing. . .*

The data used for this exercise have been collected in Pakistan and in France. In each country, a set of eight biscuits, composed of 4 Pakistani biscuits and 4 French biscuits, has been assessed from a hedonic point of view by a panel of consumers. Unfortunately, some consumers didn't taste all 8 products. The main idea of this exercise is to show how missing data can be handled, then how results from MDPref issued from different panels can be compared.

- Import the *biscuits_hedo.csv* file from the book website.

- For each country separately (*Nationality*), rearrange the data and create the matrix crossing the products in rows and the consumers in columns, and including the raw hedonic scores.

- Remove from each data set the consumers who have not expressed any preferences (same liking scores for all products).

- As can be seen, the two blocks of data include missing data. Estimate these missing data using the **imputePCA** function of the missMDA package. In this case, we recommend to use 2 components.

- Create the preference map for each country separately. Compare the results.

- Merge the two data sets together, and run an MFA using the settings of your choice. Justify your settings and conclude.

Exercise 8.4 *Tell me who you are and I will tell you what you like*

As a follow up to Exercise 8.3, the idea of this exercise is to identify the right ANOVA model in order to compare the liking from one country to the other.

- Import the *biscuits_hedo.csv* file from the book website.

- Without making any distinction between products or consumers, do French (*resp.* Pakistani) consumers prefer French (*resp.* Pakistani) biscuits? To evaluate this, define the right ANOVA model before running the analysis. The **AovSum** function of the FactoMineR package can be used here.

- Following the same idea, run another analysis by including the product differences within each country, and including the consumer effect. In this case, the *Origin* effect is ignored, and the **aov** function is used. Conclude.

Exercise 8.5 *Defining and characterizing homogeneous clusters of consumers*

Let's consider the same situation as in Exercise 8.3 and in Exercise 8.4, but let's ignore the fact that the panels come from two different countries.

- Import the *biscuits_hedo.csv* file from the book website.

- Create the matrix crossing the products in rows, and the consumers in columns, and including the raw hedonic scores.

- Remove the consumers who did not express any preferences (no variability).

- Using the **imputePCA** function of the missMDA package, estimate the remaining missing data.

- Perform a PCA on the resulting matrix after transposing it.

- Apply the **HCPC** function of the FactoMineR package to the results of your PCA. Define the number of clusters that seems suitable for you.

- Compare numerically and graphically the difference in terms of preference between clusters.

- Can these differences be explained by external information? Conclude and compare to the results obtained in Exercises 8.3 and 8.4.

8.5   Recommended readings

- Couronne, T. (1996). Application de l'analyse factorielle multiple à la mise en relation de donnes sensorielles et de données de consommateurs. *Sciences des Aliments*, 16, 23-35.

- de Mendiburu, F. (2013). agricolae: statistical procedures for agricultural research. R package version 1.1-4, `http://CRAN.R-project.org/package= agricolae`.

- Ennis, D.M. (2005). Analytic approaches to accounting for individual ideal point. IFPress, 82, 2-3.

- Ennis, D. M., & Ennis, J. M. (2013). Mapping hedonic data: a process perspective. *Journal of Sensory Studies*, 28, 324-334.

- Husson, F., Lê, S., & Pagès, J. (2011). Exploratory Multivariate Analysis by Example Using R, Chapman & Hall/CRC Computer Science & Data Analysis.

- Lê, S., Husson, F. & Pagès, J. (2006). Another look at sensory data: How to have your salmon and eat it, too!. *Food Quality and Preference*, 17, 658-668.

- Lê, S., Pagès, J., & Husson, F. (2008). Methodology for the comparison of sensory profiles provided by several panels: application to a cross-cultural study. *Food Quality and Preference*, 19, (2), 179-184.

- Lim, J. (2011). Hedonic scaling: A review of methods and theory. *Food Quality and Preference*, 22, 733-747.

- Moskowitz, H., & Krieger, B. (1993). The contribution of sensory liking to overall liking: an analysis of six food categories. *Food Quality and Preference*, 6, 83-90.

- Pagès, J., Bertrand, C., Ali, R., Husson, F., & Lê, S. (2007). Sensory analysis comparison of eight biscuits by French and Pakistani panels. *Journal of Sensory Studies*, 22, (6), 665-686.

- Rousseau, B., Ennis, D. M., & Rossi, F. (2012). Internal preference mapping and the issue of satiety. *Food Quality and Preference*, 24, 67-74.

- Steel, R., Torry, G., & Dickey, D. (1997). *Principles and procedures of statistics: a biometrical approach*. McGraw-Hill Inc.,US. Third Edition.

- Thomson, D. M. H., Crocker, C., & Marketo, C. G. (2010). Linking sensory characteristics to emotions: An example using dark chocolate. *Food Quality and Preference*, 21, 1117-1125.

9

When products are described by both liking and external information "independently"

CONTENTS

"... the two information could be analysed separately. However, it is once combined through the use of statistics, that they take all their interest. Indeed, if we can explain, through sensory characteristics, why products are liked or disliked (defining the drivers of liking and disliking), we can guide product developers in improving the products. In fine, we can even provide the location, on the sensory space, of the potential optimum product. This is the aim of the so-called external preference mapping *(PrefMap), that can be performed using the* **carto** *function of the* **SensoMineR** *package."*

9.1   Data, sensory issues, and notations

In the previous chapters, the focus every time was on one particular type data, whether it was the sensory description of the products (as in Chapter 2) or the acceptance of the products by the consumers (through the hedonic scores, as in Chapter 8). If both sets of information are of utmost importance, they take their full power once combined. What if we could explain the differences in the preferences with the sensory properties? If such information were available, the Research and Development would be in position to create better products, *i.e.*, products with a higher chance of being successful.

However, and as it has been shown in the previous chapters, the different information regarding the products are provided by different panels. The sensory profiles are usually generated with trained panelists, whereas the hedonic scores are provided by consumers. In this situation, the only point in common is the products: it is then natural to consider each product as a statistical unit.

Since the products are described according to different types of data, the variables are defined in multiple matrices. The first matrix noted H is related to the consumers' data and contains the liking scores. This matrix usually considers one consumer as one variable. Hence, at the intersection of row i and column j, the value h_{ij} corresponds to the hedonic score provided by consumer j to product i[1]. As the information is more dense, we consider here the entire matrix of individual liking scores.

The second matrix is related to the expert panel and contains the sensory profile of the products. This matrix X hence contains the mean scores by products on the different sensory attributes. In other words, at the intersection of row i and column j, the value x_{ij} corresponds to the intensity of product i on the sensory attribute j.

The main aim of this analysis is hence to combine the two matrices H and X by using the products as anchor points, and to assess the relationship between the two sets. In other words, we expect from the analysis that it explains as much as possible the differences in preferences between products (matrix H) by using the sensory characteristics of the products (matrix X). To some further extent, we expect to have information on what is the most important sensory characteristics that can explain liking at the panel of consumers level, for each cluster of consumers, and at the individual level. Additionally, if links can be made, we expect to get information on a potential optimum product, *i.e.*, a product that would be appreciated by a maximum of consumers. Such optimal product (often referred to as *ideal* product) is then used as reference to match in the optimization procedure.

It is based on these expectations that the following chapter is organized. First, the focus is on the hedonic data, and we attempt to explain the differences in the products preferences using the sensory characteristics of the products. This can be done by projecting as supplementary variables the sensory profiles of the products within the preference space. This is the core of the *internal preference mapping* technique, also known as MDPref, which is tackled in Section 9.2.1. Since in MDPref, only linear relationships are measured between the sensory characteristics and the hedonic scores, simple and

---

[1]In some cases, this matrix is summarized into one or few variables, which then correspond to the mean liking scores by product across all consumers, or across consumers who belong to a homogeneous cluster.

quadratic regressions are computed. This procedure allows defining the so-called drivers of liking and disliking, *i.e.*, the characteristics that seem to play a role in the appreciation of the products. The definition of drivers of liking is presented in Section 9.2.2.

Finally, we attempt to predict an optimal product through individual models. In this case, the focus is on the sensory characteristics of the products; and for each potential point of the product space, the way each consumer would appreciate that product if it happens to exist is estimated. Based on these individual models, a response surface with density lines showing the percentage of consumers that would accept the product is created. This is the core of the so-called external preference map (or PrefMap) that is presented in Section 9.2.3.

As explained in the following sections, both the MDPref and PrefMap have the drawback that the focus is on one matrix of data only (hedonic for MD-Pref, sensory for PrefMap), the second matrix being then "projected" on the first one. For that reason, we propose another methodology called PrefMFA in Section 9.3 (*For experienced users*), which considers the best common configuration between the hedonic and the sensory one, before performing the PrefMap regression routine.

9.2 In practice

9.2.1 How can I explain the differences in preferences using sensory data?

The data sets used to illustrate this methodology are the same as in Chapter 8 (hedonic data) and in Chapter 2 (sensory data), as they correspond to the hedonic and sensory evaluation of the same 12 perfumes. Since a small modification in the product names was required, use the *perfumes_liking_conso.csv* file for the hedonic data, and the *perfumes_qda_experts2.csv* file for the sensory data. Both files can be downloaded from the book website.

Import the two data sets in your R session, and arrange the data so that they are suitable.

```
> sensory <- read.table("perfumes_qda_experts2.csv",header=TRUE,sep=",",
+ dec=".",quote="\"")
> sensory$Session <- as.factor(sensory$Session)
> sensory$Rank <- as.factor(sensory$Rank)
> head(sensory)
  Panelist Session Rank       Product Spicy Heady Fruity Green Vanilla
1       S0       1    1 Coco Mademoiselle  0.6   0.7    7.1   0.8     2.0
2       S0       1    2 Lolita Lempicka   1.4   1.5    3.2   1.3     5.3
3       S0       1    3           Angel   3.8   9.7    1.0   0.6     1.9
4       S0       1    4     Pure Poison   1.1   1.2    7.4   0.2     2.5
5       S0       1    5       Chanel N5   4.9   8.4    3.0   0.3     0.1
```

```
6        S0       1    6  Aromatics Elixir   8.6  10.0    0.0   0.0    3.1
  Floral Woody Citrus Marine Greedy Oriental Wrapping
1     8.6   0.7    0.3    3.4    0.8    0.9      3.8
2     4.4   1.1    1.8    1.0    9.5    1.0      7.9
3     3.6   0.6    0.7    0.5    9.8    0.7      7.3
4     9.5   0.8    1.0    0.8    3.8    7.0      8.9
5     4.9   3.3    0.0    0.0    1.9    9.0      9.5
6     5.0   1.2    0.0    0.0    0.0    6.0      8.5
> hedonic <- read.table("perfumes_liking_conso.csv",header=TRUE,sep=",",
+ dec=".",quote="\"")
> hedonic$consumer <- as.factor(hedonic$consumer)
> head(hedonic)
  consumer           product liking
1      171             Angel      3
2      171  Aromatics Elixir      3
3      171         Chanel N5      7
4      171            Cinéma      6
5      171 Coco Mademoiselle      6
6      171          Jadore EP      5
```

As highlighted by the **head** function, both the *sensory* and *hedonic* data sets are not in the format expected. Indeed, as explained in Section 9.1, the hedonic matrix should have the products in rows and the consumers in columns, whereas the sensory matrix should correspond to the sensory profiles, *i.e.*, a matrix with the products in rows and the sensory attributes in columns.

The transformation of the hedonic matrix in the right format can be done manually within Excel by copying and pasting each subject one next to the other. However, such a manual procedure is tedious and easily subject to mistakes (for instance, pasting the wrong products together if the data set is not sorted correctly). For this reason, we propose to do it automatically in R. This short line code requires to perform a `for` loop on the *consumer* variable:

```
> conso <- levels(hedonic$consumer)
> nbconso <- length(conso)
> product <- levels(hedonic$product)
> hedonic.c <- matrix(0,length(product),0)
> rownames(hedonic.c) <- product
> for (j in 1:nbconso){
+     data.c <- as.matrix(hedonic[hedonic$consumer==conso[j],3])
+     rownames(data.c) <- hedonic[hedonic$consumer==conso[j],2]
+     hedonic.c <- cbind.data.frame(hedonic.c,data.c[rownames(hedonic.c),])
+                 }
> colnames(hedonic.c) <- paste("C",conso,sep="")
> head(hedonic.c[,1:5])
                  C171 C553 C991 C1344 C1661
Angel                3    3    7     5     5
Aromatics Elixir     3    5    6     6     4
Chanel N5            7    1    8     8     1
Cinéma               6    4    6     8     7
Coco Mademoiselle    6    7    7     6     5
Jadore EP            5    8    7     9     5
```

Regarding the sensory matrix, the average table by product should be computed. This can easily be done using the **averagetable** function of the SensoMineR package.

```
> library(SensoMineR)
> sensory <- averagetable(sensory,formul="~Product+Panelist",firstvar=5)
> head(sensory)
                    Spicy    Heady    Fruity    Green   Vanilla   Floral
Angel             3.900000 7.841667 1.9208333 0.1125000 7.1833333 2.491667
Aromatics Elixir  6.304167 8.308333 0.6125000 0.5166667 1.8208333 4.295833
Chanel N5         3.733333 8.212500 0.9666667 0.4375000 1.7875000 6.150000
Cinéma            1.083333 2.195833 5.1250000 0.2125000 4.8625000 5.550000
Coco Mademoiselle 0.912500 1.141667 5.0625000 0.7791667 1.9500000 7.975000
Jadore EP         0.262500 1.179167 6.4041667 1.5625000 0.4666667 8.400000
                      Woody    Citrus    Marine     Greedy Oriental Wrapping
Angel             1.1750000 0.4083333 0.14166667 7.8875000 4.758333 7.550000
Aromatics Elixir  2.6375000 0.6041667 0.09166667 0.3416667 7.450000 7.720833
Chanel N5         0.9500000 0.9291667 0.15000000 0.6250000 6.379167 7.845833
Cinéma            1.0166667 1.0500000 0.58750000 4.3750000 2.870833 5.570833
Coco Mademoiselle 0.8041667 1.2416667 0.66250000 2.9375000 3.087500 4.795833
Jadore EP         0.9125000 2.1666667 1.02500000 1.3000000 1.137500 3.566667
```

The first point of view adopted on these data consists in evaluating the variability between products in terms of preferences. This is done by performing PCA on the *hedonic.c* data set, as proposed in Chapter 8. Since we are interested in explaining the differences in preferences between products using the sensory description of the products, we propose to project the sensory profiles of the products as supplementary variables within this hedonic space.

Note that with this procedure, the main focus is on the liking scores, and the sensory description of the products is not involved in the product separation (*i.e.*, in the construction of the dimensions), as they are only projected as supplementary. *De facto*, only *true* linear relationships between the main differences in preferences and the sensory characteristics are observed here.

To do so, the two tables *hedonic.c* and *sensory* should be combined using the **cbind** function. The PCA is then performed on this resulting matrix named here *mdpref.data*. Since the sensory matrix is projected as illustrative, the quanti.sup parameter should be properly informed.

```
> mdpref.data <- cbind(hedonic.c,sensory[rownames(hedonic.c),])
> mdpref <- PCA(mdpref.data,quanti.sup=(ncol(hedonic.c)+1):ncol(mdpref.data),
+ graph=FALSE)
> plot.PCA(mdpref,choix="ind")
> plot.PCA(mdpref,choix="var",label="quanti.sup")
```

As shown previously in Chapter 8, the first dimensions separated *Angel*, *Shalimar*, and *Aromatics Elixir* to *Jadore ET*, *Jadore EP*, and *Coco Mademoiselle* (*cf.* Figure 9.1) in terms of preferences. In other words, the consumers who tend to like *Jadore EP* also tend to like *Coco Mademoiselle*, and tend to dislike *Shalimar*.

Still, it would be good if we could explain these differences between products by using the sensory description of the products. A closer look at the

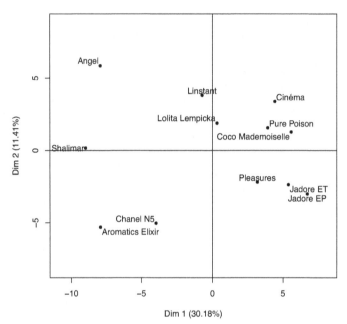

FIGURE 9.1
Representation of the perfumes on the first two dimensions resulting from a
PCA on the hedonic scores (MDPref), using the **PCA** function (*mdpref.data*
data set).

projection as supplementary variables of the sensory attributes helps interpreting these differences. Indeed, as shown in Figure 9.2, the differences in
preferences on the first dimension can be explained by the attributes *Fruity,
Marine, Wrapping, Oriental, etc.* Indeed, the consumers who preferred the
products *J'adore EP* and *Coco Mademoiselle* responded positively to *Fruity*
and *Marine*, and responded negatively to *Wrapping* and *Oriental*. In other
words, the stronger the intensity in *Fruity* and *Marine*, the more they liked
the product. Similarly, the weaker the intensity in *Wrapping* and *Oriental*,
and the more they liked the product.

By using the **dimdesc** function, the sensory attributes related to each dimension can be shown numerically. This procedure is particularly useful when
numerous sensory attributes are projected as supplementary variables. Since
the **dimdesc** function returns all the variables (*i.e.*, consumer and sensory
attributes) related to each dimension, an automatic selection of the sensory
attribute is performed. Such selection is made using the following code:

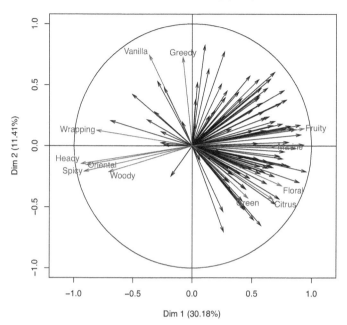

FIGURE 9.2

Representation of the sensory attributes on the first two dimensions resulting from a PCA on the hedonic scores (MDPref), using the **PCA** function
(*mdpref.data* data set).

```
> res.dimdesc <- dimdesc(mdpref)
> select.supp <- which(rownames(res.dimdesc$'Dim.1'$quanti) %in%
+ colnames(sensory))
> res.dimdesc$'Dim.1'$quanti[select.supp,]
          correlation       p.value
Fruity      0.9351164 8.116539e-06
Floral      0.7460137 5.333904e-03
Marine      0.7007259 1.113696e-02
Citrus      0.6750551 1.601156e-02
Woody      -0.7052946 1.040116e-02
Wrapping   -0.8068787 1.514421e-03
Oriental   -0.8941778 8.728868e-05
Spicy      -0.9139439 3.212391e-05
Heady      -0.9398604 5.597798e-06
```

This output shows that the positive side of the first dimension is related
to *Fruity*, *Floral*, and *Marine*, whereas the negative side is related to *Heady*,
Oriental, and *Spicy*. These sensory attributes help to explain the opposition
of the products in terms of preferences along the first dimension.

9.2.2 How can I evaluate the relationship between each sensory attribute and the hedonic scores, at different levels?

To evaluate the relationship between sensory attributes and the liking scores, the PCA uses the correlation coefficient. In other words, linear relationships are considered here. However, in PCA, statistical tests are not performed and the relationship is only visual.

For that reason, we propose to complement the results of the PCA with simple linear regressions. These simple linear regressions are performed on each attribute, and aim to explain the liking scores by using each attribute separately. If the attribute can explain the liking scores, and the relationship is positive, the given attribute is defined as a driver of liking. If the attribute can explain the liking scores, and the relationship is negative, the given attribute is a driver of disliking. Finally, if no relationship between the given attribute and the liking scores is found, the given attribute is not driving liking (or disliking)[1].

Depending on the hedonic scores considered, different drivers of liking can be estimated: if the liking scores correspond to the mean liking scores by product across all consumers, the drivers of liking are the global drivers of liking. If the liking scores considered correspond to the mean liking scores by product across consumers belonging to a homogeneous cluster, the drivers of liking are the ones defined for that particular cluster. Finally, if the liking scores considered are the raw hedonic scores provided by one consumer of interest, the drivers of liking are the ones for that particular consumer.

To define the global linear drivers of liking, the **lm** function is used. This function is the classical regression function in R that we use to express the overall liking scores (average over all consumers) in function of each attribute. Let's consider *Vanilla* and *Oriental*, and let's evaluate whether these attributes are drivers of (dis)liking. To do so, we first need to compute the average liking scores for each product (over all consumers), and add this column to the *sensory* data set.

```
> hedonic.means <- as.matrix(apply(hedonic.c,1,mean))
> rownames(hedonic.means) <- rownames(hedonic.c)
> data.dol <- cbind(sensory,hedonic.means[rownames(sensory),])
> colnames(data.dol)[ncol(data.dol)] <- "Liking"
```

A simple regression expressing the liking scores in function of the *Vanilla* is performed. The **lm** function is used:

```
> vanilla.reg <- lm(Liking~Vanilla,data=data.dol)
> summary(vanilla.reg)
```

---

[1]This is actually true within the product space studied: an attribute can be a driver of liking within a category of product and not being defined as such if the product space studied is too narrow regarding that attribute.

```
Call:
lm(formula = Liking ~ Vanilla, data = data.dol)

Residuals:
    Min      1Q  Median      3Q     Max
-1.4878 -0.6857  0.3703  0.6112  0.8700

Coefficients:
            Estimate Std. Error t value Pr(>|t|)
(Intercept)  5.95452    0.39579  15.045  3.4e-08 ***
Vanilla     -0.08570    0.09947  -0.862    0.409
---
Signif. codes:  0 '***' 0.001 '**' 0.01 '*' 0.05 '.' 0.1 ' ' 1

Residual standard error: 0.8539 on 10 degrees of freedom
Multiple R-squared:  0.06911,    Adjusted R-squared:  -0.02398
F-statistic: 0.7424 on 1 and 10 DF,  p-value: 0.4091
```

The same procedure is applied for the attribute *Oriental*:

```
> oriental.reg <- lm(Liking~Oriental,data=data.dol)
> summary(oriental.reg)

Call:
lm(formula = Liking ~ Oriental, data = data.dol)

Residuals:
     Min       1Q   Median       3Q      Max
-0.82458 -0.21410  0.05758  0.25309  0.58817

Coefficients:
            Estimate Std. Error t value Pr(>|t|)
(Intercept)  6.83743    0.23454  29.152 5.26e-11 ***
Oriental    -0.31080    0.05398  -5.757 0.000183 ***
---
Signif. codes:  0 '***' 0.001 '**' 0.01 '*' 0.05 '.' 0.1 ' ' 1

Residual standard error: 0.4261 on 10 degrees of freedom
Multiple R-squared:  0.7682,     Adjusted R-squared:  0.7451
F-statistic: 33.15 on 1 and 10 DF,  p-value: 0.0001833
```

The test not being significant (*p-value*=0.41), *Vanilla* is not a linear driver of liking, as opposed to *Oriental* (*p-value*=0.0002). Since the regression coefficient is negative (*estimate*=-0.31) for *Oriental*, the lower the intensity of *Oriental*, the higher the liking score: *Oriental* is a driver of disliking.

These relationships between the attributes and the liking scores can also be evaluated graphically. This can be done by simply representing in a scatter plot the intensity of the attribute *versus* the liking scores of the different products. Such graphic can be obtained using the following code:

```
> layout(matrix(1:2,1,2))
> plot(data.dol$Vanilla,data.dol$Liking,xlab="Vanilla",ylab="Liking",
+ type="p",pch=20,main="Simple Regression")
> abline(a=summary(vanilla.reg)$coefficients[1],
+ b=summary(vanilla.reg)$coefficients[2],lwd=2)
```

```
> plot(data.dol$Oriental,data.dol$Liking,xlab="Oriental",ylab="Liking",
+ type="p",pch=20,main="Simple Regression")
> abline(a=summary(oriental.reg)$coefficients[1],
+ b=summary(oriental.reg)$coefficients[2],lwd=2)
```

This code generates Figure 9.3. As can be seen, *Vanilla* is not related to *Liking* as the points are scattered all around the graphs, whereas for *Oriental*, a structure is observed.

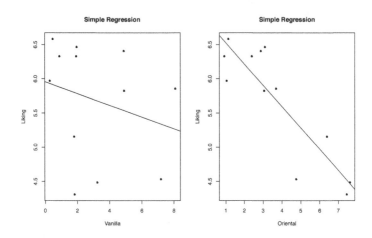

FIGURE 9.3
Representation of the linear relationship between the average liking scores and the attributes *Vanilla* (left) and *Oriental* (right), using the **lm** function (*sensory* and *hedonic* data sets).

Since PCA links the sensory description with the hedonic scores through correlations, only linear relationships between the attributes and liking scores are evaluated. In practice, although linear relationships are generally sufficient, it can be a limitation in some particular cases. Indeed, some attributes might show a saturation point, *i.e.*, a point after which increasing (or decreasing) further the intensity of that attribute does not affect positively liking anymore, but negatively. To evaluate such a relationship, quadratic effects need to be considered as well within the models.

By following the same procedure, the quadratic effect of an attribute can also be tested within the model. In such case, the quadratic version of the attribute first needs to be computed. Let's consider here the attribute *Wrapping*. After computing its squared level, let's run a regression by expressing *Liking* in function of *Wrapping* and *Wrapping2*:

```
> data.dol2 <- cbind.data.frame(data.dol$Wrapping,data.dol$Wrapping^2,
+ data.dol$Liking)
> rownames(data.dol2) <- rownames(data.dol)
```

```
> colnames(data.dol2) <- c("Wrapping","Wrapping2","Liking")
> wrapping.reg <- lm(Liking~Wrapping+Wrapping2,data=data.dol2)
> summary(wrapping.reg)

Call:
lm(formula = Liking ~ Wrapping + Wrapping2, data = data.dol2)

Residuals:
    Min       1Q   Median       3Q      Max
-0.65023 -0.33385  0.02432  0.21339  0.97435

Coefficients:
            Estimate Std. Error t value Pr(>|t|)
(Intercept)  4.28201    1.58977   2.693   0.0247 *
Wrapping     1.03519    0.62956   1.644   0.1345
Wrapping2   -0.12529    0.05713  -2.193   0.0560 .
---
Signif. codes:  0 '***' 0.001 '**' 0.01 '*' 0.05 '.' 0.1 ' ' 1

Residual standard error: 0.4996 on 9 degrees of freedom
Multiple R-squared:  0.7133,    Adjusted R-squared:  0.6495
F-statistic: 11.19 on 2 and 9 DF,  p-value: 0.00362
```

The global test is significant, meaning that there is a relationship between *Liking* and *Wrapping*. More precisely, a look at the coefficients shows that the quadratic effect (*p-value* for *Wrapping2*=0.056) is very close to being significant, whereas the linear effect is not significant at all (*p-value* for *Wrapping2* is equal to 0.1345). Since the *estimate* for *Wrapping2* is negative, this attribute seems to show some saturation, hence highlighting a maximum or optimum level. Such a maximum can be computed mathematically, or estimated graphically.

```
> plot(data.dol$Wrapping,data.dol$Liking,xlab="Wrapping",ylab="Liking",
+ type="p",pch=20,main="Quadratic Regression")
> xseq <- seq(min(data.dol$Wrapping),max(data.dol$Wrapping),0.05)
> lines(xseq,y=(summary(wrapping.reg)$coefficients[1]+
+ summary(wrapping.reg)$coefficients[2]*xseq+
+ summary(wrapping.reg)$coefficients[3]*xseq^2),lwd=2)
```

The code presented here generates Figure 9.4, which highlights the quadratic relationship between *Wrapping* and *Liking*. From this figure, it seems that the optimal intensity level for *Wrapping* is slightly above 4.

The drivers of liking considered here are global drivers of liking, as they are calculated for the entire panel of consumers. This is possible here since the data set highlights a high agreement amongst consumers in terms of preferences. When the panel is more segmented (*i.e.*, different preferences behavior) it is of utmost importance to split the panel into different clusters, and to define the drivers of liking for each cluster. Still at this stage, individual differences between consumers can be observed. To evaluate the individual differences between consumers, the raw liking scores provided by the different consumers are required.

Let's consider the consumers *C1755* and *C10815*, and let's consider the

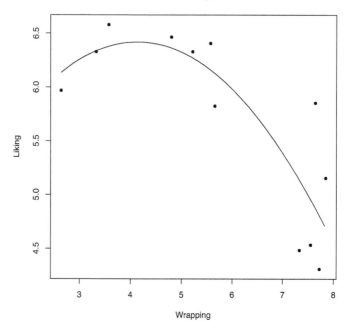

FIGURE 9.4
Representation of the quadratic relationship between the average liking scores and the attribute *Wrapping*, using the **lm** function (*sensory* and *hedonic* data sets).

attribute *Fruity*. By using a similar code as previously, we estimate the impact of the perception of this attribute on *Liking* (here, only the linear effect is considered):

```
> Fruity.1755 <- lm(C1755~Fruity,data=mdpref.data)
> summary(Fruity.1755)

Call:
lm(formula = C1755 ~ Fruity, data = mdpref.data)

Residuals:
     Min       1Q   Median       3Q      Max
-1.15524 -0.65123 -0.04382  0.60655  1.24827

Coefficients:
            Estimate Std. Error t value Pr(>|t|)
(Intercept)   9.5373     0.5287  18.040 5.87e-09 ***
Fruity       -0.4648     0.1333  -3.488  0.00584 **
```

```
---
Signif. codes:  0 '***' 0.001 '**' 0.01 '*' 0.05 '.' 0.1 ' ' 1

Residual standard error: 0.8736 on 10 degrees of freedom
Multiple R-squared:  0.5488,     Adjusted R-squared:  0.5037
F-statistic: 12.17 on 1 and 10 DF,  p-value: 0.005844

> Fruity.10815 <- lm(C10815~Fruity,data=mdpref.data)
> summary(Fruity.10815)

Call:
lm(formula = C10815 ~ Fruity, data = mdpref.data)

Residuals:
    Min      1Q  Median      3Q     Max
-1.1376 -0.9072 -0.3434  0.6926  1.8206

Coefficients:
            Estimate Std. Error t value Pr(>|t|)
(Intercept)   2.3852     0.6963   3.426 0.006487 **
Fruity        0.8216     0.1755   4.681 0.000866 ***
---
Signif. codes:  0 '***' 0.001 '**' 0.01 '*' 0.05 '.' 0.1 ' ' 1

Residual standard error: 1.151 on 10 degrees of freedom
Multiple R-squared:  0.6867,     Adjusted R-squared:  0.6553
F-statistic: 21.91 on 1 and 10 DF,  p-value: 0.0008659
```

For both consumers, *Fruity* influences *Liking*. However, a closer look at the results shows that for *C10815*, *Fruity* is a driver of liking (*estimate*=0.82), whereas for *C1755*, it is a driver of disliking (*estimate*=-0.46).

9.2.3 How can I locate an optimum product within the product space?

Combining the hedonic and sensory description of a set of products would not be complete if it would not lead to the definition of an optimum product, *i.e.*, a product that would be accepted by a maximum of consumers. Previously, in Section 9.2.2, we saw how to detect which attributes play an important role on liking (positively and negatively), and which attributes do not.

However, it is well known that sensory attributes are highly correlated together, meaning that modifying the perception of one attribute affects the perception of many others. For instance, it is well known that modifying the perception of sweetness also affects the perception of bitterness and sourness.

To take such aspect into consideration, an alternative to simple regression is proposed. This alternative consists in considering all the attributes in one unique model. This cannot be done directly by adding all the attributes in one multiple regression, and this is for two reasons: first, the number of degrees of freedom is often insufficient to estimate all the parameters, and second, the regression procedure is unstable with multi-collinear data.

To overcome these two limitations, alternative methodologies exist. The most famous alternatives are the Partial Least Square (PLS) regression and the regression on Principal Component (PCR). Here, we focus our attention on PCR, which consists in practice in summarizing the explanatory matrix into into a smaller number of orthogonal variables. These particular variables usually correspond to the first principal component of a PCA performed on the original data set. Indeed, as we have seen previously, each principal component is a linear combination of the raw variables (through the loadings). By substituting within the regression model the attributes by the first principal components, the problem of multi-collinearity and of degrees of freedom is solved.

Since the information to summarize lies in the *sensory* data set, we apply the **PCA** function on this data set.

```
> sensory.pca <- PCA(sensory,graph=FALSE)
> names(sensory.pca)
[1] "eig"  "var"  "ind"  "svd"  "call"
> sensory.pca$var$coord
              Dim.1       Dim.2        Dim.3        Dim.4        Dim.5
Spicy     0.8716443  0.45963687  0.107856675  0.08976533  0.033406069
Heady     0.9246736  0.22330561  0.198242266 -0.12056791 -0.126629461
Fruity   -0.8985310 -0.31098223 -0.111280152  0.24776600 -0.129826609
Green    -0.7409308  0.31449573  0.544549889 -0.11199439  0.149363058
Vanilla   0.5518335 -0.80242160  0.064000738  0.10689898  0.117028889
Floral   -0.8727855  0.33940701 -0.295290558 -0.13800792 -0.031290877
Woody     0.6414953  0.60021583 -0.050048095  0.43595209  0.159966247
Citrus   -0.8138241  0.22734986  0.258793055  0.20388280 -0.381013940
Marine   -0.8659626  0.05604599  0.037178995  0.03779279  0.351476971
Greedy    0.2787301 -0.93036189  0.195788226  0.09349221 -0.005996179
Oriental  0.9157147  0.33647512 -0.005797401 -0.01476709 -0.042191781
Wrapping  0.9637400 -0.11909239 -0.042781294 -0.10386606 -0.080858506
```

The object `res.pca$var$coord` shows the relationship between the sensory attributes and the principal components: in this case, the first principal component is positively linked to *Wrapping, Heady, Oriental, etc.*, and is negatively linked to *Fruity, Floral, Marine, etc.*

From these results, let's extract the three first dimensions (present in the object `res.pca$ind$coord`, as we are interested in the coordinates of the products here), and let's combine them to the overall liking scores (averaged over all consumers present in the object `liking.means`). A regression model is then performed by expressing *Liking* in function of the first three dimensions of the PCA.

```
> data.pcr <- cbind.data.frame(hedonic.means,
+ sensory.pca$ind$coord[rownames(hedonic.means),1:3])
> colnames(data.pcr)[1] <- c("Liking")
> res.pcr <- lm(Liking~Dim.1+Dim.2+Dim.3,data=data.pcr)
> summary(res.pcr)

Call:
lm(formula = Liking ~ Dim.1 + Dim.2 + Dim.3, data = data.pcr)
```

```
Residuals:
     Min        1Q    Median        3Q       Max
-0.48311 -0.07683   0.07510   0.15256   0.36949

Coefficients:
            Estimate Std. Error t value Pr(>|t|)
(Intercept)  5.68770    0.08347  68.138 2.40e-12 ***
Dim.1       -0.24987    0.03007  -8.310 3.32e-05 ***
Dim.2       -0.15382    0.05153  -2.985   0.0175 *
Dim.3       -0.30940    0.11133  -2.779   0.0240 *
---
Signif. codes:  0 '***' 0.001 '**' 0.01 '*' 0.05 '.' 0.1 ' ' 1

Residual standard error: 0.2892 on 8 degrees of freedom
Multiple R-squared:  0.9146,    Adjusted R-squared:  0.8826
F-statistic: 28.56 on 3 and 8 DF,  p-value: 0.0001263
```

The test being significant, the three first dimensions are affecting *Liking*. Additionally, a closer look at the results shows that the coefficient related to the first dimension is negative (*estimate*=-0.25). Hence, the attributes that are negatively linked to the first dimension (*i.e.*, *Fruity*, *Floral*, *Marine*, *etc.*) are positive drivers of liking.

As we have seen previously, in some cases, linear effects are not sufficient and quadratic effects should be considered. Interaction between attributes could also be considered. Based on this observation, different PCR models can be considered:

- linear or vector model: only linear effects are considered, and the model considered is $Liking = a + b_1 * F_1 + b_2 * F_2$;

- circular model: the quadratic effects on the dimensions are also considered, and the model is $Liking = a + b_1 * F_1 + b_2 * F_2 + b_3 * (F_1^2 + F_2^2)$;

- elliptic model: different quadratic effects are considered on the dimensions, and the model is $Liking = a + b_1 * F_1 + b_2 * F_2 + b_{11} * F_1^2 + b_{22} * F_2^2$;

- quadratic model (or Danzart model): the linear, quadratic, and two-way interaction between dimensions are considered, and the model is $Liking = a + b_1 * F_1 + b_2 * F_2 + b_{11} * F_1^2 + b_{22} * F_2^2 + b_{12} * F_1 F_2$.

Increasing the complexity of the model increases the accuracy of the estimation, but also requires more degrees of freedom. For instance, for the quadratic model, a minimum of 7 degrees of freedom (*i.e.*, test with at least 7 products) is required[2].

Finally, and as it has been shown previously, each consumer has their own drivers of liking. In order to keep such information, the models defined here are applied to each consumer separately. This is the core of the external preference mapping technique (PrefMap).

---

[2]For technical reasons (degrees of freedom), we only recommend the use of quadratic models in studies involving a minimum of 10 products.

Once these individual models are defined, the liking score is estimated on each point of the product space (*i.e.*, for each pair of coordinates on the two dimensions of interest), for each consumer separately. Based on these estimations, individual areas of acceptance are defined. These areas of acceptance are zones within the product space that are associated with an estimated liking score that is larger than a certain threshold value (by default, this corresponds to the average liking score provided by the consumer of interest). The overlay of all the individual areas of acceptance generates the response surface. This representation is often completed with contour lines, which highlight the percentage of consumers accepting the products in each area.

Such procedure is the one performed in the **carto** function of the SensoMineR package. This function takes as inputs the coordinates of the product in the sensory space on the two dimensions of interest (here, `Mat=sensory.pca$ind$coord[,1:2]`), the matrix of raw hedonic scores crossing the products in rows and the consumers in columns (`MatH=hedonic.c`), the model to consider (by default, `regmod=1` corresponds to the quadratic model), and the threshold of acceptance (by default, `level=0` corresponds to the average liking score provided by each consumer, and is expressed in standard deviation). On our data sets, the **carto** function returns the following results:

```
> prefmap <- carto(sensory.pca$ind$coord[,1:2],hedonic.c,level=0,regmod=1)
```

The surface response plot (*cf.* Figure 9.5) highlights a zone of maximum liking close to *Jadore ET* and *Coco Mademoiselle*. Such a product would be accepted by around 70% of the consumers.

To complete the optimization, the sensory profile of such optimum product should be estimated. Such profile can either be obtained by using the data reconstitution formula of the PCA, or reverse regressions. The latter methodology is not presented here, but proposed in Exercise 9.5.

9.3   For experienced users: Finding the best correspondence between the sensory and hedonic matrices, using PrefMFA

Both internal and external preference mapping are techniques that aim to combine the hedonic evaluation of a set of products to external characteristics of the products (*i.e.*, sensory, instrumental, *etc.*). The aim of these techniques is hence to find which external characteristics of the products explain the differences in liking.

Although the data sets considered are the same, and although the objectives are very similar, these two techniques adopt two very different points of

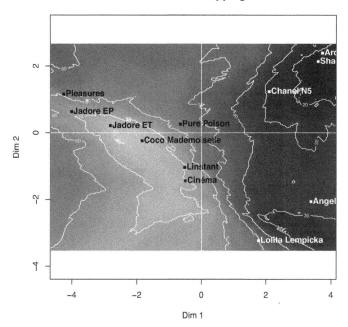

FIGURE 9.5
Representation of the perfumes and the surface response on the first two dimensions resulting from a PCA (PrefMap), using the **carto** function (*sensory* and *hedonic.c* data sets).

view. In each case, the two points of view adopted are directly related to the data set the attention is focused on:

- In the MDPref, the attention is focused on the liking data, *i.e.*, the product separation is directly linked to the differences of preferences between consumers. In this case, the external information only plays a secondary role, as it aims at explaining the differences in preferences between products.

- In the PrefMap, the attention is focused on the external data, *i.e.*, the product separation considered usually corresponds to the first two dimensions of variability defined within the external data. On the resulting external space, the hedonic scores are then regressed in order to define the location that would be accepted by a maximum of consumers.

Despite the two different points of view, MDPref and PrefMap suffer the same critics, *i.e.*, that each technique is focusing on one matrix of data only. Hence, finding relationships between the two matrices might become difficult.

Indeed, with MDPref, two products that are very different from a sensory point of view can be appreciated equally (*e.g.*, apples and pears). With PrefMap, only two dimensions are considered in the regression model, and some relevant information might be discarded, which can lead to "irrelevant" model for a non-negligible number of consumers. Moreover, still in the case of PrefMap, the choice of the two external dimensions to consider is crucial: logically, the first two dimensions are often considered, as they contain the maximum of information. However, in some cases, it is worth using other dimensions, as they would bring more "relevant" information regarding the purpose of the analysis. Another solution consists in using more than two dimensions in the individual regression model, but such an addition of information might lead to over-fitting the hedonic scores, and with more complex models (*e.g.*, circular, elliptic, and quadratic), individual models are quickly relying on a very low number of degrees of freedom.

Since the main criticism against MDPref and PrefMap lies on the *unidimensionality*[1] of the product space, an improvement of these techniques can be obtained by considering a more "relevant" product space, *i.e.*, a space that is linked to both sensory and hedonic data. Such point of view is the one adopted in the PLS-based procedure. Indeed, the core of PLS consists in best explaining a variable (or group of variables) Y (here the hedonic matrix) from a group of variable X (here the sensory matrix). To do so, "the PLS model will try to find the multidimensional direction in the X space that explains the maximum multidimensional variance direction in the Y space" (source: `http://en.wikipedia.org/wiki/Partial_least_squares_regression`).

However, PLS-based methods also have some drawbacks. First, they are usually performed on average liking scores. In other words, the individual variability is "lost." Alternatively, successive PLS-regression on each individual consumer can be performed, but this procedure does not solve the problem, since for each consumer, the sensory configuration used to explain the liking scores is changed. Hence, with the PLS-based method, a surface response plot similar to the one in PrefMap cannot be produced.

By following the same path, analyses of multiple data table in which external descriptions of the products are considered on the one hand, and the individual hedonic scores are considered on the other hand, can be used. The maximum part in common in the different matrices is then highlighted. In this case, multiple data methods, such as Generalized Procrustes Analysis, STATIS, or Multiple Factor Analysis (MFA), can be used.

As discussed in Chapter 3, and thanks to its numerous valuable features, MFA appears to be a good solution to find the sensory space that explains the most of the liking data. Indeed, MFA is an exploratory multivariate method dedicated to the analysis of so-called multiple data tables, in the sense that one set of statistical individuals is described by several groups of variables.

---

[1]Here, unidimensional refers to the unique nature of the product space (*i.e.*, only related to hedonic data in the case of MDPref, and only related to sensory data in the case of PrefMap) rather than to its fundamental structure.

When two groups of variables are considered (as it is the case here with the sensory and the hedonic groups), MFA balances them in the analysis (using a particular weighting procedure), and the consensual space defined corresponds to the maximum common information shared between the two groups.

In the perfume example presented in Section 9.2, let's evaluate the relationship between the hedonic data, structured with the products in rows and the consumers in columns, and the sensory profiles of the products. To do so, let's first import the *perfumes_qda_experts2.csv* and the *perfumes_liking_conso.csv* files.

```
> sensory <- read.table("perfumes_qda_experts2.csv",header=TRUE,sep=",",
+ dec=".",quote="\"")
> sensory$Session <- as.factor(sensory$Session)
> sensory$Rank <- as.factor(sensory$Rank)
> hedonic <- read.table("perfumes_liking_conso.csv",header=TRUE,sep=",",
+ dec=".",quote="\"")
> hedonic$consumer <- as.factor(hedonic$consumer)
```

Once imported, the sensory profiles of the products are computed from the *sensory* data using the **averagetable** from the SensoMineR package. Additionally, the *hedonic* data are transformed into a *Product × Consumer* matrix of hedonic scores. To do so, the code previously defined is used:

```
> library(SensoMineR)
> sensory <- averagetable(sensory,formul="~Product+Panelist",firstvar=5)
> conso <- levels(hedonic$consumer)
> nbconso <- length(conso)
> product <- levels(hedonic$product)
> hedonic.c <- matrix(0,length(product),0)
> rownames(hedonic.c) <- product
> for (j in 1:nbconso){
+     data.c <- as.matrix(hedonic[hedonic$consumer==conso[j],3])
+     rownames(data.c) <- hedonic[hedonic$consumer==conso[j],2]
+     hedonic.c <- cbind.data.frame(hedonic.c,data.c[rownames(hedonic.c),])
+                 }
> colnames(hedonic.c) <- paste("C",conso,sep="")
```

Once generated, the two tables are combined using the **cbind** function. The common part in the two matrices of data is then assessed by MFA. As usual, the **MFA** function of the FactoMineR package is used, and the two groups are scaled (**type=rep("s",2)**).

```
> data.mfa <- cbind(sensory,hedonic.c[rownames(sensory),])
> res.mfa <- MFA(data.mfa,group=c(ncol(sensory),ncol(hedonic.c)),
+ type=rep("s",2),name.group=c("Sensory","Hedonic"),graph=FALSE)
> names(res.mfa)
 [1] "separate.analyses" "eig"               "group"
 [4] "inertia.ratio"     "ind"               "summary.quanti"
 [7] "summary.quali"     "quanti.var"        "partial.axes"
[10] "call"              "global.pca"
```

Before evaluating and interpreting the product space, let's first have a look at some coefficients of the MFA that assess the strength of the relationship between the two tables. The first numerical criterion to look at is the first eigenvalue of the MFA. As the number of active groups equals 2, let's remark that the first eigenvalue lies between 1 (the groups are perfectly orthogonal) and 2 (the first dimension of both groups are homothetic).

```
> res.mfa$eig[1,1]
[1] 1.917092
```

The first eigenvalue of the MFA being 1.92, the main dimension of variability of the sensory space seems to coincide well with the main differences in terms of preferences between products.

The next criterion to evaluate is the RV coefficient. Let's recall that this coefficient measures the link between two groups of variables. This coefficient lies between 0 (all the variables of the first group are orthogonal to all the variables of the second group) and 1 (the two groups are homothetic). This coefficient is stored in **res.mfa$group$RV**.

```
> res.mfa$group$RV
          Sensory   Hedonic       MFA
Sensory 1.000000 0.7525320 0.9236500
Hedonic 0.752532 1.0000000 0.9474592
MFA     0.923650 0.9474592 1.0000000
```

The RV coefficient measured between the sensory and the hedonic group is equal to 0.75: a strong link exists between the two matrices of data. However, the two configurations are not homothetic: a part of the information present in one of the two data sets (at least) cannot be explained by the other group.

The last criterion used to evaluate the relationship between the two groups is a criterion, specific to MFA: the N_g or the L_g coefficient (the N_g coefficient corresponds to the L_g coefficient when applied to a single group). For each group separately, the N_g coefficient tells us about the dimensionality of the group itself. For a group in which the information is well balanced along the first S dimensions, the N_g coefficient is larger (between 1 and S) than for a group in which most of the inertia is explained along the first dimension only. When measured between pairs of groups, the L_g coefficient measures the richness of the common structure between the two groups: the larger the L_g coefficient, the larger the common structure. This coefficient is stored in the object **res.mfa$group$Lg**.

```
> res.mfa$group$Lg
          Sensory  Hedonic       MFA
Sensory 1.127031 1.016121 1.117918
Hedonic 1.016121 1.617728 1.373877
MFA     1.117918 1.373877 1.299778
```

By looking at the coefficient within each group, *Hedonic* appears to be more multidimensional than *Sensory* ($N_g(Sensory)$ = 1.13 *versus* $N_g(Hedonic) = 1.62$). Additionally, the L_g coefficient measured between both groups ($L_g(Hedonic, Sensory) = 1.01$) highlights the fact that *Sensory* can be largely used to explain *Hedonic*, but due to the differences in the structure of the groups, a part of *Hedonic* remains particular to that group.

Graphically, the relationship between the groups and the common space provided by the MFA is evaluated through the partial axes representation. This representation consists in projecting as supplementary variables the dimension of the separate analysis performed on each group within the MFA space. Such representation highlights how the MFA space is linked to the different groups. To generate such a graphic, the **plot.MFA** function is used:

```
> plot.MFA(res.mfa,choix="axes",habillage="none")
```

Figure 9.6 shows that the first plane of the MFA is highly correlated with the first plane of both the *Sensory* and *Hedonic* groups (the sign of the correlation should not be interpreted here). This result confirms the strong link between groups, and confirms the use of dimensions 1 and 2 of *Sensory* in the PrefMap routine performed in Section 9.2.3. Indeed, in this case, the first plane of the sensory space is the best possible two-dimensional solution to explain the liking data.

Finally, let's evaluate the common product space provided by the MFA. To fully illustrate the differences between both groups, let's represent on this product space the partial points representation. This partial points representation is obtained by projecting as supplementary, on the MFA solution, each group separately (the second matrix of data being composed of 0 values only). To obtain such representation, the **plot.MFA** function is used. In this case, both `choix="ind"` and `partial="all"` parameters should be set.

```
> plot.MFA(res.mfa,choix="ind",partial="all",habillage="none")
```

Thanks to the partial points representation (*cf.* Figure 9.7), it can be seen that the products most common to the two groups are *Jadore ET* and *L'instant*, whereas the products less common to the two groups are *Lolita Lempicka* and *Pleasures*.

By construction, the partial point configuration related to the *Sensory* group, within the MFA common space, constitutes the sensory configuration of the product that explains the most liking. Let's extract the coordinates of these partial points, and let's use this configuration as a starting point for the regression routine of the PrefMap. Such procedure generates a surface response plot, which is the core of the methodology also known as PrefMFA.

The partial points are stored in the object `res.mfa$ind$coord.partiel`. To select the points directly related to the *Sensory* group, the **grep** function is used:

```
> senso.select <- grep(".Sensory",rownames(res.mfa$ind$coord.partiel))
> coord.partial <- res.mfa$ind$coord.partiel[senso.select,1:2]
> coord.partial
```

Partial axes

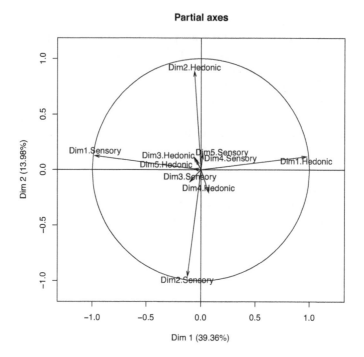

FIGURE 9.6
Representation of the dimensions resulting from the separate analyses on the first two dimensions resulting from an MFA, using the **MFA** function (*sensory* and *hedonic* data sets).

	Dim.1	Dim.2
Angel.Sensory	-1.6587026	1.36466166
Aromatics Elixir.Sensory	-1.9902251	-0.74664287
Chanel N5.Sensory	-1.1095586	-0.42861015
Cinéma.Sensory	0.3180935	0.67645160
Coco Mademoiselle.Sensory	0.9568817	-0.07239563
Jadore EP.Sensory	2.0176622	-0.74901544
Jadore ET.Sensory	1.4330424	-0.42821176
Linstant.Sensory	0.3125909	0.48356617
Lolita Lempicka.Sensory	-0.7908025	1.74566509
Pleasures.Sensory	2.0913552	-1.06380148
Pure Poison.Sensory	0.3360602	-0.15998343
Shalimar.Sensory	-1.9163973	-0.62168376

Finally, after renaming the rows of `coord.partiel` so that they match *hedonic.c* (make sure that the order of the rows match), the **carto** function of the SensoMineR package is used to generate the surface response plot. Like

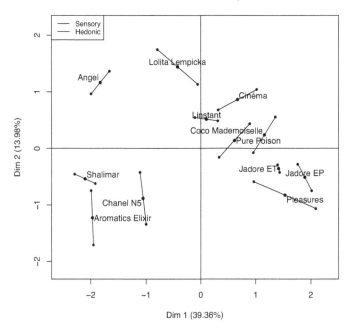

Individual factor map

FIGURE 9.7
Representation of the perfumes and their partial points on the first two dimensions resulting from an MFA, using the **plot.MFA** function (*sensory* and *hedonic* data sets).

previously, the quadratic model is considered (**regmod**=1), and the level of acceptance is set to the average liking score provided by the consumer (**level**=0).

```
> rownames(coord.partial) <- rownames(hedonic.c)
> PrefMFA <- carto(coord.partial,hedonic.c,regmod=1,level=0,
+ main="PrefMFA solution")
```

The PrefMFA solution obtained is presented in Figure 9.8. This solution is very similar to the PrefMap presented in Figure 9.5. This result was expected, as the first plane of the *Sensory* group corresponds to the best two-dimensional solution to explain the liking data.

Although the two techniques show similar results, the methodology of PrefMFA shows three advantages over PrefMap. First, by selecting the external space that explains the most liking, the individual models obtained are better, in terms of predictions. Second, by construction, the MFA automatically searches for the best representation of the sensory space to explain the

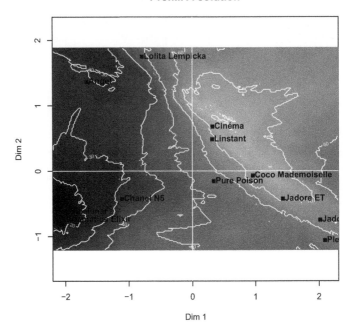

FIGURE 9.8
Representation of the perfumes and the surface response on the first two dimensions resulting from an MFA (PrefMFA), using the **MFA** and the **carto** functions (*sensory* and *hedonic* data sets).

liking data. This configuration does not need to be the first plane of the sensory space, but can be dispersed over multiple dimensions. Finally, thanks to the MFA outputs, the strength of the relationship between the two matrices of data can be assessed.

9.4 Exercises

Exercise 9.1 *Introduction to Consumer's Preference Analysis*

As described in Section 9.2.1, MDPref usually consists in performing a PCA on the product × consumer matrix of raw hedonic scores. In this case, the product separation is defined through the differences between products in terms of preference: two products are close if they have been liked by the same consumers. Alternatively, the PCA can be performed on this matrix after being transposed (*e.g.*, consumer × product matrix). To see the interest of such analysis, a data set involving 16 different cocktails is used.

- Import the data set *cocktail_hedo* from the book website.

- Run a PCA on this matrix after transposing it (using the **t** function). Interpret the results.

- Run the same analysis after standardizing the consumers' hedonic scores (the scale function can be used). In this case, we suggest you to perform the PCA on the covariance matrix. Interpret the results.

- Since the sensory profiles of the products are available, we propose to use them to help interpreting the consumer space. Import the data set *cocktail_senso* from the book website.

- Add the matrix of sensory profiles to the matrix of hedonic scores (standardized by consumer), and run the same analysis by adjusting the parameter appropriately. Interpret the results: how would you interpret the attribute "odor mango"?

- The analysis just performed is called Consumer Preference Analysis (CPA). Run the same analysis using the **cpa** function from the SensoMineR package, and interpret the results.

Exercise 9.2 *Defining linear and quadratic drivers of liking*

The aim of this exercise is to understand the relationship between the average liking score and the perception of each attribute. Which attributes are linear drivers of liking, linear drivers of disliking? Which attribute show an effect of saturation?

- Import the data set *cocktail_hedo* from the book website. Compute the average overall liking score. The **apply** function combined with the **mean** function can be used.

- Import the *cocktail_senso* data set from the book website. Combine these two blocks of information together (*i.e.*, sensory and hedonic blocks).

- Regress the average liking scores on each sensory attribute separately. Which attributes are drivers of liking? Which attributes are drivers of disliking?

- Repeat the same process by adding the quadratic effect for each sensory attribute. Which attributes show saturation?

- Represent the results graphically.

Exercise 9.3 *Defining drivers of (dis)liking through PCR and PLS regression*

In the previous exercise, we defined each driver of liking separately, without taking care of the collinearity between attribute. To some extent, the previous models considered could be made more complex by evaluating the impact of 2, 3, or more sensory attributes simultaneously. However, quickly, the problem of lacking degrees of freedom and the problem of multi-collinearity arise. For that reason, we need to consider alternatives, such as PCR and PLS regression.

- Import the data set *cocktail_hedo* from the book website. Compute the average overall liking score. The **apply** function combined with the **mean** function can be used.

- Import the *cocktail_senso* data set from the book website. Combine these two blocks of information together (*i.e.*, sensory and hedonic blocks).

- Using the **oscorespls.fit** function of the pls package, run a PLS regression expressing the liking scores in function of the sensory attributes. We suggest here to scale the table before running the **oscorespls.fit** function.

- Using the **PCA** function of the FactoMineR package, reduce the sensory space to 3 dimensions.

- Perform a multiple regression evaluating the impact of each dimension on liking.

- Using the loadings and the regression weights, estimate the impact of each attribute on liking.

- Compare the results between the PLS on one component and the PCR.

Exercise 9.4 *Comparing the results of PrefMap and PrefMFA*

In this example, the results of the PrefMap and of PrefMFA are compared.

- Import the *cocktail_hedo* and *cocktail_senso* data sets from the book website.

- Reduce the sensory data to its first two dimensions using the **PCA** function of the FactoMineR package.

- Use this configuration together with the matrix of hedonic scores to generate the surface response with the **carto** function of the SensoMineR package.

- Combine the original sensory profiles to the raw hedonic scores using the **cbind** function.

- Perform an MFA (using the **MFA** function of the FactoMineR package) on these two groups.

- By looking at some of the MFA outputs, how strong is the relationship between the sensory and the hedonic description of the products?

- From the MFA outputs, extract the coordinates of the products associated with partial points related to the sensory group. Readapt the names of the products so that it matches with the one of the hedonic data.

- Use this configuration of the products in the **carto** function. Compare the two preference maps obtained and conclude.

Exercise 9.5 *Estimating the sensory profile of an optimum product*

In the previous exercise, we showed you how to generate the response surface in the case of PrefMap and PrefMFA. Once such response surfaces are obtained, the location of an eventual optimum product can be defined. In the PrefMap example, it could be for instance at the coordinates $(-3; 0.8)$ on dimension 1 and dimension 2 respectively. If the product space is obtained by PCA, it is possible to estimate from these coordinates the sensory profile of the products by using the so-called *data reconstitution* formula. Indeed, let us remind you that PCA is simply a geometrical transformation that relocates a clouds of points into another better structured reference. Hence, from the set of coordinates, it is possible to go back to the original data. Another solution consists in using a different series of regression. Similarly to the PrefMap regression routine, we can express each sensory attribute in function of the product configuration used. The regression formula thus obtained can then be used to estimate the intensity of each sensory attribute at a certain pair of coordinates. This is what we are proposing here:

- Import the *cocktail_senso* data set from the book website.

- Create the product space by using the **PCA** function of the FactoMineR package.

- Extract the coordinates of the products on the first 2 dimensions.

- Combine these 2 dimensions to the original sensory profile data.

- Regress each sensory attribute separately on the first two dimensions of the PCA.

- Using the **predict** function, estimate the intensity level for each sensory attribute of a product position at the coordinates $(-3, 0.8)$. Save the predictions in a matrix.

- Add this optimum product to your sensory profile data, and run a PCA again by projecting it as supplementary entity. Conclude.

9.5 Recommended readings

- Danzart, M. (1998). Quadratic model in preference mapping. 4th Sensometric meeting, Copenhagen, Denmark, August 1998.

- Danzart, M. (2009a). Cartographie des préférences. In Evaluation sensorielle. Manuel méthodologique (SSHA, 3e édition) (pp. 443-450). Edited by Tec & Doc. Paris: Lavoisier.

- Danzart, M. (2009b). Recherche d'un point optimal. In Evaluation sensorielle. Manuel méthodologique (SSHA, 3e édition) (pp. 431-439). Edited by Tec & Doc. Paris: Lavoisier.

- Faber, N. M., Mojet, J., & Poelman, A. A. M. (2003). Simple improvement of consumer fit in external preference mapping. *Food Quality and Preference*, 14, (5), 455-461.

- Gower, J., Groenen, P. J. F., Van de Welden, M., & Vines, K. (2014). Better perceptual maps: Introducing explanatory icons to facilitate interpretation. *Food Quality and Preference*, 36, 61-69.

- Greenhoff, K., & MacFie, H. J. H. (1994). Preference mapping in practice. In Measurement of food preferences (MacFie, H. J. H., & Thompson, D. M. H. eds.). London: Blackie Academic & Professionals (pp. 137-166).

- Lengard, V., & Kermit, M. (2006). 3-Way and 3-block PLS regressions in consumer preference analysis. *Food Quality and Preference*, 17, (3), 234-242.

- Mage, I., Menichelli, E., & Naes, T. (2012). Preference mapping by PO-PLS: Separating common and unique information in several data blocks. *Food Quality and Preference*, 24, (1), 8-16.

- Mao, M., & Danzart, M. (2008). Multi-response optimization strategies for targeting a profile of product attributes with an application on food data. *Food Quality and Preference*, 19, (2), 162-173.

- McEwan, J. A. (1996). Preference Mapping for product optimization. In Multivariate analysis of data in sensory science (Naes, T., & Risvik, E. eds.). New York: Elsevier (pp. 71-102).

- Menichelli, E., Hersleth, M., Almoy, T., & Naes, T. (2014). Alternative methods for combining information about products, consumers and consumers' acceptance based on path modelling. *Food Quality and Preference*, 31, 143-155.

- Meullenet, J.F., Lovely, C., Threlfall, R., Morris, J.R. & Striegler, R.K. (2008). An ideal point density plot method for determining an optimal sensory profile for Muscadine grape juice. *Food Quality and Preference*, 19, (2), 210-219.

- Mevik, B. H., & Wehrens, R. (2007). The pls package: principal component and partial least squares regression in R. *Journal of Statistical Software*, 13.

- Pagès, J., & Tenenhaus, M. (2001). Multiple factor analysis combined with PLS path modelling. Application to the analysis of relationships between physicochemical variables, sensory profiles and hedonic judgments. *Chemometrics and Intelligent Laboratory Systems*, 58, (2), 261-273.

- Pagès, J., & Tenenhaus, M. (2002). Analyse factorielle multiple et approche PLS. *Revue de Statistique Appliquée*, 50, 5-33.

- Tenenhaus, M., Pagès, J., Ambroisine, L., & Guinot, C. (2005). PLS methodology to study relationships between hedonic judgments and product characteristics. *Food Quality and Preference*, 16, (4), 315-325.

- Van Kleef, E., Van Trijp, H. C. M., & Luning, P. (2006). Internal versus external preference mapping: An exploratory study on end-user evaluation. *Food Quality and Preference*, 17, (5), 387-399.

- Worch, T. (2013). PrefMFA, a solution taking the best of both internal and external preference mapping techniques. *Food Quality and Preference*, 30, (2), 181-191.

- Yenket, R., Chambers IV, E., & Adhikari, K (2011). A comparison of seven preference mapping techniques using four software programs. *Journal of Sensory Studies*, 26, (2), 135-150.

10

When products are described by a mix of liking and external information

CONTENTS

"... consumers are more involved in the sensory task. Indeed, in such tasks, and additionally to liking ratings, consumers are also asked to provide information on how they perceive the products, and how they would like them. Such information is either acquired directly (Ideal Profile Method [IPM]), or as deviation from the ideal point (Just About Right [JAR] scale). Although this "ideal" information is of utmost importance for product optimization, it is questionable, as it comes from consumers who are rating fictive products. Before analyzing such data, when possible, it is recommended to check for their consistency. This can be done by using the **ConsistencyIdeal** *function of the* SensoMineR *package."*

10.1   Data, sensory issues, and notations

In the previous chapter, we saw that consumers are required in the product development procedure, since they are the final decider of the product success. If, historically, consumers were only used for hedonic tasks, they are nowadays more and more solicited. For example, consumers are often in the heart of "*new*" fast sensory tasks, such as with sorting task and Napping (*cf.* Chapter 6 and Chapter 7).

Consumers can also be used for profiling products, and two strategies can be adopted. The first strategy consists in avoiding misinterpretation of the sensory terms by consumers by giving them the freedom to use their own attributes or words. This strategy is the one adopted in Free Choice Profile and Flash Profile, and word elicitation task, for which the analyses were presented in Chapter 3 and Chapter 4. The second strategy consists in getting similar profiles as with experts (QDA®), but from consumers. A predefined list of attributes is then provided to consumers. Due to the absence of training sessions, such strategy requires using attributes that are not too technical to be understandable by naive consumers.

In practice, it is well known that all consumers are not able to perform sensory tasks. From our experience, around 20% of the consumers are giving random scores. These consumers are often considered as random noise. As a direct consequence, the random noise in consumer tests is larger than in tests involving experts. To counterbalance, larger panel sizes are used in consumer tests (usually 10 to 12 experts *versus* a minimum of 100 consumers).

Using consumers to profile products shows advantages. For example, having consumers in an early stage of the process reinforces the relationship between the perception of the products and their appreciation (both information comes from the same consumers). Moreover, the use of consumers in sensory description tasks allows seeking for additional useful information. By assuming that the majority of consumers knows what they like, knows what they want, and can describe it, it is our task to extract this valuable information from them. The notion of "*ideal*" product is then introduced implicitly or explicitly within questionnaires.

By definition, an ideal is "a conception representing an abstract or hypothetical optimum. It is satisfying one's conception of what is perfect, or most suitable. It is also regarded as a standard to be aimed at" (source: Oxford Dictionary; http://oxforddictionaries.com/). By extension to sensory science, the ideal product is defined as "a product, with a particular sensory characteristic, and that would maximize liking. The ideal product is hence defined with two main components: its sensory profile (reflecting the notion of standard, goal to achieve) and its high expected liking score (reflecting the notion of optimum)."

It is fair to say that ideals are personal, and that it takes all kinds to make a world. The representation of the ideal product can differ largely from one consumer to another. Based on this reality, the statistical analysis of ideal data is reversed compared to classical sensory data. First, in terms of goal, the analysis of ideal data is not directed to find differences between products, but to find the largest consensus between consumers' ideal. This "consensual ideal" is then used as reference to match in the optimization process. Second, in terms of data itself, a large variability between consumers (*i.e.*, consumers' ideals can differ largely) and a small variability between products (*i.e.*, a consumer should associate a set of *relatively* similar products to the same ideal product) are suspected, as opposed to the large variability within panelist (*i.e.*, a panelist should discriminate the products) and the small variability within product (*i.e.*, panelists should agree on the description of a product) expected with classical sensory data.

As both the assumption about the data and the aim of the analysis are reversed, the statistical units of interest are also different. In the classical profiling task, as we have seen previously, the statistical unit of interest is often product-related. When ideal information is involved, one statistical unit of interest is consumer-related, whether it corresponds to the representation of the ideal product (as in the Ideal Profile Method [IPM]) or the deviation of a product from the ideal of a consumer (as in Just-About-Right task [JAR]). However, the structure of the data corresponds to the one used for ANOVA, with as many rows as there are evaluations ($I \times K$, I being the number of products and K the number of consumers), and $2 + J + 1$ columns (J being the number of "sensory" questions asked to the consumer): two columns informing about the product and the consumer information, followed by J sensory variables[1] and, finally, the overall liking response.

As we will see, the two tasks presented here (*i.e.*, JAR and IPM) have a different use of the notion of ideal. Hence, the corresponding analyses and their outcomes differ. But since in both tasks the notion of ideal is implied, the objectives of the corresponding analyses are similar, *i.e.*, optimizing the products tested. In that matter, both JAR and IPM can be seen as diagnostic methods, as they aim at understanding how products differ from their ideals, and how they can hence be improved.

Since the two tasks (JAR and IPM) have different inputs and adopt different points of view (*i.e.*, implicit *versus* explicit measure of the ideal), it is naturally based on this principle that the next section is divided.

Tips: Correction of the use of the scale in IPM

In consumer tests, it is well known that the major source of variability between consumers is related to a different use of the scale. Usually, such differences in the use of the scale between panelists is not important, as the data are averaged across panelists (and eventually standardized). In consumer tests, such differences are corrected by centering within consumer the hedonic scores (*cf.* Chapter 8). In such case, the data are called "preference" rather than "liking scores."

Since, with ideal data obtained from the IPM, most analyses aim at comparing the ideal scores between consumers, such differences in the use of the scales strongly noises the data. Hence, such differences in the scale usage needs to be corrected. To do so, the average perceived intensity of each consumer

---

[1]Depending on the task, J either corresponds to the number of JAR questions, or to the block of $2 \times J$ variables corresponding to the perceived and ideal intensity asked for each sensory attribute in IPM.

and each attribute is used as reference point. In other words, the corrected ideal scores between consumers is obtained by subtracting, for each consumer, the average perceived intensity to the raw ideal scores.

10.2   In practice

Sensory tasks, in which ideal questions are asked, can be of two forms: consumers can either be asked to rate their perception of products on attributes relatively to their ideal, or can be asked to rate directly both perceived and ideal intensities for each attribute and each product.

In the first situation, called JAR task, consumers are asked to rate products on each sensory attribute by mentioning whether it is much too weak, too weak, too strong, much too strong, or just about right. Here, the notion of ideal is implicit, as the deviation from the ideal is measured.

In the second situation, called IPM, consumers are asked to rate products on both their perceived and ideal intensity, for each attribute. In practice, this is done alternatively, for each product and each attribute: after rating their perception of an attribute for a product, consumers are asked to rate the intensity of this attribute for this product if it was at its ideal level. Here, the notion of ideal is explicit within the questionnaire.

As described in Section 10.1, the data set obtained in both cases is structured in a similar way, according to the data set structure used for ANOVA. Let's recall that, usually, the first two columns correspond to the consumer and product information, followed by the different attributes as they are asked in the questionnaire. In the JAR task, one qualitative variable per JAR attribute is recorded. This JAR variable indicates the deviation from the ideal (too weak, too strong, or JAR) the consumer defined for the product and attribute involved. In the IPM task, this information is split into two quantitative variables, one presenting the perceived intensity for that attribute and one presenting the ideal intensity for that attribute. Finally, since both tasks are performed with consumers, overall liking is also asked. Hence, an additional column with the hedonic ratings is also present in the data set.

These two tasks are illustrated through a real case study involving 12 luxurious women perfumes, which were evaluated by 103 Dutch consumers according to the IPM. The test was conducted by OP&P Product Research, Utrecht, The Netherlands (http://www.opp.nl)[1]. Without loss of generality, JAR scores were generated from these data. For each consumer and each attribute, the deviation between perceived and ideal intensity is computed.

---

[1]The aim of this book is to present sensory methods, their corresponding analyses and how to perform them in R. In other words, the origin of the data set is only important to a lesser extent.

Based on the deviations thus obtained (and to thresholds arbitrarily defined), the products are categorized as either much too weak, too weak, just about right, too strong, or much too strong for each consumer and each attribute.

Remark. Like for any optimization procedure, it is highly recommended to define homogeneous clusters of consumers, and to optimize the products for each cluster separately (to define cluster, *cf.* Chapter 8). For brevity and without loss of generality, the data set used for illustration is treated as if it corresponds to one unique cluster.

10.2.1 How can I optimize products based on *Just About Right* data?

The analysis of JAR data consists in evaluating, for each product, the impact on the liking scores of not being at the optimum level (*i.e.*, just about right) for an attribute. Intrinsically, these data are implicitly self-reported ideal data. The loss in liking thus observed is also called *penalty* (of not being just about right), and the global analysis is called *penalty analysis*. This analysis is assimilated to a diagnostic of the products, since its objective is to understand which attributes are at an acceptable/optimum level and which attributes are not. For attributes that are not at their optimal levels, the direction of improvement is defined: the intensity of an attribute must be increased (*resp.* decreased) if the intensity of that attribute is diagnosed to be too low (*resp.* too high).

The data used to illustrate the analysis are stored in the *perfumes_jar.csv* file, which can be downloaded from the book website. After importing the data, transform the *consumer* column to a categorical variable, and print the summary of this table.

```
> jar <- read.table("perfume_jar.csv",header=TRUE,sep=",",dec=".",quote="\"")
> jar$consumer <- as.factor(jar$consumer)
> summary(jar)
    consumer                   product        intensity            freshness
 171    :  12   Angel             :103   Min.   :-2.0000   Min.   :-2.0000
 553    :  12   Aromatics Elixir  :103   1st Qu.: 0.0000   1st Qu.:-1.0000
 991    :  12   Chanel N5         :103   Median : 0.0000   Median : 0.0000
 1344   :  12   Cinéma            :103   Mean   : 0.2864   Mean   :-0.4895
 1661   :  12   Coco Mademoiselle :103   3rd Qu.: 1.0000   3rd Qu.: 0.0000
 1679   :  12   J'adore EP        :103   Max.   : 2.0000   Max.   : 2.0000
 (Other):1164   (Other)           :618
     jasmin             rose             camomille          fresh.lemon
 Min.   :-2.0000   Min.   :-2.0000   Min.   :-2.0000   Min.   :-2.0000
 1st Qu.:-1.0000   1st Qu.:-1.0000   1st Qu.:-1.0000   1st Qu.:-1.0000
 Median : 0.0000   Median : 0.0000   Median : 0.0000   Median : 0.0000
 Mean   :-0.1974   Mean   :-0.3244   Mean   :-0.2112   Mean   :-0.2799
 3rd Qu.: 0.0000   3rd Qu.: 0.0000   3rd Qu.: 0.0000   3rd Qu.: 0.0000
 Max.   : 2.0000   Max.   : 2.0000   Max.   : 2.0000   Max.   : 2.0000

     vanilla            citrus             anis             sweet.fruit
 Min.   :-2.0000   Min.   :-2.0000   Min.   :-2.00000   Min.   :-2.0000
```

```
1st Qu.:-1.0000    1st Qu.:-1.0000    1st Qu.: 0.00000   1st Qu.:-1.0000
Median : 0.0000    Median : 0.0000    Median : 0.00000   Median : 0.0000
Mean   :-0.3058    Mean   :-0.2638    Mean   :-0.07443   Mean   :-0.2346
3rd Qu.: 0.0000    3rd Qu.: 0.0000    3rd Qu.: 0.00000   3rd Qu.: 0.0000
Max.   : 2.0000    Max.   : 2.0000    Max.   : 2.00000   Max.   : 2.0000

      honey              caramel             spicy              woody
Min.   :-2.0000    Min.   :-2.0000    Min.   :-2.0000    Min.   :-2.00000
1st Qu.:-1.0000    1st Qu.: 0.0000    1st Qu.:-1.0000    1st Qu.: 0.00000
Median : 0.0000    Median : 0.0000    Median : 0.0000    Median : 0.00000
Mean   :-0.1869    Mean   :-0.1553    Mean   :-0.0801    Mean   :-0.03722
3rd Qu.: 0.0000    3rd Qu.: 0.0000    3rd Qu.: 0.0000    3rd Qu.: 0.00000
Max.   : 2.0000    Max.   : 2.0000    Max.   : 2.0000    Max.   : 2.00000

     leather              nutty               musk              animal
Min.   :-2.00000   Min.   :-2.000     Min.   :-2.00000   Min.   :-2.00000
1st Qu.: 0.00000   1st Qu.: 0.000     1st Qu.: 0.00000   1st Qu.: 0.00000
Median : 0.00000   Median : 0.000     Median : 0.00000   Median : 0.00000
Mean   : 0.06472   Mean   :-0.106     Mean   : 0.05906   Mean   : 0.09142
3rd Qu.: 0.00000   3rd Qu.: 0.000     3rd Qu.: 1.00000   3rd Qu.: 0.00000
Max.   : 2.00000   Max.   : 2.000     Max.   : 2.00000   Max.   : 2.00000

     earthy             incense             green              liking
Min.   :-2.0000    Min.   :-2.0000    Min.   :-2.0000    Min.   :1.000
1st Qu.: 0.0000    1st Qu.: 0.0000    1st Qu.:-1.0000    1st Qu.:5.000
Median : 0.0000    Median : 0.0000    Median : 0.0000    Median :6.000
Mean   : 0.2597    Mean   : 0.1464    Mean   :-0.3964    Mean   :5.688
3rd Qu.: 0.0000    3rd Qu.: 0.0000    3rd Qu.: 0.0000    3rd Qu.:7.000
Max.   : 2.0000    Max.   : 2.0000    Max.   : 2.0000    Max.   :9.000
```

The JAR variables have values comprised between -2 and +2. This is often the case in a JAR task. In fact, JAR scale is a particular scale that always includes an uneven number of levels, in which the middle point (here 0) is the optimum (*i.e.*, just about right), negative scores the "too weak" levels, and positive scores the "too strong" levels.

For each JAR attribute, it is important to assess the proportion of consumers that selected each of the levels. This is done using the **table** function. To transform the raw proportions to percentage, the **prop.table** function is used. Since the percentage should be computed for each product (hence in rows), the `margin` argument is set to 1. For the intensity attribute we then have:

```
> intensity.freq <- table(jar$product,jar$intensity)
> intensity.freq

                   -2 -1  0  1  2
Angel               6  6 27 21 43
Aromatics Elixir    5  9 26 17 46
Chanel N5           6  6 42 22 27
Cinéma             20 17 45 11 10
Coco Mademoiselle  12 14 43 20 14
J'adore EP         11  9 52 17 14
J'adore ET         11  7 56 16 13
L'instant          12 10 43 17 21
```

```
   Lolita Lempicka   16 12 41 15 19
   Pleasures         16 14 45 12 16
   Pure Poison       10 13 53 11 16
   Shalimar           8  9 32 19 35
> intensity.pct <- round(100*prop.table(intensity.freq,margin=1),2)
> intensity.pct
```

	-2	-1	0	1	2
Angel	5.83	5.83	26.21	20.39	41.75
Aromatics Elixir	4.85	8.74	25.24	16.50	44.66
Chanel N5	5.83	5.83	40.78	21.36	26.21
Cinéma	19.42	16.50	43.69	10.68	9.71
Coco Mademoiselle	11.65	13.59	41.75	19.42	13.59
J'adore EP	10.68	8.74	50.49	16.50	13.59
J'adore ET	10.68	6.80	54.37	15.53	12.62
L'instant	11.65	9.71	41.75	16.50	20.39
Lolita Lempicka	15.53	11.65	39.81	14.56	18.45
Pleasures	15.53	13.59	43.69	11.65	15.53
Pure Poison	9.71	12.62	51.46	10.68	15.53
Shalimar	7.77	8.74	31.07	18.45	33.98

The products presented here seem to be very different in terms of intensity since around 36% of the consumers considered *Cinéma* as not intense enough, while 62% considered *Angel* as too intense; 54% of the consumers considered *J'adore ET* as just about right in terms of intensity.

Let's automatically generate this table for each one of the 21 attributes, that we store in a list of objects called percentage.

```
> percentage <- vector("list",21)
> names(percentage) <- colnames(jar)[3:23]
> for (i in 1:21){
+     attribute.freq <- table(jar$product,jar[,i+2])
+     percentage[[i]] <- as.matrix(100*prop.table(attribute.freq,margin=1))
+ }
```

In practice, JAR experts only consider an attribute as relevant to improve, if a minimum proportion of consumers, who agree on this attribute not being just about right (towards one of the directions), is reached. In the literature, the proportion of 20% is often used.

To visualize graphically the frequency of one particular attribute (say, *vanilla*), the **barplot** function is used. This generates the barplots presented in Figure 10.1.

```
> barplot(t(percentage$vanilla),main='Vanilla',beside=TRUE,
+ legend.text=c('Much Too Little','Too Little','JAR','Too Much',
+ 'Much Too Much'))
```

As said previously, the penalty analysis consists in computing the difference between the average liking score of each non-JAR level with the average liking score of the JAR level, for each attribute. The drop in liking measured here corresponds to the penalty for that attribute. Note that for each attribute, two penalties are calculated: one for the "weak" levels and one for the "strong" levels.

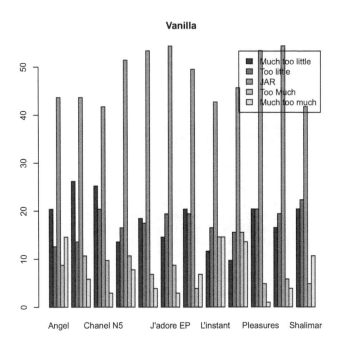

FIGURE 10.1
Representation of the proportion of consumers associated with each JAR and non-JAR category for *Vanilla*, for each product, using the **barplot** function (*jar* data set).

Intrinsically, such means deviation corresponds to the results of a one-way ANOVA, in which the hedonic scores are expressed in function of each JAR attribute separately. By setting the contrast in a way that the coefficient $\alpha_{jar} = 0$ for the JAR level, the α_i related to the too weak and too high levels directly provide the different penalties.

Prior to performing the ANOVA, the data set requires a slight transformation. As no distinction between the too weak and much too weak levels (*resp.* too strong and much too strong levels) is made, they are combined in one unique level arbitrarily called "low" (*resp.* "high"). The JAR level is kept as "just about right," and is set as the reference level for each jar variable. To do so, the **relevel** function is used. To generate such table, the **recode** function from the car package is required.

```
> library(car)
> jar2 <- jar
> for (i in 1:21){
```

```
+    jar2[,i+2] <- recode(jar[,i+2],recodes="c(-2,-1)='low';0='jar';
+    c(1,2)='high'",as.factor.result=TRUE)
+    jar2[,i+2] <- relevel(jar2[,i+2],ref="jar")
+ }
> summary(jar2)
    consumer                  product     intensity  freshness   jasmin
  171   :  12  Angel             :103   jar :505   jar :589   jar :618
  553   :  12  Aromatics Elixir  :103   high:472   high:144   high:227
  991   :  12  Chanel N5         :103   low :259   low :503   low :391
  1344  :  12  Cinéma            :103
  1661  :  12  Coco Mademoiselle :103
  1679  :  12  J'adore EP        :103
 (Other):1164 (Other)           :618
    rose      camomille  fresh.lemon vanilla      citrus      anis     sweet.fruit
 jar :602   jar :717   jar :567    jar :593   jar :701   jar :815   jar :686
 high:188   high:168   high:231    high:199   high:167   high:184   high:181
 low :446   low :351   low :438    low :444   low :368   low :237   low :369



    honey      caramel     spicy      woody      leather     nutty       musk
 jar :731   jar :806   jar :616    jar :748   jar :874   jar :738   jar :643
 high:179   high:144   high:280    high:222   high:207   high:200   high:324
 low :326   low :286   low :340    low :266   low :155   low :298   low :269



    animal     earthy     incense     green       liking
 jar :936   jar :932   jar :776    jar :673   Min.    :1.000
 high:184   high:251   high:282    high:141   1st Qu.:5.000
 low :116   low : 53   low :178    low :422   Median :6.000
                                             Mean    :5.688
                                             3rd Qu.:7.000
                                             Max.    :9.000
```

For each product and each attribute, an ANOVA is performed. After setting up the contrast in such a way that $\alpha_{jar} = 0$ for the JAR level, the subset of data that is related to each product is selected. An ANOVA is then performed on each attribute for each subset, using the **aov** function. The results of the ANOVA are stored in a list of vector arbitrarily called **penalty**. This list has two levels, the first level being related to the product, the second level to the attributes. To obtain the penalties, the **coefficients** element of the **summary.lm** function is called.

```
> options(contrasts=c("contr.treatment","contr.poly"))
> product <- levels(jar2[,2])
> penalty <- vector("list",12)
> names(penalty) <- product
> for (p in 1:length(product)){
+    penalty[[p]] <- vector("list",21)
+    names(penalty[[p]]) <- colnames(jar2)[3:23]
+    jar2.p <- jar2[jar2[,2]==product[p],]
+    for (i in 1:21){
```

```
+            penalty[[p]][[i]] <- summary.lm(aov(jar2.p$liking~
+            jar2.p[,i+2]))$coefficients
+        }
+ }
> round(penalty$Angel$intensity,3)
                      Estimate Std. Error t value Pr(>|t|)
(Intercept)             5.852      0.388   15.099   0.000
jar2.p[, i + 2]high    -2.117      0.462   -4.582   0.000
jar2.p[, i + 2]low     -0.019      0.699   -0.027   0.979
```

For *Angel*, and the attribute *intensity*, the JAR level is associated with an average liking score of 5.852. The penalty (*i.e.*, loss in liking) for the *low* level is -0.019 (not significantly different from 0, with a *p-value* of 0.979), whereas the loss in liking for the *high* level is -2.117 (significantly different from 0). For *Angel*, being perceived as too intense is penalizing liking of more than 2 points.

To facilitate the visualization (hence the interpretation) of the results, current practice consists in representing the penalties in function of the proportion of consumers associated with each non-JAR level. Since the optimization is done product by product, one graph per product is generated. Such a graphic can be generated using this code (here for *Pleasures*):

```
> attribute <- colnames(jar)[3:23]
> plot(0,0,type="n",xlim=c(0,50),ylim=c(-0.2,2),
+ xlab="Proportion (%)",ylab="Penalty",main="Pleasures")
> for (i in 1:21){
+        points(sum(percentage[[i]][rownames(percentage[[i]])=="Pleasures",4:5]),
+        abs(penalty$Pleasures[[i]][2,1]),pch=20,col="black")
+        if (penalty$Pleasures[[i]][2,4]<=0.05){
+          text(sum(percentage[[i]][rownames(percentage[[i]])=="Pleasures",4:5]),
+          abs(penalty$Pleasures[[i]][2,1])+0.05,paste(attribute[i],"*",sep=""),
+          col="black",cex=0.8)
+        } else {
+          text(sum(percentage[[i]][rownames(percentage[[i]])=="Pleasures",4:5]),
+          abs(penalty$Pleasures[[i]][2,1])+0.05,attribute[i],col="black",cex=0.8)
+        }
+        points(sum(percentage[[i]][rownames(percentage[[i]])=="Pleasures",1:2]),
+        abs(penalty$Pleasures[[i]][3,1]),pch=20,col="grey60")
+        if (penalty$Pleasures[[i]][3,4]<=0.05){
+          text(sum(percentage[[i]][rownames(percentage[[i]])=="Pleasures",1:2]),
+          abs(penalty$Pleasures[[i]][3,1])+0.05,paste(attribute[i],"*",sep=""),
+            col="grey60",cex=0.8)
+        } else {
+          text(sum(percentage[[i]][rownames(percentage[[i]])=="Pleasures",1:2]),
+          abs(penalty$Pleasures[[i]][3,1])+0.05,attribute[i],col="grey60",
+            cex=0.8)
+        }
+ }
> abline(v=20,lwd=2,lty=2)
> legend("topleft",bty="n",legend=c("Not enough","Too much"),col=
+ c("grey60","black"),pch=20,cex=0.8)
```

The graphic obtained is shown Figure 10.2. On this graphic, the two non-JAR levels are printed in black (high) and gray (low). Based on this, it can easily be seen that, although a smaller proportion tends to be involved (in average), having too much of an attribute is more penalizing than having not enough (in average).

Additionally, the significant non-JAR attributes are highlighted with a "*." Here, it appears that considering only attributes associated with large proportions of consumers (right side of the vertical line) is misleading. Indeed, some attributes that involve more than 20% of the consumers are associated with non-significant penalty (*e.g.*, *caramel*-not enough), whereas attributes that involve less than 20% of the consumers are associated with a significant penalty (*e.g.*, *leather*-too much).

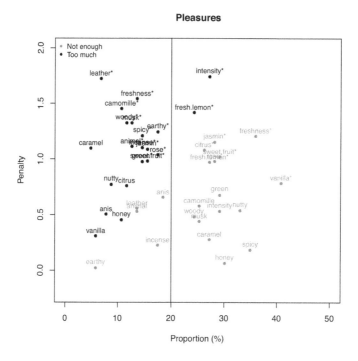

FIGURE 10.2
Representation of the penalties in function of the proportion of consumers for each non-JAR level, for *Pleasures*, using the **plot** function (*jar* data set).

In this example, the optimization procedure for *Pleasures* should focus on reducing the attributes *intensity*, *fresh.lemon* and increasing the attribute *freshness* as these sensory characteristics are associated with large penalties, and affect a large proportion of consumers.

10.2.2   How can I optimize products based on *Ideal Profile Method* data?

In the IPM task, each consumer is asked to rate both the perceived and ideal intensity, as well as the liking scores, for each product. In that sense, IPM data are explicitly self-reported ideal data. The data set hence obtained is richer than in the JAR task. For an overview of a data set obtained with IPM, import the file entitled *perfumes_ideal.csv*, which can be downloaded from the book website. This data set corresponds to the perfume example presented in Section 10.1.

```
> ideal <- read.table("perfumes_ideal.csv",header=TRUE,sep=",",
+ dec=".",quote = "\"")
> ideal[,1] <- as.factor(ideal[,1])
> summary(ideal[,1:6])
     consumer                    product         intensity          id_int
 171     :  12   Angel              :103   Min.   :  2.70   Min.   : 1.80
 553     :  12   Aromatics Elixir   :103   1st Qu.: 47.60   1st Qu.:47.90
 991     :  12   Chanel N5          :103   Median : 61.50   Median :54.80
 1344    :  12   Cinéma             :103   Mean   : 60.45   Mean   :56.27
 1661    :  12   Coco Mademoiselle  :103   3rd Qu.: 75.20   3rd Qu.:65.50
 1679    :  12   J'adore EP         :103   Max.   :100.00   Max.   :97.00
 (Other):1164    (Other)            :618
    freshness         id_fresh
 Min.   : 0.30   Min.   : 1.80
 1st Qu.:35.80   1st Qu.:48.50
 Median :51.80   Median :59.10
 Mean   :51.21   Mean   :59.01
 3rd Qu.:67.90   3rd Qu.:69.78
 Max.   :97.90   Max.   :97.90
```

As can be seen, the data set presented here is a succession of blocks of two variables, one corresponding to the perceived intensity of an attribute, followed by its ideal intensity. The ideal variables can easily be detected, as they all start with the same sequence, here "id_". These particularities (*i.e.*, order of variables and names of the ideal attribute) are of utmost importance since, as we will see, the functions related to the analysis of Ideal Profile data proposed in the SensoMineR package use this information to detect the ideal variables from the perceived intensity one.

As mentioned at the end of Section 10.1, any optimization technique requires first to define homogeneous clusters of consumers, and to optimize the products for each of them. The clustering procedure used here being the same as the one presented in Chapter 8, it is not described here. To simplify, and without loss of generality, we suppose here that the panel of consumers consists in one homogeneous cluster.

Besides the notion of homogeneous cluster of consumers, another concept arises with ideal data, and particularly in the case where ideal intensities are measured for each product separately. Is it legitimate to associate the product space to one unique ideal? Or do consumers associate some products to different ideals? To evaluate such a concept, we need to evaluate whether the ideal intensities provided from the consumers shift when associated with some particular products. This is the principle of the **MultiIdeal** function from the SensoMineR package, which generates fictive panels of consumers from the original one by re-sampling with replacement. Such a procedure generates confidence ellipses around the average ideal products computed over the consumers. If two ellipses overlap, no systematic shift is observed and the two products of interest are associated with a similar ideal. Inversely, if two ellipses are clearly separated, a shift in the ideal ratings is observed between these two products, and both products are not strictly associated with one unique ideal.

To evaluate whether the products tested should be associated with one or multiple ideals, the **MultiIdeal** function of the SensoMineR package is used. Like any of the functions related to the Ideal Profile Analysis, this function presents the `id.recogn` argument, which takes the string sequence that allows recognizing ideal attributes from the other variables. All the variables directly preceding an ideal attribute are considered as the corresponding sensory attribute (hence the necessity of arranging correctly the data set).

```
> library(SensoMineR)
> res.MultiId <- MultiIdeal(ideal,col.p=2,col.j=1,id.recogn="id_")
[1] "The attribute camomille is removed from the analysis."
> round(res.MultiId,3)
```

	Angel	Aromatics Elixir	Chanel N5	Cinéma	Coco Mademoiselle
Angel	1.000	0.273	0.067	0.264	0.028
Aromatics Elixir	0.273	1.000	0.354	0.331	0.295
Chanel N5	0.067	0.354	1.000	0.657	0.873
Cinéma	0.264	0.331	0.657	1.000	0.370
Coco Mademoiselle	0.028	0.295	0.873	0.370	1.000
J'adore EP	0.002	0.014	0.282	0.091	0.324
J'adore ET	0.000	0.001	0.024	0.004	0.039
L'instant	0.086	0.207	0.889	0.756	0.648
Lolita Lempicka	0.466	0.437	0.526	0.909	0.287
Pleasures	0.001	0.006	0.159	0.041	0.204
Pure Poison	0.020	0.247	0.795	0.302	0.985
Shalimar	0.276	0.778	0.097	0.106	0.078

	J'adore EP	J'adore ET	L'instant	Lolita Lempicka	Pleasures
Angel	0.002	0.000	0.086	0.466	0.001
Aromatics Elixir	0.014	0.001	0.207	0.437	0.006
Chanel N5	0.282	0.024	0.889	0.526	0.159
Cinéma	0.091	0.004	0.756	0.909	0.041
Coco Mademoiselle	0.324	0.039	0.648	0.287	0.204
J'adore EP	1.000	0.471	0.402	0.045	0.955
J'adore ET	0.471	1.000	0.048	0.002	0.631
L'instant	0.402	0.048	1.000	0.560	0.254
Lolita Lempicka	0.045	0.002	0.560	1.000	0.019
Pleasures	0.955	0.631	0.254	0.019	1.000
Pure Poison	0.368	0.053	0.589	0.227	0.246
Shalimar	0.001	0.000	0.048	0.176	0.000

	Pure Poison	Shalimar
Angel	0.020	0.276
Aromatics Elixir	0.247	0.778
Chanel N5	0.795	0.097
Cinéma	0.302	0.106

Coco Mademoiselle	0.985	0.078
J'adore EP	0.368	0.001
J'adore ET	0.053	0.000
L'instant	0.589	0.048
Lolita Lempicka	0.227	0.176
Pleasures	0.246	0.000
Pure Poison	1.000	0.064
Shalimar	0.064	1.000

The analysis generates the graphical output presented in Figure 10.3. Additionally, the Hotelling T^2 test is performed for each pair of products, and the resulting *p-values* are provided in a table. Two products are associated with a similar ideal product if the corresponding *p-value* is larger than a certain threshold.

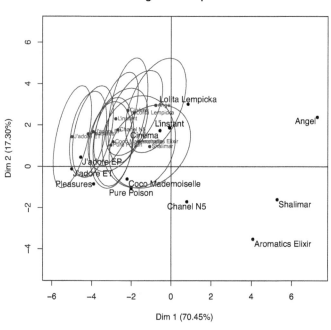

FIGURE 10.3
Assessment of the heterogeneity of the product space through the variability of the ideal ratings, using the **MultiIdeal** function (*ideal* data set).

In this example, it appears that all the products are projected in a similar area of the product space. Still, a closer look at the confidence ellipses highlights the existence of a systematic shift between some pairs of products. Since the systematic shift is not so evident (some products would belong to multiple subsets of products), and without loss of generality, we consider here that all the products belong to the same subset.

Remark. The interpretation of both the Hotelling T^2 table and the confidence ellipses should not be done in a *strict* way, and the analyst should also consider their own judgment in the final decision (as in the example presented here). Do the differences between ideal justify considering multiple ideals? In case of doubts, we suggest that each product should be optimized according to its own ideal, *i.e.*, the ideal product that is defined with respect to that product.

Once homogeneous groups of consumers and homogeneous groups of products are defined (here, homogeneous in terms of ideal product they are linked to), the next step consists in defining the sensory profile of the ideal product used as reference, and this for each cluster of consumers and each homogeneous subset of products. To define such ideal products of reference, two points of view can be adopted:

- the ideal product of reference corresponds to the average ideal product defined within each cluster of consumers;

- the ideal product of reference corresponds to the ideal product common to a maximum of consumers within each cluster.

In the first case, the ideal product of reference is obtained by computing the average ideal score for each attribute, across products, within each cluster of consumers. In the second case, the ideal product of reference is obtained by defining which part of the product space corresponds to the ideal product of a maximum of consumers. To do so, each consumer is associated with an ideal area (*i.e.*, the area of the space in which their ideal belongs to), and the point of the sensory space that is shared by a maximum of consumers (in practice, where a maximum of individual ideal areas overlap) defines the ideal product of reference. Such individual ideal areas are obtained by re-sampling techniques with replacement, which define confidence ellipses around the average ideal product provided by each consumer.

Remark. Compared to the different procedures generating confidence ellipses described throughout the book, it can be noted that in this particular case, the consumer is the center of attention. Hence, the re-sampling technique (with replacement) is done on the set of products. In other words, the question we try to answer here is: where would be the average ideal product of that consumer if the product space were defined with these *new* products? This procedure is hence reversed compared to the one described before, which justifies the remark in Section 10.1.

Once all the individual ideal areas are defined on the product space, the area shared by a maximum proportion of consumers defines the ideal product to consider as reference. This procedure is very similar to the one defining the optimum product in PrefMap (*cf.* Section 9.2.3 from Chapter 9), although no regression models are involved here. Such product space, which includes the individual ideal areas, defines an Ideal Map (or IdMap).

Let's consider the second point of view here. To obtain the IdMap for the *ideal* data set, the **IdMap** function of the FactoMineR package is used. To zoom in a particular location of the IdMap, the `xlim` and `ylim` arguments of the **plot.IdMap** function are used.

```
> res.IdMap <- IdMap(ideal,col.p=2,col.j=1,col.lik=ncol(ideal),id.recogn='id_')
> plot.IdMap(res.IdMap,xlim=c(-10,10),ylim=c(-10,10),color=FALSE,inverse=TRUE)
```

The IdMap obtained is shown in Figure 10.4. In this map, as we set the `color=FALSE` and `inverse=TRUE`, the darker area corresponds to larger proportions of consumers sharing a common ideal product.

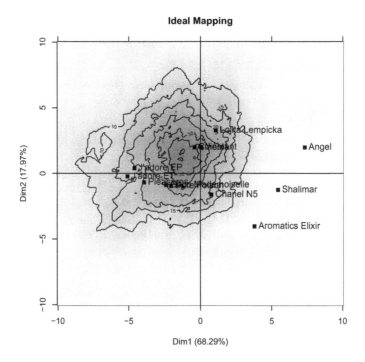

FIGURE 10.4
Representation of the Ideal Map on the first two dimensions resulting from a PCA, using the **IdMap** and the **plot.IdMap** function (*ideal* data set).

Now the ideal product of reference, it is worth getting its profile and getting an idea of the proportion of consumers sharing it. All this information are outputs of the **IdMap** function, which are stored in `res.IdMap$ideal`. After printing its profile, let's save the ideal profile of reference in an object called `ideal.ref`:

```
> names(res.IdMap)
[1] "PCA"   "idmap" "ideal"
> res.IdMap$ideal
$profiles
        intensity freshness    jasmin    rose camomille fresh.lemon vanilla
Ideal   55.70271   58.7438  42.71279 41.1093  31.77093     40.2843 37.4436
           citrus     anis sweet.fruit   honey  caramel     spicy   woody
Ideal   33.17074 27.50484    36.9876 30.64806 28.03934 37.56512 25.59419
          leather    nutty      musk  animal    earthy  incense    green
Ideal   20.04593 30.23391  31.89109 20.46996 17.43876 24.92636 33.14748

$pct.conso
[1] 41.7
> ideal.ref <- round(res.IdMap$ideal$profiles,2)
```

Based on these outputs, it can be noted that the ideal profile used as reference is common to approximately 42% of the consumers.

The sensory profile of the ideal product of reference is then used to optimize the actual products. To do so, the deviation between the perceived intensity of a product and the ideal level of reference is used. After computing the product profiles using the **averagetable** function from the SensoMineR package, let's calculate the deviation from the ideal for *Chanel N5*, for example:

```
> perc.att <- grep("id_",colnames(ideal))-1
> senso.profile <- averagetable(ideal[,c(1,2,perc.att)],
+ formul="~product+consumer",firstvar=3)
> deviation <- round(ideal.ref-senso.profile["Chanel N5",],2)
> deviation
          intensity freshness jasmin rose camomille fresh.lemon vanilla citrus
Chanel N5    -10.42     10.76   4.41 1.17       2.8        5.47    7.75   6.04
          anis sweet.fruit honey caramel spicy woody leather nutty  musk animal
Chanel N5 4.06       8.34  5.88    3.79 -2.83 -1.57   -1.29  2.33 -3.29   1.05
          earthy incense green
Chanel N5  -3.18   -4.49   9.8
```

The deviation from the ideal can be represented in barplots by using the **barplot** function. By recoding the deviations based on their sign (**recode** function of the car package), we color the bars in black if the deviation is positive (the attribute needs more), and in gray if the deviation is negative (the attribute needs less).

```
> library(car)
> col.dev <- as.numeric(sign(deviation))
> col.dev <- as.character(recode(col.dev,"1='black'; else='grey60'",
+ as.factor.result=TRUE))
> pos <- barplot(t(deviation),xlim=c(0,32),ylim=c(-12,12),
+ ylab="Deviation from Ideal",beside=TRUE,space=0.5,col=col.dev)
> text(pos-0.8,-1.5,colnames(deviation),srt=45,cex=0.85)
```

The code generates Figure 10.5. To improve *Chanel N5*, it is advised to decrease *intensity* and increase *freshness, etc.* But this comment is especially true if *intensity* and *freshness* have a strong impact on liking (to estimate the impact of an attribute on liking, please refer to Section 9.2.2 in Chapter 9).

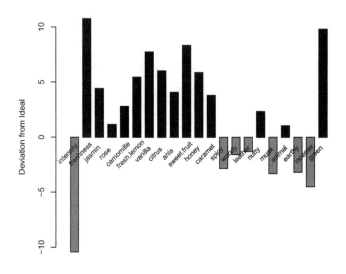

FIGURE 10.5
Barplots representing the deviations of the perceived intensity from the ideal intensity, for *Chanel N5*, using the **barplot** function (*ideal* data set).

10.3 For experienced users: Assessing the consistency of ideal data in IPM

The JAR task, as well as the IPM, are potentially powerful since the data acquired from these tasks are directly actionable to optimize products. However, since these data are provided from consumers who are asked to rate a fictive product, their quality is questionable.

Since the data acquisition in the IPM is very rich (both the perceived and ideal intensity, as well as the hedonic ratings, are measured), the consistency of the ideal data can be assessed. Unfortunately, to our knowledge, such validation process of the data is not possible with JAR data.

By definition, the ideal product is a product with particular sensory characteristics that maximize liking. From this definition, it appears that the ideal product has two components: one related to its sensory profile and one related to its liking. For that reason, the consistency of the ideal data is assessed in two steps: the sensory consistency and the hedonic consistency.

The sensory consistency consists in ensuring that the ideal product of a consumer has similar sensory characteristics than the most liked products. In other words, if a consumer likes more the products perceived as sweeter, we expect their ideal to be rather sweet. The hedonic consistency consists in ensuring that a consumer would like more their ideal product than the existing products, if it happened to exist.

For the sensory consistency, the analysis evaluates whether the ideal information provided from the consumer is making the link between the sensory profiles of the products and their liking. The panel is consistent (from a sensory point of view) if a link exists between the sensory and hedonic information, within the ideal product space. For the hedonic consistency, the liking score (also called liking potential) of the ideal product of a consumer is estimated through Principal Component Regression (PCR), and is compared to the liking scores provided to the product space. Consumers providing consistent ideals (from a hedonic point of view) have provided ideal products for which the liking potential is relatively high compared to the liking scores provided to the products.

To assess the sensory consistency of the ideal data, the **ConsistencyIdeal** function of the SensoMineR package is used. In this case, the `type` argument is set to `sensory`, and we are particularly interested in the `$Senso` element of the results.

```
> senso.consist <- ConsistencyIdeal(ideal,col.p=2,col.j=1,col.lik=ncol(ideal),
+ id.recogn="id_",type="sensory")$Senso
> names(senso.consist)
[1] "panel" "conso"
> plot.PCA(senso.consist$panel$PCA.ideal_senso,choix="ind",
+ label="ind.sup",col.ind.sup="grey40")
> plot.PCA(senso.consist$panel$PCA.ideal_hedo,choix="var",
+ col.quanti.sup="grey40")
```

At the panel level, the evaluation of the sensory consistency is done through PCA on the ideal profiles (corrected from the use of the scale). On this ideal space, the sensory profiles are projected as supplementary entities. The rationale behind this projection is to consider each product as a particular consumer who has that product as ideal (*cf.* Figure 10.6). Additionally, the raw hedonic scores are projected as supplementary variables (*cf.* Figure 10.7).

From this PCA, the sensory consistency is evaluated through the ideal data as a link between the sensory and hedonic information about the product. Such link is assessed through the direct correspondence between the same products seen through the sensory and/or hedonic description. In this example, since the consumers who rated their ideal with similar characteristics than *J'adore ET* (*resp. Shalimar*) also preferred *J'adore ET* (*resp. Shalimar*), the panel is considered as consistent, from a sensory point of view.

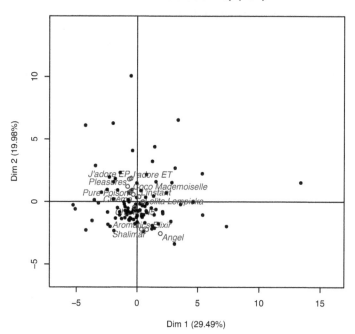

Individuals factor map (PCA)

FIGURE 10.6

Representation of the ideal product space for the assessment of the sensory consistency of the consumer, using the **ConsistencyIdeal** function (*ideal* data set).

Numerically, the sensory consistency can be evaluated at the panel level through the correlation coefficients measured between the scores of the supplementary entities and the loadings of the supplementary variables (in both cases, it corresponds to the products). These coefficients are expected to be high and positive for the first corresponding dimensions. The correlation coefficients are stored in the object senso.consist$panel$correlation.

```
> round(senso.consist$panel$correlation[1:3,1:3],3)
                  Dim.1_ideal.hedo Dim.2_ideal.hedo Dim.3_ideal.hedo
Dim.1_ideal.senso            0.814           -0.756           -0.071
Dim.2_ideal.senso           -0.845            0.870           -0.022
Dim.3_ideal.senso           -0.240            0.239            0.618
```

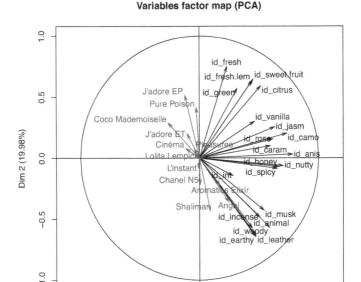

FIGURE 10.7
Representation of the ideal and hedonic variables for the assessment of the sensory consistency of the consumer, using the **ConsistencyIdeal** function (*ideal data set*).

From this output, the correlation coefficients between dimension 1, dimension 2, and dimension 3 are, respectively, 0.81, 0.87, and 0.62. This confirms the consistency of the panel of consumers, from a sensory point of view.

At the consumer level, the consistency is evaluated through the link between the ideal scores on one hand, and the individual drivers of liking on the other hand. Indeed, for a given consumer, if an attribute positively drives liking, we expect that consumer to provide an ideal rating rather high for that attribute. Inversely, if an attribute negatively drives liking, they should provide an ideal rating rather low for that attribute.

Such link is evaluated through the correlation coefficient between the corrected ideal data on one hand, and the drivers of liking on the other hand. This correlation coefficient is expected to be high and positive for consistent consumers.

For each consumer, the correlation coefficient between their ideal ratings and their individual drivers of liking is stored in the object senso.consist$conso$correlations. Since it can be tedious to go through the 103 correlation coefficients, let's represent them graphically. To do so, the **plot** and the **density** functions are combined.

```
> plot(density(senso.consist$conso$correlations[,1]),
+ main="Sensory consistency evaluated at the consumer level")
> abline(v=0.37,lwd=2,lty=2)
```

In the perfume example, the distribution of the individual correlation coefficients are shown in Figure 10.8. The threshold value for the correlation coefficient at 5% being around 0.37 (19 degrees of freedom), it can be concluded here that the large majority of consumers is consistent from a sensory point of view.

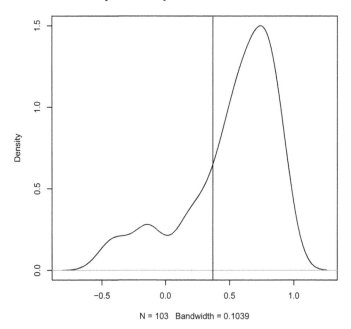

FIGURE 10.8
Distribution of the individual correlation coefficients assessing the sensory consistency of the ideal data at the consumer level, using the **ConsistencyIdeal** function (*ideal* data set).

As mentioned above, to be consistent from a hedonic point of view, the consumers should provide ideal products that would be more liked than the products tested if they happened to exist. Since the ideal potential (*i.e.*, the liking score of the ideal product of a given consumer) cannot be measured directly from the consumer, it needs to be estimated. To do so, individual models explaining the liking in function of the sensory attributes are computed for each consumer. We then apply each individual model on the ideal profile provided by the consumer of interest, in order to estimate the liking potential of their ideal product.

Such procedure is the core of the hedonic consistency, as developed in the **ConsistencyIdeal** function of the SensoMineR package. However, in this case, the `type` argument needs to be set to `hedonic`. Here, our interest is in the `$Hedo` element of the results.

Besides estimating the liking potential of the individual ideal products, a test based on permutation can be performed. This test evaluates the significance of these liking potentials. To run this test, the number of `nbsim` argument should be different from 0 (value by default). In practice, 200 to 500 simulations seem to be sufficient.

```
> hedo.consist <- ConsistencyIdeal(ideal,col.p=2,col.j=1,col.lik=ncol(ideal),
+ id.recogn="id_",type="hedonic",nbsim=200,graph=FALSE)$Hedo
```

The overall distribution of the liking scores and of the liking potentials are provided in Figure 10.9. This box plot gives an overall view whether the ideal data are consistent from a hedonic point of view. Here, it seems to be the case as the liking potentials (ideal products) have higher means, median, and quartiles than the hedonic scores provided to the products.

Although the ideal products seem to be consistent from a hedonic point of view at the panel level, an assessment at the consumer level is required. To do so, the individual liking potentials should be compared to the hedonic scores provided by the consumer of interest to the products. To facilitate the interpretation of the results, rather than using the raw estimates, the liking potentials are made relative to the hedonic scores provided by the consumers. To do so, values similar as *z-scores* are computed, and for each consumer, the relative liking potential associated with their ideal product is obtained by subtracting to the ideal potential the average liking score. This difference is then divided by the standard deviation of the liking scores.

By definition, the relative liking potentials are expected to be large, at least positive, for consistent consumers. Since the liking potentials are estimated using models, the quality of the individual models (the *adjusted* R^2 is used here) is also of utmost importance, and should be taken into consideration in the interpretation of the results. For that reason, the relative liking potentials are plotted in function of the *adjusted* R^2 of the corresponding individual models (*cf.* Figure 10.10).

When permutations are performed, additional outputs are generated.

Let's recall that the aim of these permutations is to evaluate the significance of the liking potential estimated for each consumer. More precisely, the permutations are performed to generate the distribution of the liking po-

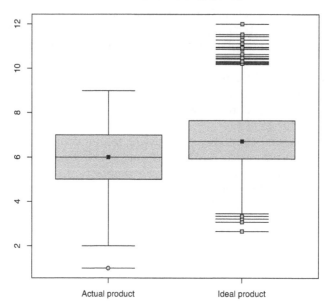

FIGURE 10.9

Distribution of the liking scores (*Actual product*) and of the estimated liking potential (*Ideal product*) for the assessment of the hedonic consistency, using the **ConsistencyIdeal** function (*ideal* data set).

tential of each consumer under H_0[1]. The liking potential observed is then positioned on the corresponding distribution according to the usual approach of hypothesis testing in statistics. Specifically, we count (in percentage) how many times the liking potential obtained in random situation is higher than the liking potential observed. This percentage is used as a *p-value*.

To facilitate the interpretation of such results, we represent graphically the distribution of the individual *p-values*. To do so, the **plot** and the **density** functions are applied on the hedo.consist$simulation$pvalue object.

```
> plot(density(na.omit(hedo.consist$simulation$pvalue)),
+ main="Distribution of the p-values (liking potential)")
> abline(v=0.05,lwd=2,lty=3)
> abline(v=0.10,lwd=2)
```

---

[1] H_0: the ideal profile is defined randomly, and no structure is observed in the ideal profile; H_1: the ideal profile is not defined randomly, and a structure is observed in the ideal profile.

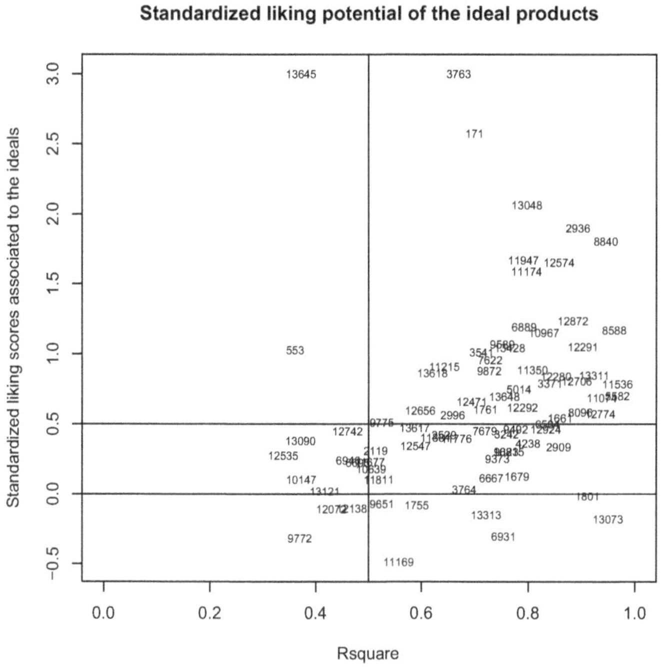

FIGURE 10.10
Representation of the standardized liking potential in function of the quality of the individual models for the assessment of the hedonic consistency, using the **ConsistencyIdeal** function (*ideal* data set).

The distribution of the *p-values* shows that the large majority of the consumers is providing ideal products that are associated with liking potential that cannot be obtained randomly. Indeed, the large majority of *p-values* is below 5%, and almost all of them are below 10% (*cf.* Figure 10.11). In other words, most of the consumers do not provide random ideal profiles.

Finally, as the significance of the liking potential is provided here, it seems interesting to add this information on Figure 10.10. Indeed, in this representation, which consumers are providing random ideal profiles?

To answer this question, we propose to reproduce Figure 10.10, in which a color code corresponding to the *p-value* associated with the liking potential is added. To do so, three main outputs are required: the individual relative liking potential, the *adjusted* R^2 associated with the individual models, and the *p-values* associated with the liking potential. These three outputs are, respectively, stored in `hedo.consist$hedo$relativ.ideal`, in `hedo.consist$R2aj`, and in `hedo.consist$simulation$pvalue`.

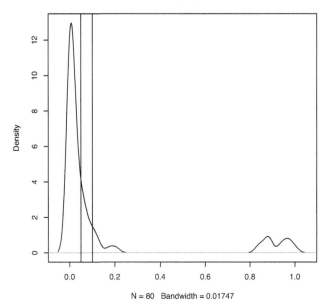

Distribution of the p–values (liking potential)

N = 80   Bandwidth = 0.01747

FIGURE 10.11
Distribution of the *p-values* associated to the significance of the estimated
liking potential for the assessment of the hedonic consistency, using the **ConsistencyIdeal** function (*ideal* data set).

To produce such a graphic, the following code is used:

```
> plot(0,0,type="n",xlim=c(0,1),ylim=c(-1,3),
+ xlab="Adjusted R2",ylab="Relative Liking Potential")
> for (i in 1:nrow(hedo.consist$R2aj)){
+     if (!is.na(hedo.consist$simulation$pvalue[i,1])){
+         if (hedo.consist$simulation$pvalue[i,1]<0.05){
+             color="black"
+             point=16
+         } else if (hedo.consist$simulation$pvalue[i,1]<0.10){
+             color="grey40"
+             point=17
+         } else {
+             color="grey70"
+             point=18
+         }
+         points(hedo.consist$R2aj[i,1],hedo.consist$hedo$relativ.ideal[i,1],
+         pch=point,col=color)
+         text(hedo.consist$R2aj[i,1],hedo.consist$hedo$relativ.ideal[i,1]+0.1,
+         labels=rownames(hedo.consist$R2aj)[i],cex=0.7,col=color)
+     }
+ }
```

```
> legend("bottomleft",legend=c("P<5%","P<10%","P>10%"),bty="n",
+ col=c("black","grey40","grey70"),text.col=c("black","grey40","grey70"),
+ horiz=TRUE,cex=0.8,pch=c(16,17,18))
> abline(v=0.5,lwd=2)
> abline(h=0,lwd=2)
> abline(h=0.5,lty=3)
```

Thanks to the color code used, it appears that the non-consistent (from the hedonic point of view) consumers (*i.e.*, consumers associated with low relative liking potential) are, in fact, associated with liking potentials that are not significant. Hence, the consumers most probably provided random ideal scores (*cf.* Figure 10.12).

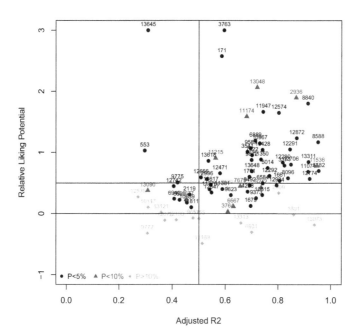

FIGURE 10.12
Representation of the standardized liking potential (color-coded based on their significance) In function of the quality of the individual models for the assessment of the hedonic consistency, using the **ConsistencyIdeal** function (*ideal* data set).

JAR and IPM are two tasks that are based on a similar assumption (*i.e.*, consumers have a clear representation of their ideal products) and that have similar objectives (*i.e.*, optimize products). However, if one of the tasks seems simpler in practice (*i.e.*, JAR), the other one (*i.e.*, IPM) gathers more information, which allows a deeper understanding of the consumers and a broader analysis of the data (*e.g.*, assessment of the consistency of the data, single *versus* multiple ideals, IdMap).

10.4   Exercises

Exercise 10.1 *Penalty analysis on data based on emotions: the Sennheiser case study*

In the Exercise 4.3 and Exercise 4.4, we introduced you to a study on sound branding. 27 variants of the Sennheiser sound logo were generated according to three factors with three modalities. Amongst the data collected, 8 JAR questions on emotional state were asked. In the Exercise 4.4, we showed you how to analyze such data in a textual way. The aim of this exercise is now to analyze this data in the classical way, *i.e.* using penalty analysis.

- Import the *Sennheiser_JAR_og.csv* file from the book website.

- Recode each JAR variable into a factor with 3 levels: *Low*, *JAR*, and *High*. The **recode** function of the car package can be used.

- Select the data for one of the variants of your choice.

- Using the **table** function, create a contingency table defining the proportion of consumers in each JAR and non-JAR category.

- After adjusting the contrast, run a 1-way ANOVA expressing the liking score in function of the JAR level of each attribute. Extract from this analysis the mean drop related to the non-JAR level for each attribute.

- Create the graph representing the penalty in function of the proportion of consumers in each non-JAR level for the product selected. Different color code can be used for the Low and High categories, and a "*" can be added for significant mean drops.

Exercise 10.2 *Alternative approach of the Penalty analysis*

Generally, and as we have seen in the previous exercise, the penalty analysis is performed for each product separately. Moreover, within each product, the impact (*i.e.*, the mean drop) of each attribute is measured separately. Such observation raises two major concerns:

1. What does a penalty associated with one product really mean?

2. By analyzing each attribute separately, the underlying structure of the data set in terms of collinearity is ignored. If a product is considered as not sweet enough, is it also too sour?

As the penalty analysis can be performed through ANOVA, different models can be considered. Additionally, the relationship between JAR variables can be assessed through multivariate analysis. In this case, since the variables are qualitative, MCA is used to understand the product space.

To illustrate this new approach, the same case study as in Exercise 10.1 is used.

- Import the *Sennheiser_JAR_og.csv* file from the book website.

- Recode each JAR variable into a factor with 3 levels. The **recode** function of the car package can be used. For interpretation reasons, recode the variables by indicating the attribute involved (*e.g.*, sweet_low, swet_JAR, sweet_high).

- After adjusting the contrast, run a 2-way ANOVA with interaction explaining the liking score in function of the JAR level of each attribute separately and the product effect. Extract from this analysis the mean drop related to the non-JAR level for each attribute. Interpret the results.

- Perform MCA on the raw recoded data, and project the consumer and the product information as supplementary qualitative variables.

- Represent the results by focusing on the modalities of the JAR variables.

Exercise 10.3 *Comparing the optimum product from PrefMap and IdMap*

In this study, 12 perfumes were evaluated by 103 consumers using the Ideal Profile Method. The products were rated on both perceived and ideal intensity for 21 attributes, and on overall liking. Since the consumers are provided both the sensory profiles of the products and their ideal profile, we are interested in comparing the optimum products obtained from PrefMap and from IdMap.

- Import the *perfumes_ideal.csv* file from the book website.

- After selecting the adequate columns, compute the sensory profile of the products using the **averagetable** function of the SensoMineR package. The selection of the perceived intensity column can be done using the **grep** function.

- Create the product space using the **PCA** function of the FactoMineR package. Extract the coordinates of the products on the first two dimensions.

- Transform the column of liking scores into a matrix crossing the products in rows and the consumers in columns.

- Generate the response surface of the PrefMap using the **carto** function of the SensoMineR package.

- Using the reverse regression procedure (see Exercise 9.5) estimate the sensory profile of the optimum profile previously selected.

- Define the optimum product based on the IdMap procedure. To do so, use the **IdMap** function of the SensoMineR package.

- Compare the sensory profile of the optima obtained through the two methodologies. Such comparison can be done in a spider plot, for instance.

Exercise 10.4 *Optimization of skin creams*

Eight skin creams were formulated using an experimental design involving 4 factors, each factor including 2 levels. These 8 creams were then tested by 72 consumers following the IPM procedure involving 13 sensory attributes and 4 liking questions. These data were collected by Amandine Crine and Adeline Gruel during their master's degree studies.

- Import the *cream_ideal1.csv* file from the book website.

- After selecting the adequate columns, compute the sensory profile of the products using the **averagetable** function of the SensoMineR package. The selection of the perceived intensity column can be done using the **grep** function.

- Create the product space using the **PCA** function of the FactoMineR package. Extract the coordinates of the products on the first two dimensions.

- Transform the column of liking scores into a matrix crossing the products in rows and the consumers in columns.

- Generate the response surface of the PrefMap using the **carto** function of the SensoMineR package.

- Using the reverse regression procedure (see Exercise 9.5) estimate the sensory profile of the optimum profile previously selected.

- Define the optimum product based on the IdMap procedure. To do so, use the **IdMap** function of the SensoMineR package.

- Compare the sensory profile of the optima obtained through the two methodologies. Such comparison can be done in a spider plot, for instance.

Exercise 10.5 *Validation of the IPM*

This exercise is a follow up of Exercise 10.4. By applying the guidance on improvement proposed earlier, two new variants of skin cream were formulated. From our impression, one should show a slight improvement (*product 9*) whereas the second one should show a larger improvement (*product 2*). To make sure that the improvements are real, the 2 new prototypes were evaluated in the same condition as in the previous study. To do so, we substitute 2 of the 8 previous products by these 2 new variants, and performed the same test with 65 of the 72 previous consumers on the new set of 8 products.

- Import the *cream_ideal2.csv* file from the book website.

- After selecting the adequate variables, construct the table of sensory profiles using the **averagetable** function of the SensoMineR package. Then, create the product space using the **PCA** function of the FactoMineR package. Is the product space similar to the one defined in the first test?

- Using the ideal map procedure, define the zone of the optimum product. Are the new products closer to that zone?

- Using the liking questions, compare the different products. To do so, the **decat** function of the SensoMineR package can be used. Alternatively, the **HSD.test** function of the agricolae package can be used. Can we conclude that the new prototypes are showing an improvement?

10.5   Recommended readings

- Cooper, H.R., Earle, M.D., & Triggs, C.M. (1989). Ratios of Ideals ? A New Twist to an Old Idea, in Product Testing with Consumers for Research Guidance, Philadelphia, ASTM STP 1035, 54-63.

- Gacula, M., Rutenbeck, S., Pollack, L., Resurreccion, A. V. A., & Moskowitz, H. R. (2007). The Just About Right intensity scale: functional analyses and relation to hedonics. *Journal of Sensory Studies*, 22, 194-211.

- Hoggan, J. (1975). New Product Development, MBAA Technical Quarterly, 12, 81-86.

- Li, B., Hayes, J. E., & Ziegler, G. R. (2014). Just-about-right and ideal scaling provide similar insights into the influence of sensory attributes on liking. *Food Quality and Preference*, 37, 71-78.

- Meullenet, J.F., Xiong, R., & Findlay, C.J. (2007). Multivariate and probabilistic analyses of sensory science problems, Ames, Blackwell Publishing, 208-210.

- Moskowitz, H.R. (1972). Subjective ideals and sensory optimization in evaluating perceptual dimensions in food. *Journal of Applied Psychology*, 56, (1), 60-66.

- Moskowitz, H.R., Stanley, D.W., & Chandler, J.W. (1977). The Eclipse method: Optimizing product formulation through a consumer generated ideal sensory profile. *Canadian Institute of Food Science Technology Journal*, 10, (3), 161-168.

- Pagès, J., Berthelo, S., Brossier, M., & Gourret, D. (2014). Statistical penalty analysis. *Food Quality and Preference*, 32, Part A, 16-23.

- Rothman, L., & Parker, M. (2009). Just-About-Right JAR scales: Design, usage, benefits and risks. ASTM International, Manual MNL-63-EB, USA.

- Szczesniak, A., Loew, B.J., & Skinner, E.Z. (1975). Consumer texture profile technique. *Journal of food science*, 40, (6), 1253-1256.

- Van Trijp, H.C., Punter, P.H., Mickartz, F., & Kruithof, L. (2007). The quest for the ideal product: comparing different methods and approaches. *Food Quality and Preference*, 18, (5), 729-740.

- Worch, T., Crine, A., Gruel, A., & Lê, S. Analysis and validation of the Ideal Profile Method: Application to a skin cream study. *Food Quality and Preference*, 32, Part A, 132-144.

- Worch, T., Dooley, L., Meullenet, J.F., & Punter, P. (2010). Comparison of PLS dummy variables and Fishbone method to determine optimal product characteristics from ideal profiles. *Food Quality and Preference*, 21, 1077-1087.

- Worch, T., & Ennis, J.M. (2013) Investigating the single ideal assumption using Ideal Profile Method. *Food Quality and Preference*, 29, (1), 40-47.

- Worch, T., Lê, S., & Punter, P. (2010). How reliable are the consumers? Comparison of sensory profiles from consumers and experts. *Food Quality and Preference*, 21, (3), 309-318.

- Worch, T., Lê, S., Punter, P., & Pagès, J. (2012). Assessment of the consistency of ideal profiles according to non-ideal data for IPM. *Food Quality and Preference*, 24, 99-110.

- Worch, T., Lê, S., Punter, P., & Pagès, J. (2012). Construction of an Ideal Map (IdMap) based on the ideal profiles obtained directly from consumers. *Food Quality and Preference*, 26, 93-104.

- Worch, T., Lê, S., Punter, P., & Pagès, J. (2012). Extension of the consistency of the data obtained with the Ideal Profile Method: Would the ideal products be more liked than the tested products? *Food Quality and Preference*, 26, 74-80.

- Worch, T., Lê, S., Punter, P., & Pagès, J. (2013). Ideal Profile Method: the ins and outs. *Food Quality and Preference*, 28, 45-59.

A

The R survival guide

CONTENTS

A.1 What is R?

Quæ sunt Cæsaris, Cæsari: the first section of this appendix describing the R project is directly borrowed from "The R Project for Statistical Computing" website: http://www.r-project.org/.

A.1.1 Introduction to R

R is a language and environment for statistical computing and graphics. It is a GNU project which is similar to the S language and environment which was developed at Bell Laboratories (formerly AT&T, now Lucent Technologies) by John Chambers and colleagues. R can be considered as a different implementation of S. There are some important differences, but much code written for S runs unaltered under R.

R provides a wide variety of statistical (linear and nonlinear modelling, classical statistical tests, time-series analysis, classification, clustering,...) and graphical techniques, and is highly extensible. The S language is often the vehicle of choice for research in statistical methodology, and R provides an Open Source route to participation in that activity.

One of R's strengths is the ease with which well-designed publication-quality plots can be produced, including mathematical symbols and formulae where needed. Great care has been taken over the defaults for the minor design choices in graphics, but the user retains full control.

R is available as Free Software under the terms of the Free Software Foundation's GNU General Public License in source code form. It compiles and runs on a wide variety of UNIX platforms and similar systems (including FreeBSD and Linux), Windows and MacOS.

A.1.2 The R environment

R is an integrated suite of software facilities for data manipulation, calculation and graphical display. It includes

- an effective data handling and storage facility,

- a suite of operators for calculations on arrays, in particular matrices,

- a large, coherent, integrated collection of intermediate tools for data analysis,

- graphical facilities for data analysis and display either on-screen or on hardcopy, and

- a well-developed, simple and effective programming language which includes conditionals, loops, user-defined recursive functions and input and output facilities.

The term "environment" is intended to characterize it as a fully planned and coherent system, rather than an incremental accretion of very specific and inflexible tools, as is frequently the case with other data analysis software.

R, like S, is designed around a true computer language, and it allows users to add additional functionality by defining new functions. Much of the system is itself written in the R dialect of S, which makes it easy for users to follow the algorithmic choices made. For computationally-intensive tasks, C, C++ and Fortran code can be linked and called at run time. Advanced users can write C code to manipulate R objects directly.

Many users think of R as a statistics system. We prefer to think of it of an environment within which statistical techniques are implemented. R can be extended (easily) via packages. There are about eight packages supplied with the R distribution and many more are available through the CRAN family of Internet sites covering a very wide range of modern statistics.

R has its own LaTeX-like documentation format, which is used to supply comprehensive documentation, both on-line in a number of formats and in hard-copy.

A.2 Installing R

To install R, first visit the adequate page (depending on your operating system) and download the installer version of your choice. For Windows, go to `http://cran.at.r-project.org/`, click on "Download R for Windows", then "Base", then "Download R x.x.x for Windows". For Mac OS X, go to `http://cran.at.r-project.org/bin/macosx/`. For Linux, go to `http://cran.at.r-project.org/bin/linux/`.

On Windows, after downloading the installer, double-click on "R-x.x.x-win.exe" and follow the instructions. If you have an account with Administrator privileges, you will be able to install R in the Program Files area, and to set all the optional registry entries; otherwise you will only be able to install R in your own file area. You may need to confirm that you want to proceed with installing a program from an "unknown" or "unidentified" publisher. With the recent versions of R (R-3.0.0 and posterior), after installation, you should choose a working directory for R. You will have a shortcut to "Rgui.exe" on your desktop and/or somewhere on the Start menu file tree, and perhaps also in the Quick Launch part of the taskbar (Windows Vista and earlier). Right-click each shortcut, select Properties... and change the "Start in" field to your working directory. On some systems you will have two shortcuts, one for 32-bit with a label starting R i386 and one for 64-bit starting R x64.

For additional information on how to install R, please refer to `http://cran.r-project.org/doc/manuals/R-admin.html`.

A.3 Running my first R function

A.3.1 Import/export data

In order to analyze a data set with R, the data set has to be loaded into R. This operation can be done in at least two different ways: 1) by entering the value manually in R, or 2) by importing the data set directly from an existing file. Entering the data manually is tedious and is subject to many typo errors. For that reason, we advise you to always use the second option which is importing the data set from a file. Many types of files can be imported; amongst them text files and csv files are the most used. To import the data set, you can use the functions **read.table** or **read.csv**. These functions take as arguments the location of the file on your computer, as well as several options needed to read the file, the main options being the header (TRUE if the first row of your data set contains the header), the separator between cells ("\t" being used for tabulation, "," in csv files), the decimal that is used in the data set ("." for anglo-saxon countries, "," for latin countries). By default, and unless it is specified, R does not consider the first column of the data set imported as the row names. If one of the columns (let's say the first one) corresponds to the row names of the data set, the option row.names=1 can be used.

In R, it is important to save the outputs of a function in an object. For instance, in order to use the data set that we import in R, it is important to save it in an object that we can call mydataset.

```
> mydataset <- read.table("myfile.csv",header=TRUE,sep=";",dec=".")
```

The data set is imported and saved in the object called mydataset that can be used within R. To check if the data set is imported correctly in R, or to have some first global information about the data set, we can use the function **summary** to summarize the data.

```
> summary(mydataset)
```

From the summary, some important information can be visualized: first, if a variable is quantitative (summarized with minimum, maximum, mean, median, first and third quartiles) or qualitative (the six most used levels are printed on the screen with their respective frequency). If missing entries are detected, the notation NA's with the amount of missing entries is added as a seventh row.

If problems during the importation procedure are encountered, they can be caused by different reasons. The most common factor causing errors is the use of special characters not supported by R, or that have another meaning in R. For example, the apostrophe (') is considered by R as a quote for comment. If a name contains an apostrophe, the rest of the line can be considered as a comment. To change that, either remove all apostrophes from your data set, or mention to R (in the **read.table** function) that quotes are only specified by "\"" (using the quote parameter). Warnings should also be given: a data set can be imported successfully without being imported correctly. If the decimal is not specified properly, all the variables that are supposed to be quantitative will be qualitative. In this case, analyses will not be supported the same way. Moreover, qualitative variables that are using numbers as labels (it happens with the assessors or the products) will be considered as quantitative variables. An example of a way to transform this variable from quantitative to qualitative is shown in Chapter 1. In some software, missing data are written with some particular character such as ".". In R, missing data should be considered as empty cells. If one cell containing missing data actually contains "." or " " (*space*), the importation process will consider this cell as a value. Such variable is considered as qualitative with an additional level ("." or " "). For those reasons, it is important to always check that 1) the importation is successful (otherwise R returns an error message) and 2) that the importation is correct (R does not validate the importation, you have to do it manually).

It is important to have the possibility to export data or results from R to other software. Similar to the function **read.table**, the functions **write.table** and **write.csv** exist, for instance. This function takes as argument the object (value, vector, matrix, or data frame) to export, the physical address of where the result should be written in, and some option such as the separator to use (sep="\t" for tabulation, sep="," for *csv* files, *etc.*), the decimal to use (dec="." or dec=","), and if the file to export includes row names (row.names=TRUE/FALSE).

```
> write.table(x=myresult,file="location of my file",
+ sep=";",dec=".",row.names=TRUE)
```

Alternatively, the **write.infile** function of the SensoMineR package allows exporting list of numerical tables into one file automatically.

A.3.2 Objects in R

In R, there are four main objects. The simplest object is a value or string. It consists in one entry unit.

```
> X=2
> Y="word"
```

X and Y are values or strings that count for one entry unit.

A succession of entry units forms a vector. This is defined by the function **c** that stands for "combine."

```
> V <- c(9,8,7,6,5,4,3,2,1)
```

To get a particular entry from that vector, the "[]" indicating the position of the entry to show is needed. For instance, to print the second entry of V, we would use:

```
> V[2]
```

If many vectors are added one on top of each other, we have a matrix. A matrix is defined by its number of rows and number of columns. It can be created using the function matrix which takes the entries, as well as the number of rows and columns of the matrix. Here, we enter the values going from 1 to 12 in a matrix M with three rows and four columns.

```
> M <- matrix(data=1:12,nrow=3,ncol=4,byrow=TRUE)
```

A matrix is indexed with the element present on a row i and a column j. To retrieve the entry presents in the second row and third column, the following code is used:

```
> M[2,3]
```

To print the entire second row or the entire first and third columns, the following code is used, respectively:

```
> M[2,]
> M[,c(1,3)]
```

Similarly to matrix, data frame also exists. Data frames are matrices with additional structural options such as unique names for the rows and columns. The data frames are the main elements supported by most R's modeling software.

In the case where a matrix has more than two dimensions, it is then called an array. To create an array in three dimensions (three elements on the first dimension, four on the second, and five on the third), the following code is used:

```
> A <- array(data=c(1:60),dim=c(3,4,5))
```

The same syntax as for the matrix is used except that in this case, the coordinates on the third component should also be mentioned. The entry present on the coordinate (2,4,1) is obtained using the following code:

```
> A[2,4,1]
```

The last object that is importantly used in R is the list of elements. A list of elements is a list of objects that can be of different nature. Such an object is often created by functions that return a large number of outputs from a different nature. Let's create a list L that will include all the objects created before.

```
> L <- list(X,V,M,A)
> names(L) <- c("Value","Vector","Matrix","Array")
```

To retrieve a name, two syntaxes can be used: either the name of the object by using the character "$" or "[[ ]]." Hence, to retrieve the object called matrix, the two following forms are equivalent:

```
> L$Matrix
> L[[3]]
```

To retrieve an entry from one of the objects, the same syntax as previously is used. For example, if we want to retrieve the entry from the second row and third column of the matrix element present in L, the following commands can be used equivalently:

```
> L$Matrix[2,3]
> L[[3]][2,3]
```

This operation has been widely used throughout the examples presented in the book.

A.3.3 First operations on a data set

In R, data set (matrix or data frame) are flexible. An advantage of this flexibility relies on the fact that the same data set can be modeled in all the possible ways to suit the analysis to perform. Simple but major transformations are presented here. If we are only interested in a subset of the data set, a selection of the columns of interest can be done. Inversely, if additional columns/rows are needed, they can be added. Let's import the data set *perfumes_ideal_small.csv* which is a subset of the data set *perfumes_ideal.csv* used in Section 10.2.2. In this data set, let's say we are interested in the perceived intensities only. To get this subset, there are two possibilities: 1) select the column of interest only or 2) remove the columns that are not of interest. Since we want to keep the information concerning the consumers and products, we will keep the columns 1, 2, 3, 5, 7, 9 or remove columns 4, 6, 8, 10, and 11.

```
> perfumes_ideal_small <- read.table("perfumes_ideal_small.csv",header=TRUE,
+ sep=",",dec=".",quote="\"")
> intensity1 <- perfumes_ideal_small[,c(1,2,3,5,7,9)]
> intensity2 <- perfumes_ideal_small[,-c(4,6,8,10,11)]
```

To remove columns or rows, the character – (minus) is used in front of the number of columns to remove. To add a column (*resp.* a row), the function **cbind** (*resp.* **rbind**) is used. Let's consider we are interested in the interaction between attribute *intensity* and *freshness*; the following code can be added:

```
> intensity1.interact <- cbind(intensity1,
+ as.matrix(intensity1[,3]*intensity1[,4]))
```

Note that this column has a name that is not the one we would like to have. To change it, we will call the function **colnames** that corresponds to the names of the columns (similarly, **rownames** deals with the names of the rows). This can either be done by relocating the names to all the columns of *intensity1.interact*, or by only giving a name to the last element. The second option is adopted here:

```
> colnames(intensity1.interact)[ncol(intensity1.interact)] <-
+ "intensity:freshness"
```

Here, the function **ncol** provides the number of columns in *intensity1.interact*.

When running the summary of the data set, it can be seen that the *user* column that corresponds to the consumer is quantitative. This happens because the consumers are coded with numbers. In order to change a factor from quantitative to qualitative, the function **is.factor** is used. In this case, in order to avoid duplicating the column, a solution consists in overwriting the existing *user* column with the same column set as qualitative. The following code is performing this:

```
> summary(perfume_ideal_small)
> perfume_ideal_small[,1] <- as.factor(perfume_ideal_small[,1])
```

In order to see the levels of a qualitative variable, the function **levels** can be used.

```
> products <- levels(perfume_ideal_small[,2])
```

The number of products that are existing in the data set can then either be obtained using the **nlevels** function or by looking at the length of the object named *products*, using the **length** function:

```
> nbprod <- nlevels(perfume_ideal_small[,2])
> nbprod <- length(products)
```

To make sure that the change from quantitative to qualitative has been done properly, we advise you to look at the summary again.

```
> summary(perfume_ideal_small)
```

As can be seen here, the first column (*user*) is now considered as qualitative.

The last check that we propose concerning the data consists in detecting outliers. To do so, a graphical procedure will be used. In this case, we propose to create a box plot for each quantitative variable by product. The function **boxprod** from the SensoMineR package is used:

```
> library(SensoMineR)
> res.boxprod <- boxprod(perfume_ideal_small,col.p=2,firstvar=3)
```

For the attribute *jasmine*, outliers seem to be detected for the products *CocoMelle*, *PurePoison*, and *Shalimar*.

A.4   Functions in R

A.4.1   Help

All the functions available in R come with some help file. The structure of the help file is standard across R. On the top of the help, the package in which the function belongs is given. Then a short explanation of the function is given in Description. Usage gives the functions the way it should be used as well as all the available options (and their values by default). A description of the options is given in Arguments. Information about what the function does internally can be given in Details. The outputs of the function are given in Value. Some extra information such as Notes, Author(s), and References are also given. Finally, all the help files end up with Examples. These examples are including code that can be run directly in R (by copying and pasting the lines of codes in R). These examples are very useful since they show how to use some functions. To call the help of a function, there are different ways. If you don't know the name of a function, the **help.search** function can be used. Another way of calling it is by using "??." For example, if we want to know how to perform Principal Component Analysis, you could add in R:

```
> ??PCA
> help.search("PCA")
```

These two functions open a long list of functions that are available and that are performing or using PCA. You could focus your attention on three major functions: **PCA** from the FactoMineR package, **princomp** and **prcomp** both from the stats package. Let's use the function **PCA** from the FactoMineR package.

If the name of the function is known, and after making sure that the package it belongs to is loaded, open the help file using "?."

```
> ?PCA
```

An alternative way to open the help file of a function when the name is unknown is by going into R>Help>Html help>Packages. The list of all the packages included on your computer opens. By clicking on the package of interest (FactoMineR for example), a list including the names and description of the functions included in the selected packages opens.

A.4.2 Package

By default, R already provides various functions to perform analyses. However, these functions might not be enough and some additional analyses are required. Two solutions: 1) the user develops his/her own functions to analyze the data. This requires good knowledge in statistics and in R to be able to perform it; 2) check if some colleagues did not already develop functions that would suit you and include them in packages that are freely available.

Let's consider we are interested in the statistical analysis of sensory data. A quick search on the Internet indicates to us that a package called SensoMineR is dedicated to sensory analysis with R (http://sensominer.free.fr). In order to have the functions of SensoMineR available for use, we need to install this package on our computer. To do so, open R, and click Packages>Install Package.

A window opens. Select any CRAN mirror to use. Another window opens. This window highlights all packages that are present on the CRAN mirror selected. Select the package(s) you would like to install and click OK. Note that this installation requires a working Internet connection. If you don't have an active Internet connection, you can still install the packages by downloading the zip file version on a stick from a computer that has access to Internet and by clicking on Packages > Install package(s) from local zip file. Alternatively, you can install the package of interest manually using the following code, in the R-GUI window:

```
> install.packages("SensoMineR")
```

Although SensoMineR is installed, its functions are not yet available on your R session. The package still needs to be loaded. This can either be done by clicking on Packages>Load package or by writing in R:

```
> library(SensoMineR)
```

Important note: once a package is installed, it is not necessary to reinstall it every time you close a session of R. However, you do need to load the package every time you start a new session of R.

If new versions of packages exist, it is not necessary to uninstall them to reinstall the newer versions. Instead, you can update your list of packages by either clicking on Packages>Update Packages..., or alternatively by writing in your R-GUI window:

```
> update.packages(ask='graphics',checkBuilt=TRUE)
```

A list of packages to update opens in a new window. Select the packages to update and click OK. The option checkBuilt=TRUE is particularly important here since it updates the packages based on the version of R installed on your machine. By setting up this option, you avoid errors of packages due to incompatibilities between R versions.

A.5   Non-exhaustive list of useful functions in R

Without loading any specific package, R already provides a very large amount of functions that are either directly linked to statistical analyses, to mathematics, to graphics, or simply to data management. A non-exhaustive list of useful functions is provided below. These functions are classified based on their purposes. These lists also contain functions that have not been used in the book.

Note that, similar to these lists, R reference cards exist. These reference cards present other (more or less complete) non-exhaustive lists of functions, also grouped based on their purposes. An example of reference card can be found at the following link: http://cran.r-project.org/doc/contrib/Short-refcard.pdf.

Import and export functions

Function	Description
getwd	gets the address of the working directory
history	finds the most recently executed lines of code

Function	Description
load	finds the objects saved using the **save** function
read.csv	imports a data table from a *.csv* file and creates a data frame
read.table	imports a data table from a file and creates a data frame (table containing quantitative and/or categorical variables as well as information such as the names of the rows and columns)
save	saves R objects in a *.Rdata* file
save.history	saves the history of the most recently executed lines of code
setwd	sets the working directory
write.table	writes a table into a file

Data management functions

Function	Description
c	combines argument
cbind	juxtaposes data frames by columns; the names of the rows from the data frames must be identical
colnames	gives the names of the columns for a data frame or a matrix
dim	gives the dimensions of an object
dimnames	gives the names of the dimensions of an object (list, matrix, data frame, *etc.*)
factor	defines a vector as a factor, *i.e.*, a categorical variable
grep	searches for a pattern within a vector of elements
is.na	tests to see if the data are available or missing
levels	gives the different levels of a categorical variable
merge	merge two data frames by common columns or by row names
names	gets or sets the names of elements in lists: particularly useful to see the outputs of most functions
ncol	gives the number of columns in a table
nlevels	gives the number of levels in a categorical variable
nrow	gives the number of rows in a table
order	sorts a table into one or more columns (or rows)
paste	concatenates strings or vectors after converting to character
rbind	juxtaposes data frames by rows; the names of the columns from the data frames must be identical
rep(1,5)	repeats the value 1 five times
round	round values with x decimals
rownames	gives the names of the rows for a data frame or a matrix

Function	Description
sample	random selection with or without replacement of elements amongst a set of elements
seq(1,5,0.5)	generates a sequence of values going from 1 to 5 and spaced by 0.5
sort	sorts a vector in ascending or descending order
which	gives the positions of a value within a vector or a table

Basic statistical functions

The following statistical functions are used to describe a quantitative variable x. For most functions, the setting `na.rm=TRUE` is used to eliminate the eventual missing data prior to calculation. If `na.rm=FALSE` and data are missing, the function often returns either an error or a missing value.

Function	Description
apply(x,MARGIN, FUN)	applies the function `FUN` to the rows or columns in table x (depending on `MARGIN`)
cor(x)	correlation matrix of x
density	estimates the distribution of a set of values
lapply	applies **apply** to all the elements of a list
max(x)	maximum of x
mean(x, na.rm=TRUE)	average of x
min(x)	minimum of x
quantile(x, probs)	quantiles of x of the type `probs`
range(x)	minimum and maximum of x
scale(x, center=TRUE, scale=TRUE)	centers (`center=TRUE`) and reduces (`scale=TRUE`) x
sd(x)	standard deviation of x
sqrt(x)	root square of x
sum(x)	sum of the elements of x
summary(x)	provides a summary of x. This function can also be applied to the results of some functions to summarize their outputs (*e.g.*, it is often used to summarize the results of models obtained with **lm** or **aov**)
table	creates contingency table between 2 ordinal or categorical variables
var(x)	variance of x if x is a vector or variance-covariance matrix if x is a matrix (unbiased variance)
xtabs	creates a contingency table from cross-classifying factors

Graphic functions

One of R's strengths is its powerful and unlimited possibilities in generating graphics. In this table, only few graphical possibilities are highlighted. For more details, we suggest you type **?par** in your R-GUI to get an overview of some parameters that can be used.

Function	Description
abline	adds a line to the current graphic
barplot	generates bar charts
bmp, jpeg, pdf, png, postscript	saves a graph in *bmp, jpeg, pdf, png,* or *postscript* format
boxplot	generates box plot
dev.new()	creates a new, empty graph window
dev.off()	closes the last graphical display
graphics.off()	closes automatically all the graphical windows opened
hist	generates histograms
legend	adds a legend to the current graphic
lines	adds a line to the current graphic
plot	generates a graphic of the type scatter plot, or line charts
points(x,y)	adds a point at the coordinates (x,y) of the current graph. The type of point to draw is defined by the parameter **pch**
segment(x1, y1, x2, y2)	adds a segment from A(x1,y1) to B(x2,y2) in the current graphic
text(x, y, labels='mytext')	adds the text *mytext* at the coordinates (x,y) in the current graph
title	adds a title to the current graphic

Statistical tests

R being a statistical software, it proposes a lot of statistical functions. Amongst the most common ones, we can mention:

Function	Description
aov	fits an analysis of variance (intrinsically, **aov** calls **lm**, but returns the results in different ways)
chisq.test	performs the chi-squared test on a contingency table
cor.test	performs the test for the correlation coefficient
density	computes kernel density estimates
df,pf,qf	density, distribution function, and quantile function for the F distribution
glm	fits a generalized linear model
lm	fits a linear model

Function	Description
predict	predicts values from the results of various model fitting functions
t.test	performs Student's *t*-test

Functions of particular packages: **FactoMineR** and **SensoMineR**

Since the FactoMineR and SensoMineR packages are intensively used throughout this book, it seems important for us to provide a list of the major functions present in each of them.

The FactoMineR package is a package dedicated to multivariate analysis in R. The website associated with this package can be found at the following link: `http://factominer.free.fr`.

Function	Description
AovSum	performs ANOVA with $\sum \alpha_i = 0$
CA	performs Correspondence Analysis
catdes	describes the categories of a factor from qualitative and/or quantitative variables
coeffRV	computes the RV coefficients between pairs of tables, and tests its significance
condes	describes continuous variables from qualitative and/or quantitative variables
descfreq	describes the rows or groups of rows in a contingency table
dimdesc	characterizes the dimensions of a factorial analysis by the variables and/or categories that are specific to them
estim_ncp	estimates the number of components in Principal Component Analysis
GPA	performs Generalized Procrustes Analysis
HCPC	performs Hierarchical Cluster analysis on Principal Components
HMFA	performs Hierarchical Multiple Factor Analysis
MCA	performs Multiple Correspondence Analysis
MFA	performs Multiple Factor Analysis
PCA	performs Principal Component Analysis
textual	computes the number of occurrences of each word and generates the contingency table
write.infile	exports the results of a factorial analysis in a *.csv* or *.txt* file

The SensoMineR package is a package dedicated to the sensory analysis in R. The website associated with this package can be found at the following link: `http://sensominer.free.fr`

Function	Description
averagetable	computes the average table crossing the products in rows and the descriptors in columns
boxprod	generates box plots per category with respect to a categorical variable and a set of quantitative variables
carto	generates the external preference map
cartoconsumer	combines cluster analysis and external preference mapping technique by generating one map for each cluster
ConsistencyIdeal	evaluates the consistency of ideal data, both from a sensory and hedonic point of view
cpa	performs the Consumers' Preferences Analysis, which is a derivative of the internal preference map
decat	describes the categories (*i.e.*, the products) from the set of quantitative variables (*i.e.*, the descriptors)
fahst	performs the factorial approach for Hierarchical Sorting Task data
fasnt	performs the factorial approach for Sorted Napping data
fast	performs the factorial approach for Sorting Task data
histprod	generates histograms for each descriptor
IdMap	generates Ideal Map and weighted Ideal Map
indscal	constructs the Indscal model for Napping data
interact	computes the interaction coefficients between two categorical variables for a set of quantitative variables
magicsort	sorts the rows and columns of a matrix
MultiIdeal	evaluates whether the product space is associated with one or multiple ideals
optimaldesign	generates an optimal design balancing for first order and carry-over effects
paneliperf	evaluates the performance of the panelists
panellipse	generates confidence ellipses around the products in Principal Component Analysis
panelmatch	compares the product space obtained from different panels using confidence ellipses
panelperf	evaluates the performance of the panel
pmfa	performs Procrustean Multiple Factor Analysis
search.desc	searches for discriminating descriptors
triangle.design	generates the experimental design for the triangle test
triangle.test	computes the results for a triangle test (according to the Guessing model)
WilliamsDesign	generates an experimental design of Williams

A.6 Recommended packages

Early in 2014, the CRAN package repository featured 5087 available packages. These packages are either general (stats), for graphical purposes (RGraphics, ggplot2), specific to some statistical applications (FactoMineR dedicated to multivariate analysis, deal dedicated to bayesian networks), specific to some data (SensoMineR for sensory data, sensR for discrimination tests), or even for generating and solving puzzles (sudoku)! Besides the fact that we probably do not even know 1% of the existing packages, it is clearly not possible to list all of them. Still this large number suggests the possibility of performing almost any statistical analyses within R. If not, it is still possible to write your own code/functions, build your own package, and spread it to the R-community by uploading it on the CRAN.

Although all the packages cannot be presented, some of them still deserve special attention. As we have seen throughout this book, the SensoMineR and FactoMineR packages are useful tools for the analyses of sensory data. For analyzing data from discrimination tests, the sensR, BradleyTerry2, and prefmod packages are good companions.

Finally, last but not least, another particular package needs attention, especially for beginners. This package is called Rcmdr, and adds a graphical interface to R. This interface makes R more "user-friendly", by adding a scroll-down menu. Moreover, this package is didactic, as it helps users learning R syntax by providing the lines of code generated by each analysis performed. However, the Rcmdr interface neither contains all the functions available in R, nor contains all of the options of the different functions. As with any package, it only needs to be installed once, and then loaded when needed using:

```
> library(Rcmdr)
```

Once the package is installed and loaded, a new R-interface window opens. This interface has a scroll-down menu, a script section, and an output section. When the scroll-down menu is used, the analysis is launched and the lines of code used to generate the analysis appear in the script section.

To import data with Rcmdr, the simplest option is to import a .txt or .csv file:
Data>Import data>from text file, clipboard or URL...
The column separator (field separator) and the decimal separator (a "." or a ",") must then be specified.

To verify that the data set has been successfully imported:
Statistics>Summaries>Active data set

When importing a data set in the csv format, which contains the individuals' logins, it is not possible to specify in Rcmdr's scroll-down menu that the first column contains the login. The data set should be imported by considering the login as a variable. The line of code is therefore modified in the script window by adding the argument row.names=1, and then by clicking Submit.

In order to change the active data set, click on the `Data set` box. If the active data set is modified (for example, by converting a variable), this modification must be validated (=refresh) by:

`Data>Active data set>Refresh active data set`

Note that the output window writes the lines of code in red and the results in blue.

At the end of a Rcmdr session, the script window can be saved. This includes all the instructions as well as the outputs. Both R and Rmcdr can be closed simultaneously by going to `File>Exit>From Commander and R`.

Some packages have been converted into an "Rcmdr plugin," *i.e.*, additional menu with some functions of these packages can be added into the Rcmdr scroll-down menu. Fortunately, it is the case for the FactoMineR and the SensoMineR packages. To install permanently these two additional plugins, write the following lines in R, and click "yes" to install the different packages.

```
> source("http://factominer.free.fr/install-facto.r")
> source("http://sensominer.free.fr/install-senso.r")
```

By starting Rcmdr, one can see that the interface window presents two new buttons called SensoMineR and FactoMineR. These scroll-down menus present most (but not all) functions present in these packages. By navigating through the FactoMineR and SensoMineR menus, the reader should recognize many analyses that are described within this book, even if the names of the functions they refer to are not always explicitly mentioned.

A.7   Recommended readings

The list of functions, packages, and hence possibilities in R are much larger than what is presented here. For more information, we advise you to look at the abundance of literature available online. For example, useful documents are available in different languages on the R website (`www.r-project.org`), or at the following URL: `http://cran.r-project.org/manuals.html`. Additionally, numerous books related to R have been published. Amongst them, we can mention the two following books, which are a perfect complement to the present work:

- Cornillon, P. A., Guyader, A., Husson, F., Jegou, N., Josse, J., Kloareg, M., Matzner-Løber, E., Rouvière, L. (2012). R for Statistics., Chapman & Hall/CRC Computer Science & Data Analysis.

- Husson, F., Lê, S., Pagès, J. (2011). Exploratory Multivariate Analysis by Example Using R, Chapman & Hall/CRC Computer Science & Data Analysis.

Index